Recent Advances in Assessment and Rehabilitation of Individuals with Communication and Language Disorders

Recent Advances in Assessment and Rehabilitation of Individuals with Communication and Language Disorders

Guest Editors

Dionysios Tafiadis
Nafsika Ziavra
Louiza Voniati
Angelos Papadopoulos

MDPI

Basel • Beijing • Wuhan • Barcelona • Belgrade • Novi Sad • Cluj • Manchester

Guest Editors

Dionysios Tafiadis
Department of Speech & Language Therapy
University of Ioannina
Ioannina
Greece

Nafsika Ziavra
Department of Speech & Language Therapy
University of Ioannina
Ioannina
Greece

Louiza Voniati
Department of Health Sciences, Speech and Language Therapy
European University
Nicosia
Cyprus

Angelos Papadopoulos
Department of Speech & Language Therapy
University of Patras
Ioannina
Greece

Editorial Office
MDPI AG
Grosspeteranlage 5
4052 Basel, Switzerland

This is a reprint of the Special Issue, published open access by the journal *Brain Sciences* (ISSN 2076-3425), freely accessible at: https://www.mdpi.com/journal/brainsci/special_issues/85D7OMMX40.

For citation purposes, cite each article independently as indicated on the article page online and as indicated below:

Lastname, A.A.; Lastname, B.B. Article Title. *Journal Name* **Year**, *Volume Number*, Page Range.

ISBN 978-3-7258-5941-2 (Hbk)
ISBN 978-3-7258-5942-9 (PDF)
https://doi.org/10.3390/books978-3-7258-5942-9

Contents

About the Editors

Dionysios Tafiadis

Dionysios Tafiadis is an Assistant Professor in the Department of Speech and Language Therapy, School of Health Sciences, University of Loannina. In 2003, he received his Bachelors (Hons. B.Sc. Med.) in Speech Therapy from the TEI of Patras, Department of Speech Therapy, Greece. After his National Service as a Speech Therapist in 2008, he acquired (with a scholarship) an M.Sc. in "Speech Therapy: Oral and Written Language Disorders—Learning Disabilities", Department of Primary Education, University of Ioannina, Greece. His doctorate was in dementia studies, which was awarded by the School of Medicine at the University of Ioannina. From 2005 until July 2018, he taught at the Department of SLT (TEI of Epirus), and since 2018, he has been an Assistant Professor (full tenure) at the Department of SLT at the University of Ioannina. In 2011, he was accredited and recognized as an SLP by the American Speech Hearing Association (ASHA). His research interests include speech pathology, dementia, voice, swallowing, and language.

Nafsika Ziavra

Nafsika Ziavra is an ENT Professor and Head of the Department of Speech and Language Therapy at the School of Health Sciences, University of Ioannina, Greece, a position she has held since 2008. She has served as a faculty member in the department since 2004 and was formerly Deputy Dean of the School of Health Sciences. She received her Medical Degree (MD) in 1993 from the Faculty of Medicine at the University of Ioannina and completed her specialization in otorhinolaryngology (ENT) in 2001 at the University Hospital of Ioannina. In the same year, she was awarded a PhD from the School of Medicine, University of Ioannina, with distinction, for her thesis entitled "The Role of Otoacoustic Emissions in the Diagnosis of Sensorineural Hearing Loss." From 2001 to 2003, Dr. Ziavra was a Clinical Research Fellow in a Medical Research Council (MRC)-funded project at the Academic Department of Neuro-otology, Imperial College London, in collaboration with Charing Cross Hospital and the National Hospital for Neurology and Neurosurgery, UK. Her research interests span the fields of otorhinolaryngology, neurotology, vocology, rhinology, and speech–language pathology, with a focus on interdisciplinary approaches to diagnosis and rehabilitation.

Louiza Voniati

Louiza Voniati is an Associate Professor in the undergraduate Speech and Language Therapy Program and postgraduate Speech Pathology Programs at European University Cyprus. She was a research assistant at the Early Childhood Research Laboratory of the Department of Education of the University of Cyprus. She is the director of the iCommunicate Research Center at European University Cyprus. Her research interests include the language development of typically developing children and language-impaired children, language sample analysis procedures, the early identification of children with communication disorders, language impairment in children with learning difficulties, the scope of practice of speech–language therapists, and speech–language therapists in multidisciplinary teams. Her funding record includes leadership of and participation in national and international research projects. Dr. Voniati has presented her work in numerous seminars, workshops, and conferences on speech and language disorders, as well as publishing many scientific articles and editing many books in her field of expertise. She is also an accredited and recognized reviewer in several peer-reviewed journals.

Angelos Papadopoulos

Angelos Papadopoulos is an Assistant Professor in the Department of Speech and Language Therapy at the School of Health Sciences, University of Ioannina, since January 2025. In the past, he worked as a Speech and Language Therapist at the Children's Hospital of Patras, Greece, and as a postdoctoral researcher at the University of Patras. He earned his Ph.D. degree from the School of Medicine, University of Ioannina, and his M.S. degree from the Faculty of Medicine, University of Thessaly. During his M.S., he was awarded scholarships from the State Scholarships Foundation (I.K.Y.) for the years 2012–2014. Following his doctoral studies, he worked as a Speech and Language Therapist at the Center of Physical Medicine and Rehabilitation at the General Hospital of Rethymnon in 2021. Subsequently, in 2023, he became a Lecturer at the Department of Speech and Language Therapy, University of Patras. His research interests include speech therapy in children, pediatrics, communication disorders, voice disorders, quality of life, pragmatics, speech deficits, primary health care, early intervention, articulation disorders, phonology, and semantics.

Preface

This Reprint presents a comprehensive collection of studies dedicated to the evolving landscape of assessment and rehabilitation practices for individuals with communication speech and language disorders. The scope of this work reflects the interdisciplinary nature of the field, encompassing contributions from speech–language pathology, cognitive neuroscience, psychology, and allied health sciences. Our aim is to highlight recent evidence-based approaches, innovative methodologies, and the integration of technology in both assessment and intervention. Motivated by the need to bridge research and practice, this Reprint addresses students, clinicians, researchers, and policymakers who strive to improve the quality of care and outcomes for individuals across diverse populations and cultural contexts. As Guest Editors, we curated this collection with the intention of advancing knowledge, encouraging critical reflection, and inspiring future directions in communication sciences and disorders.

Dionysios Tafiadis, Nafsika Ziavra, Louiza Voniati, and Angelos Papadopoulos
Guest Editors

brain sciences

Editorial

Recent Advances in Assessment and Rehabilitation of Individuals with Communication and Language Disorders

Dionysios Tafiadis [1,*], Angelos Papadopoulos [1], Louiza Voniati [2] and Nafsika Ziavra [1]

1 Department of Speech & Language Therapy, School of Health Sciences, University of Ioannina, 45100 Ioannina, Greece; angelospapadopoulos@gmail.com (A.P.); nziavra@uoi.gr (N.Z.)
2 Department of Health Sciences, Speech and Language Therapy, European University, 1516 Nicosia, Cyprus; l.voniati@euc.ac.cy
* Correspondence: tafiadis@uoi.gr

1. Introduction

Currently, in the field of rehabilitation, there is a need for researchers and clinicians to stay updated on recent knowledge worldwide. Often, this is crucial for planning a treatment for individuals with communication and language disorders. Therefore, many researchers recognize the need for an update to the existing literature for various disorders, such as aphasia, autism spectrum disorders, and genetic syndromes [1–5]. Moreover, it is also beneficial for professionals, and even more so for academics, to dedicate their best and most up-to-date efforts to understanding how individuals develop, use, and recover language throughout the lifespan.

The current Editorial refers to the Special Issue "Recent Advances in Assessment and Rehabilitation of Individuals with Communication and Language Disorders". The Special Issue highlights new opportunities and challenges for advancing assessment and rehabilitation for individuals with language and communication disorders, with a specific focus on interdisciplinary research related to neurodevelopmental disorders, neurodegenerative disorders, and other related conditions. As indicated in recent years, researchers have increased the emphasis that various factors (cognitive, neurological, and environmental) have an impact on neurodevelopmental disorders, neurodegenerative disorders, and other related disorders.

The contributors of this Special Issue featured various studies (reviews, systematic reviews, and original articles). Nineteen (19) manuscripts were submitted for consideration for the Special Issue. Specifically, five (5) original articles, six (6) reviews, one (1) longitudinal case report, and one (1) brief report. One contributor study was selected as the editor's choice. The contributions are listed below.

Regarding the contributions in original articles, Sanhueza-Garrido et al. Adapted the Bogenhausen Dysarthria Scales (BoDyS) to Chilean Spanish speakers. In Greece, Zarokanellou et al. explored the narrative ability in Greek-speaking children with High-functioning autism spectrum disorders. Agbavor and Liang conducted research about multilingual prediction of cognitive impairment with large language models and speech analysis. Bachourou et al. studied cognitive vs. linguistic training in children with developmental language disorder by exploring their effectiveness on verbal short-term memory and verbal working memory. Trimmis et al. focused on creating a Greek Pediatric Word Recognition Test (GPWR) by picture identification.

Furthermore, there were six (6) reviews, one (1) brief report, and one (1) longitudinal case report. Voniati et al. published their narrative review on communication abilities, assessment procedures, and intervention approaches in Rett Syndrome. Papadopoulos

 https://doi.org/10.3390/brainsci15101095

et al. conducted a scoping review about the effectiveness of Lee Silverman Voice Treatment (LSVT LOUD) on children's speech and voice. Voity et al. proceeded with an update on how to approach a patient with Locked-In Syndrome and their communication ability. Ericson et al. conducted a systematic review of computer- and smart-tablet-based self-administered treatments in chronic post-stroke Aphasia. A systematic review was published by Georgopoulou et al. about the efficacy of rTMS combined with cognitive and language training in people living with Alzheimer's Disease. Another systematic review by Vlachou et al. about the evidence of language development using brief animated stimuli. Moreover, a Brief Report was published by Yoo on the dynamics of inferential behavior while reading expository and narrative texts. Finally, a longitudinal case report of semantic variant of primary progressive Aphasia (svPPA) was published by Strunk, Weiss, and Müller, focused on high-frequency language therapy with semantic feature analysis (SFA) and transcranial direct current stimulation (tDCS).

2. An Overview of Published Articles

The first contributor study aimed to cross-culturally adapt the Bogenhausen Dysarthria Scales (BoDyS) into Chilean Spanish and to conduct face and content validation. The adaptation process resulted in a consensus version that retained the semantic and cultural characteristics of the original scale. Both the face and content validity were evaluated by a panel of experts consisting of ten Chilean speech–language pathologists (eight women and two men), who were selected through convenience sampling. For the pilot study, 10 new experts were selected through convenience sampling, all working in different healthcare centers in Chile. Moreover, the statistical measures indicated satisfactory levels of agreement, and the pilot study demonstrated the scale's appropriateness and effectiveness for assessing dysarthria within the Chilean context. However, some experts recommended reducing task repetition for patients prone to fatigue. The Chilean version of the BoDyS seems to be a valid and valuable tool for evaluating dysarthria in Chile.

The second contributor study investigated the narrative ability of monolingual Greek-speaking HF-ASD children in comparison to that of their typically developing (TD) peers. Specifically, it explored the associations between narrative variables, ADHD symptomatology, and memory skills in the participants on the autistic spectrum. In the study, 39 children aged 7 to 12 years participated, 19 with HF-ASD and 20 age-matched, vocabulary-matched, and cognitively matched TD peers. The two groups were similar in most microstructural and macrostructural variables. Still, they differed significantly in syntactic complexity and subordination indices, indicating that the HF-ASD group exhibited syntactic delay compared to their TD peers. The HF-ASD participants showed statistically significantly higher heterogeneity in the amount of information generated for the story's main character in comparison to their TD peers. Moreover, significant associations were observed between verbal and visual memory, complex syntactic structures, and terms related to the Theory of Mind and internal state. ADHD symptomatology was negatively correlated with the generation of coordinated and straightforward clauses. Finally, complex syntax and delayed verbal short-term memory (vSTM) were correlated with retelling total scores, indicating that language ability and verbal memory compensate for narrative competence in HF-ASD children. The study's findings highlight the impact of language skills, memory ability, and ADHD symptomatology on narrative competence in children with HF-ASD, as well as the importance of narrative use in assessing language skills in populations with mild language impairment.

The third contributor study presented results from the INTERSPEECH TAUKADIAL Challenge data set, where the aim was to automatically detect mild cognitive impairment (MCI) and predict cognitive scores for English and Chinese speakers. Their approach leverages Whisper, a speech foundation model, to extract language-agnostic speech embeddings. Then, utilize ensemble models to incorporate task-specific information. The model achieved an unweighted average recall of 81.83% in an MCI classification task and a root mean squared error of 1.196 in the cognitive score prediction task, which placed the model in second and first position, respectively, in the ranking for each task. A comparison between language-agnostic and language-specific models reveals the importance of capturing language-specific nuances for the accurate prediction of cognitive impairment. This study demonstrates the effectiveness of language-specific ensemble modeling with Whisper embeddings in enabling scalable, non-invasive cognitive health assessments of Alzheimer's disease, achieving state-of-the-art results in multilingual settings.

The fourth contributor study explored comparatively the effectiveness of a cognitive (verbal short-term memory (vSTM), verbal working memory (vWM)) and of a linguistic training (10-week duration each) in the diffusion of gains in cognitive abilities (vSTM and vWM) of in-school-aged Greek-speaking children with developmental language disorder (DLD). Specifically, the researchers utilized two computerized training programs, i.e., a linguistic and a cognitive one, which were developed and applied to three groups (A, B, and C) of children with DLD (n = 49, in total). There were three assessments with two vSTM tasks (non-word repetition and forward digit span) and a vWM task (backward digit span): pre-therapeutically (time 1), where no significant between-group differences were found, post-therapeutically I (time 2), and post-therapeutically II (time 3), and two training phases. In phase I, group A received meta-syntactic training, whereas group B received vSTM/vWM training, and group C received no training. In Phase II, a reversal of treatment was performed for groups A and B: group A received vSTM/vWM, while group B received metasyntactic training. Again, group C received no training. Overall, the results indicated a significant performance improvement for the treatment groups and revealed beneficial far-transfer effects, as language therapy can affect vSTM and vWM in addition to direct and near-transfer effects. In addition, the intervention type order affected performance as follows: first, better performance on the vSTM task (non-word repetition) was shown when the linguistic treatment was delivered first; second, better performance on the vWM in Time 2 and Time 3 was shown by group B, for which the cognitive treatment was delivered first. The researchers concluded that not only the type of intervention but also the order in which it is administered can affect performance in DLD.

The fifth contributor study aimed to construct a clinically valuable closet-set WRS test with a picture identification task for young Greek-speaking children. The test material was meticulously designed in accordance with specific criteria. Then, to determine which parts of speech are used more frequently by preschool children, a spontaneous speech sample (250 words per child) was collected from 300 children aged 3 to 6 years. The researchers developed and applied two phonemically balanced 50-word lists suitable for young children, as well as created picture representations for each response set. Furthermore, all testing was conducted in an audiometric booth that exceeded the ambient noise level standards of the audiometric rooms. The speech signal was routed from a laptop computer to a GSI 61 audiometer, and all test items were delivered from the audiometer to the subject. Moreover, the results indicated that materials for a WRS test for young children are developed with high face validity and are applicable for children as young as three years old. The test satisfies the essential components needed for a WRS test. It consists of two phonemically balanced 50-word lists with low-redundancy bisyllabic words, with each list containing 227 phonemes. This closed-set WRS test presents a new and valuable tool for assessing speech perception skills in young Greek-speaking children.

The sixth contributor study aimed to synthesize the research from the last two decades regarding the verbal and nonverbal communication abilities, assessment procedures, and intervention approaches for individuals with RTT. The authors conducted a structured literature search using the Embase, Scopus, and PubMed databases, and Fifty-seven studies were selected and analyzed based on inclusion criteria. The data were categorized into four domains (verbal communication skills, nonverbal communication skills, assessment procedures, and intervention approaches). The study's findings revealed a diverse range of communicative behaviors across the RTT population, including prelinguistic signals, regression in verbal output, and preserved nonverbal communicative intent. Moreover, the results highlighted the importance of tailored assessments (Inventory of Potential Communicative Acts, eye tracking tools, and Augmentative and Alternative Communication) to facilitate functional communication. The individualized intervention approaches were found to be the most effective in improving communicative participation. This literature review provides an overview of the last two decades of evidence with an emphasis on the need for personalized and evidence-based clinical practices to improve the quality of life for individuals with RTT.

The seventh contributor study is a scoping review with a primary goal of reviewing the literature and analyzing the possible effectiveness of the LSVT LOUD approach in children with voice and speech deficits. The authors searched the Scopus and PubMed databases in May 2024, and eleven articles were obtained from the search. The standards of the PRISMA recommendations were used for scoping reviews, and the PCC framework was employed for establishing the eligibility criteria. Furthermore, the study employed the instructions outlined in the Cochrane Handbook for quality assessment. Regarding the findings, the reviewed studies employed both formal and informal measures to assess voice and speech abilities in children, with the majority of studies including children with Cerebral Palsy (CP) and those with Down Syndrome (DS). Furthermore, all the studies reported that children with CP and DS underwent a total dose of the LSVT LOUD treatment, with significant post-treatment findings indicating increased speech function and sound pressure level, as reflected in the auditory–perceptual ratings of voice and speech improvement. In many studies, parents' and expert listeners' ratings of voice, perception of vocal loudness, speech, and communication indicated improvement. In conclusion, most of the studies included provide positive evidence for the LSVT as an approach. However, the small sample size that featured in the studies, as well as their limitations, made these conclusions uncertain.

The eighth contributor study focused on the Locked-in syndrome (LIS), which is a rare and challenging condition that results in tetraplegia and cranial nerve paralysis while maintaining consciousness and variable cognitive function. This article aimed to increase knowledge of how to enhance and manage communication among patients in the LIS setting. After the acute management is completed, it is essential to collaborate with the patient in developing a plan to maintain and enhance their quality of life (QOL). A key component in increasing or maintaining QOL within this population involves establishing a functional communication system. Evaluating cognition in patients with LIS is vital for assessing patients' communication needs along with physical rehabilitation to maximize their QOL. Over the past decade or so, there has been an increase in research surrounding brain–computer interfaces aimed at improving communication abilities for patients with paralysis. This article provides an update on the available technology and protocols for finding the most effective way for patients with this condition to communicate.

In the ninth contributor study, this systematic review aimed to identify computer- and smart-tablet-based self-administered treatments and analyze the proposed interventions in terms of treatment targets, effectiveness (considering specificity, generalization, transfer, and maintenance), and clinician involvement (during and/or prior self-administered therapies). The authors used terms encompassing three main concepts (rehabilitation, self-administration, and aphasia) to search three electronic databases (Scopus, PubMed, and PsycINFO). This systematic review was conducted in accordance with the guidelines proposed by the Preferred Reporting Items for Systematic Reviews and Meta-Analyses using PRISMA guidelines2020. Regarding the process, two reviewers independently screened titles and abstracts against eligibility criteria, and three reviewers completed data extraction of the included studies. After the whole process was completed, thirty-nine studies were included in this study. In terms of treatment targets, anomia is the most treated symptom in published studies (*n* = 24). Still, the existence of promising studies for other disorders means that the targets can be broadened. Therapies are effective for trained items, and gains are maintained. There is some evidence of transfer effects for treatments targeting the sentence level. Most studies offer training sessions, previous self-administered therapy, and/or observation and monitoring sessions during therapy; more rarely, self-administered therapy is supplemented with face-to-face therapy. This systematic review is the first to focus specifically on self-administered technology-based therapies via computer or smart tablet.

The tenth contributor study focused on repetitive transcranial magnetic stimulation (rTMS) in patients with Alzheimer's disease (pwAD), similar to multidomain cognitive training (CT). The current review aims to evaluate the cognitive benefits of combining rTMS with CT, including language training, for individuals with mild to moderate AD. An extensive literature search was conducted in PubMed, Google Scholar, and the Cochrane Library using relevant terms, resulting in nine studies with a total of 290 participants (190 in the Active Group and 100 in the Control Group). The comprehensive review of the articles revealed that the combined treatment improved global cognitive function, as well as neurocognitive, neuropsychiatric, and quality of life in the AG. Nevertheless, these results should be interpreted cautiously, given the relatively small number of existing studies on this specific combination.

The eleventh contributor study is a brief report that aimed to examine changes in inferential behavior while reading different genres. The inferential behavior of 28 students with reading disabilities (RDs) and 44 students without RDs was quantified while they read expository and narrative texts. First, the average rates of inference attempts and correct inferences were measured during reading. Then, the same rates were measured separately during early and late reading to see if there was a change in inferential behavior. The results indicated that the change in inferential behavior is dependent on the genre. While reading the expository text, both groups showed no significant change in their ability to make inferences. In contrast, while reading the narrative text, both groups showed higher rates of inference attempts, and only the students without RD showed a significant increase in correct inferences.

The twelfth contributor study aimed at systematically reviewing the available literature for evidence concerning the effect of brief animation on spoken language responses (receptive—listening or expressive—speaking) in typically developing children aged 3 to 9 years. The authors searched five databases, resulting in seven studies that were included. The characteristics of animated stimuli, the manner of presentation, and the language-related tasks were recorded, and questions were posed about the effect of brief animation on children's receptive and expressive language abilities. This evidence suggested that animation may have a positive effect on the expressive language abilities of children compared to static pictures. Regarding the impact of animation on receptive language performance, the evidence is less conclusive.

The thirteenth study is a case report that aimed to investigate whether the combination of semantic feature analysis (SFA) and transcranial direct current stimulation (tDCS) is effective in treating word retrieval in the semantic variant of primary progressive aphasia (svPPA) and to determine the duration of its potential effects. Methods: A 56-year-old woman diagnosed with frontotemporal dementia (FTD) and svPPA participated in this longitudinal single-subject design. A total of four 2-week stimulation phases were conducted over 14 months, each of which began depending on the participant's language performance. Follow-up testing was conducted approximately 2 weeks and 4 weeks after the stimulation period. The authors presented results showing a significant improvement in word retrieval after SFA and tDCS therapy. Two weeks after the end of each stimulation phase, approximately 80% of the trained words could be named correctly. For the untrained words, also significantly more words were correctly named at follow-ups compared to the baseline. Furthermore, the Boston Naming Test (BNT) demonstrated a significant improvement in naming performance, showing that phonological cues facilitated word retrieval more effectively than semantic cues. The authors conclude that the combination of SFA and tDCS was able to counteract the expected language deterioration of a participant with svPPA. This effect increased until approximately 2 weeks after each intervention. Additionally, the impact on untrained words was also demonstrated.

3. Conclusions

This Special Issue included thirteen articles that reflected the new aspects in the fields of language, communication, and cognitive rehabilitation. As was presented before, the five (5) original articles focus on various conditions and disorders such as dysarthria, high-functioning autism, developmental language disorder, and cognitive decline. These studies discussed traditional methodologies and new technologies (large language models, brain stimulation, computer-based interventions). As indicated by the conclusions of all the studies, they provided vital and crucial insights for communication assessment and rehabilitation in individuals with these difficulties.

In addition, the other review studies (literature, scoping, and systematic reviews) presented data from the past two decades across various disorders like Rett syndrome, Aphasia, Locked-in syndrome, and Alzheimer's disease. A general conclusion of the majority of these studies is that it underlined the need for personalized assessment and treatment with an interdisciplinary approach. Finally, this Special Issue has delivered a message that communication remains a fundamental element of human interaction, which can have a significant impact on the quality of life. We hope that this Special Issue will be a valuable resource for professionals, clinicians, academics, and researchers, enabling them to improve outcomes for individuals facing communication and language difficulties.

Conflicts of Interest: The authors declare no conflicts of interest.

References

1. MacWhinney, B.; Fromm, D. Language sample analysis with TalkBank: An update and review. *Front. Commun.* **2022**, *7*, 865498. [CrossRef]
2. de Riesthal, M.; Ross, K.B. Patient reported outcome measures in neurologic communication disorders: An update. *Perspect. Neurophysiol. Neurogenic Speech Lang. Disord.* **2015**, *25*, 114–120. [CrossRef]
3. Kim, S.K. Recent update of autism spectrum disorders. *Korean J. Pediatr.* **2015**, *58*, 8. [CrossRef] [PubMed]
4. Johnson, R.K.; Prebor, J. Update on preservice training in augmentative and alternative communication for speech-language pathologists. *Am. J. Speech Lang. Pathol.* **2019**, *28*, 536–549. [CrossRef] [PubMed]
5. Vogels, A.; Droogmans, G.; Vergaelen, E.; Van Buggenhout, G.; Swillen, A. Recent developments in Phelan–McDermid syndrome research: An update on cognitive development, communication and psychiatric disorders. *Curr. Opin. Psychiatry* **2021**, *34*, 118–122. [CrossRef] [PubMed]

brain sciences

MDPI

Article

A Greek Pediatric Word Recognition Test by Picture Identification

Nikolaos Trimmis [1], Konstantina Chatzi [1], Vasiliki Grammatsoulia [1], Foteini Feida [1], Konstantinos Mourtzouchos [2], Angelos Papadopoulos [1] and Panagiotis Plotas [1,3,*]

1 Department of Speech and Language Therapy, University of Patras, 26504 Patras, Greece; nicktrimmis@upatras.gr (N.T.); kwnstantinaxatzh1994@gmail.com (K.C.); vasiliki.gram@gmail.com (V.G.); logotherapycenter@gmail.com (F.F.); angelospapadopoulos@gmail.com (A.P.)

2 Department of Pediatric Otorhinolaryngology, Karamandaneio Children's Hospital of Patras, 26331 Patras, Greece

3 Laboratory Primary Health Care, School of Health Rehabilitation Sciences, University of Patras, 26504 Patras, Greece

* Correspondence: pplotas@upatras.gr

Abstract: (1) Background: The study aimed to construct a clinically valuable closet-set WRS test with a picture identification task for young Greek-speaking children. (2) Methods: The test material was meticulously designed based on specific criteria. To determine which parts of speech are used more frequently by preschool children, a spontaneous speech sample (250 words per child) was acquired from three hundred children aged 3 to 6 years (M = 4.56, SD = 0.90). The study involved the development and application of two phonemically balanced 50-word lists suitable for young children, as well as the creation of picture representations for each response set. All testing was accomplished in an audiometric booth that exceeded the audiometric rooms' ambient noise level standards. The speech signal was routed from a laptop computer to a GSI 61 audiometer, and all test items were delivered from the audiometer to the subject. (3) Results: The results indicated that materials for a WRS test for young children are developed with high face validity and are applicable for children as young as three years old. The test satisfies the essential components needed for a WRS test. It consists of two phonemically balanced 50-word lists with low-redundancy bisyllabic words, with each list containing 227 phonemes. (5) Conclusions: This novel closed-set WRS test presents a valuable tool for assessing speech perception skills in young Greek-speaking children. The test results have various applications, including diagnosis, research, and (re)habilitation.

Keywords: word recognition score; WRS; children; picture identification task; speech audiometry

1. Introduction

Hearing is crucial for the development of communication skills. Even mild hearing loss can compromise speech, language, psychosocial behavior, and other developmental milestones [1]. A comprehensive audiology evaluation (CAE) is necessary if uncertainty about a child's hearing status exists. Speech audiometry is a vital part of any CAE as it provides information of greater value than pure-tone audiometry alone. It can be used throughout the auditory system to determine speech-processing abilities [2]. Two common speech audiometry procedures are the Speech-Recognition-Threshold (SRT) and the Word-Recognition-Score (WRS). The SRT measures the threshold of speech and is the lowest intensity level at which a person can correctly recognize 50% of presented speech stimuli [3,4]. The primary objective of WRS testing is to ascertain the estimated suprathreshold threshold above which an individual may accurately comprehend and reproduce a series of words. The resulting score is the proportion of words in a given list that are accurately detected at levels over the threshold [4,5]. In other words, the WRS measures speech understanding at a comfortable hearing level. Behavioral assessment of children will frequently begin with speech audiometry that has revealed hearing abnormalities, or their absence, in young children who were unable or unwilling to take pure-tone tests. This approach has resulted in substantial savings of time in initiating remedial procedures [6]. Some children will not respond to pure tones, and it appears that they can not hear. Many of these children can be conditioned to respond to speech stimuli. A child will often point to a picture or object.

Speech recognition thresholds (SRT) can frequently be obtained with picture identification by age 2. Word recognition score (WRS) test can be accomplished in the same way and is one of the most frequently used measures for speech audiometry [5,7]. Over the years, researchers have developed several WRS instruments for preschoolers as part of the hearing validation process. However, there is no generally accepted standard test for the speech assessment of young children because speech perception is an abstract construct. Therefore, it is not easy to measure in an absolute sense [8]. The most popular of these measures include the Northwestern University Children's Perception of Speech Test [9], the Word Intelligibility by Picture Identification Test [10], the Pediatric Speech Intelligibility Test [11], and the Minimal Pairs Test [12].

The assessment of WRS in young children presents a challenge to clinicians [13]. To increase reliability, the speech material of the test must be within the child's receptive vocabulary, and the response mode must be age-appropriate [14,15]. An open-set response mode means the patient must compare the stimulus item to all possible candidate words in lexical memory, whereas a closed-set response mode means the patient is provided with a choice of several possible response alternatives. In addition, it is necessary to have speech audiometry materials available in each individual's native language for reliable results.

The majority of the early evaluations of speech intelligibility, as outlined by Miller (1946), utilized open-set tests that focused on the recognition of words or syllables. Nevertheless, many researchers decided to use closed-set speech intelligibility tests due to their expediency and simplicity in terms of administration and scoring [16]. The main concept of the closed-set multiple-choice intelligibility tests is that the word recognition process remains consistent, irrespective of the test's response style. According to speech and hearing scientists, the only difference between open-set and closed-set tasks was the occurrence of chance performance [16]. In open-set tests, the chance performance is represented by 1 divided by the size of the mental lexicon (L), while in closed-set tests, it is represented by 1 divided by the number of response possibilities offered by the experimenter (N). The closed-set testing format has maintained its popularity in clinical settings due to its efficient and expedient administration and scoring procedures [16]. Closed-set tests have demonstrated their reliability even when conducted with limited trials [16]. Closed-set and open-set tasks differ in terms of information processing needs. Specifically, the level of competition between alternative responses is a fundamental distinction between these two types of tasks. In closed-set tests, the scope of competition among alternatives is constrained to the response set selected by the investigator. In the context of open-set tests for spoken word recognition, it is essential to acknowledge the presence of lexical rivalry that occurs over the entirety of the mental lexicon [16]. At this point, it is crucial to acknowledge that closed-set WRS testing has distinct advantages for implementation compared to open-set WRS testing [17].

Today, while several open-set speech audiometry tests are currently available in Greek for adults and school-aged children, there is still no valid speech test available with a picture-pointing task for pediatric evaluation, especially for children younger than six. Given the above limitations in Greek speech audiometry testing, our goal in this study was to construct a clinically valuable closet-set WRS test with a picture identification task for young Greek-speaking children. More specifically, our primary aims were to evaluate the word familiarity of list items, ensure the phonemic balance of the lists, produce digital recordings of the stimuli, select digitally illustrated pictures for each response set, and conduct a preliminary investigation of list equivalence.

2. Materials and Methods

A review of English literature regarding the significant criteria for valid speech recognition tests of the intended age group and structure of the Modern Greek language has guided us to develop a set of word lists specifically for use in preschool children's WRS measurements. The process of the study is presented in the Figure 1 flow diagram. The following seven criteria were chosen for the development of the word lists:

1. Type of Materials.
2. Familiarity of Materials.
3. Phonemic dissimilarity.
4. Number of scorable list items.
5. Phonemic Balance.
6. Lexical stress.
7. Method of response.

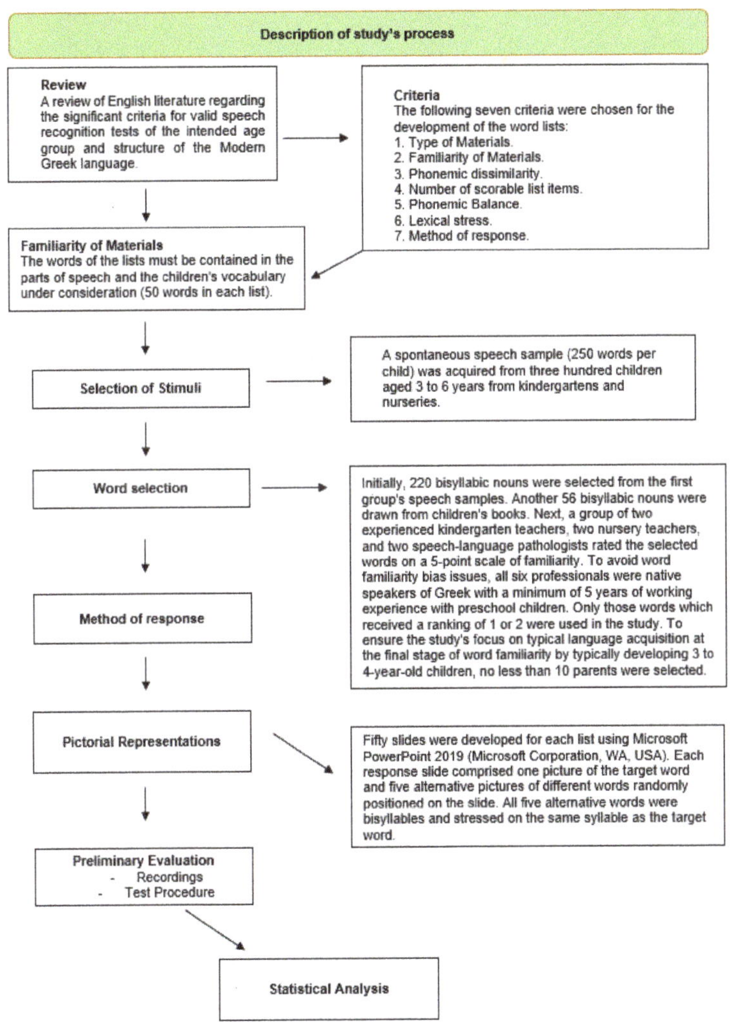

Figure 1. Flow diagram of the study process.

The criteria for five of the above, namely type of materials, phonemic dissimilarity, number of scorable list items, phonemic balance, and lexical stress, were used in previous research [4,18]. The remaining two criteria are discussed below. The institutional review board approved the present study. To avoid methodological biases, standardized procedures were consistent across participants and conditions.

2.1. Familiarity of Materials

The possibilities for constructing speech audiometry materials for young children are indeed more restricted than with older children and adults due to lower stages of language abilities development. Phonemic and linguistic requirements usually have to be adjusted, and more emphasis is placed on the familiarity and simplicity of the test words. The familiarity of a given word enhances the word's intelligibility [19]. Research has shown that highly familiar word lists offer a more precise indicator of an individual's continuous discourse perception than less familiar word lists [20].

The selection of speech materials for young children should be within the children's speech and language competence [2,13]. The words on the lists must be contained in the parts of speech and the children's vocabulary under consideration. Words fall into various categories called "parts of speech." For example, "verbs" and "nouns" are some words; "pronouns" and "adjectives" are others. Each of these categories of words represents a part of speech.

2.2. Selection of Stimuli

Parts of speech. To determine which parts of speech are used more frequently by preschool children, a spontaneous speech sample (250 words per child) was acquired from three hundred children aged 3 to 6 years (M = 4.56, SD = 0.90) from kindergartens and nurseries in the cities of Athens and Patras. All children were monolingual speakers of Greek and presented no history of speech and language problems or other developmental disorders. The children were divided into three age groups. The first group corresponded to the age of 3 to 4 years old, the second 4 to 5 years old, and the third 5 to 6 years old. Each group consisted of 100 children (50 boys and 50 girls). The speech samples were elicited through conversational strategies and narratives to produce the most complex language [21]. Each child's sample was recorded individually for later analysis. Table 1 presents the frequency of occurrence of all parts of Modern Greek speech for the 300 children.

Table 1. Mean percent scores (%) for the Modern Greek language parts of speech used by children 3–6 years old (n = 300).

Parts of Speech	Frequency of Occurrence
Article	11.98%
Noun	22.65%
Adjective	3.00%
Pronoun	12.64%
Verb	22.25%
Participle	0.16%
Adverb	10.34%
Preposition	4.69%
Conjunction	10.84%
Interjection	0.27%
Particle	1.17%

Preschoolers use more nouns followed by verbs in their production of spontaneous speech. There was no statistically significant difference between boys and girls for the nouns ($t = 0.16$, df = 298, $p > 0.05$) and verbs ($t = 0.66$, df = 298, $p > 0.05$). Comparison between the groups revealed that nouns are the most frequently used part of speech in all three groups of participants. The lower the group's age, the more nouns appeared in spontaneous speech. Specifically, the 3–4-year-old children used 24.25% of nouns, the

4–5-year-old children used 22.59%, and the 5–6-year-old group used 21.10%. A study [22] investigated vocabulary composition in Greek preschool children. Common nouns were the largest category among the fifty most frequent words. Recently, a grammatical content analysis of hard-of-hearing children's speech in Greece aged 6–14 revealed that nouns have the highest frequency [23].

2.3. Word Selection

The receptive vocabulary size of young children increases rapidly. At age 3, children recognize about 1000 words, and by age 5, they recognize at least 10,000 words [24]. Therefore, since the present study is intended to design a speech test for children of vocabulary level of age 3 through 6 years, the stimulus words must be primarily selected from the subset of word usage in the first group.

However, children's receptive vocabulary generally exceeds their expressive vocabulary [25]. Thus, children's books of this age range were also used to select simple bisyllabic nouns that could be adequately represented pictorially.

Initially, 220 bisyllabic nouns were selected from the first group's speech samples. Another 56 bisyllabic nouns were drawn from children's books.

Next, a group of two experienced kindergarten teachers, two nursery teachers, and two speech-language pathologists (six female; average age = 41.24 years; Standard Deviation = 5.77) rated the selected words on a 5-point scale of familiarity (1 = extremely familiar, 2 = very familiar, 3 = somewhat familiar, 4 = infrequently used, and 5 = rarely used). To avoid word familiarity bias issues, all six professionals were native speakers of Greek with a minimum of 5 years of working experience with preschool children. Only those words which received a ranking of 1 or 2 were used in the study. To ensure the study's focus on typical language acquisition at the final stage of word familiarity by typically developing 3 to 4-year-old children, no less than 10 parents were selected with the following inclusion and exclusion criteria:

The Inclusion criteria included: Female biological parents of children between the ages of 3 and 4 years old, native Greek speaker, graduate of Greek high school, regular interactions with their children, and no medical or developmental concerns that may impact language development.

Exclusion criteria: Parents who did not satisfy all the above criteria were excluded from the study.

To further enhance familiarity, the 214 words that were rated with 1 or 2 were judged as either familiar or unfamiliar by ten parents (ten female; average age = 32.80 years; Standard Deviation = 3.71) who had in the first group, children with normal hearing bilaterally. The final 192 words rated as familiar were used to construct the test lists.

2.4. Method of Response

The response format used to elicit accurate responses is a crucial issue to consider when designing WRS tests for young children. The type of response required must reflect the child's ability to hear, and this must not be contaminated by other factors [19]. Some children are too shy to speak in the test room environment and will not provide suitable responses to an open-set test format, where the patient repeats the word that they hear [14]. Also, an oral response is inappropriate for children with limited speech production skills, such as articulatory problems, because the tester cannot determine whether such a response is due to faulty hearing or faulty articulation. Thus, the response format must be age-appropriate and not require speech. This criterion is met with a picture identification pointing response to a set of pictures, a closed-response set. Children's speech perception can be assessed using picture-pointing tasks when their receptive language skills are 2.6 years of age or greater [2]. Fifty slides were developed for each list using Microsoft PowerPoint 2019 (Microsoft Corporation, Redmond, WA, USA). Each response slide comprised one picture of the target word and five alternative pictures of different words randomly positioned on the slide. All five alternative words were bisyllables and stressed on the same syllable as the target word. American Speech Hearing Association (ASHA) [3] recommended that if pointing responses are employed with picture cards, an excessively high number of items may be distracting and increase response time, and a deficient number of items increases the probability of chance performance. An example of a response plate is presented in Figure 2.

Figure 2. A sample response slide employed in the Picture Identification Task. All six pictures represent bisyllabic words with stress on the first syllable. The image is the property of the author.

2.5. Pictorial Representations

Items selected were simple bisyllabic words that could be easily represented pictorially. The following visual clarity criteria were adopted:

- Pictures with simple, recognizable images.
- Pictures with vibrant colors that are visually appealing to young children.
- Pictures with a strong contrast between the foreground and background.
- Pictures that are age-appropriate for preschoolers.
- Consistent visual style throughout the selection of pictures.
- Pictures do not contain any inappropriate content.

The picture selection was a lengthy process requiring scanning many online sites that provided free stock pictures. Several pictures were selected for each target word. Next, the same group of two experienced kindergarten teachers, two nursery teachers, and two speech-language pathologists working with young children were consulted as experts. The expert group was asked to rate each item for its visual clarity (easily recognizable by young children) on a 3-point scale (1 = high clarity, 2 = average clarity, 3 = low clarity). Only the pictures which received a ranking of 1 were used in the study.

2.6. Preliminary Evaluation

2.6.1. Subjects

The individuals who participated in this study were selected from kindergartens and nurseries. They included 50 children ranging in age from 3 to 6 years (25 boys and 25 girls; Mean = 4.68 years; Standard Deviation = 0.80). All participants met the following criteria:

1. Monolingual speakers of Greek.
2. No known history of auditory dysfunction, speech or language delay, or other cognitive disorders.
3. Not currently receiving speech or language therapy.
4. Hearing within normal limits bilaterally. Pure tone thresholds of $\leq$15 Decibels Hearing Level (dBHL) at all octave frequencies ranging from 250 Hertz (Hz) to 8000 Hz.
5. None of the children were on medication or had known illness on the day of testing.
6. No history of vision problems. All children passed a screening test as measured by a Snellen children's color blindness chart.

2.6.2. Recordings

Speech audiometry testing can be performed with monitored live voice or recorded material. However, the use of recorded material standardizes the test procedure. Auditory stimuli were recorded in a 40A series audiometric booth (IAC Acoustics, Winchester, UK) by a young adult female native speaker of Modern Greek. The speaker was instructed to maintain clarity, pace, and effort to produce each target word at least three times. The words were digitized with a professional unidirectional microphone at a 44.100 kilohertz (kHz) sampling frequency and 16-bit resolution. Each word was equalized to the average root mean square (RMS) amplitude. Words were presented with a 5-s interval. A 1000-Hz calibration tone of 30 s duration was synthesized and equalized to the average RMS level of the 100 test words. Two experienced judges (one audiologist and one speech pathologist) rated the repetitions of each word for the perceived quality of production, and the best production of each word was selected.

2.6.3. Test Procedure

Before data collection, each participant's parent signed an informed consent form approved by the Departmental Committee of TEI Western Greece (currently the University of Patras). All testing was accomplished in the same 40A series audiometric booth that exceeded the audiometric rooms' ambient noise level standards. The speech signal was routed from a laptop computer to a GSI 61 audiometer (Grason-Stadler, Eden Prairie, MN, USA). All test items were delivered from the audiometer to the subject via supra-aural TDH-49 headphones (Telephonics, Farming Dale, NY, USA). Although there were difficulties with younger children maintaining consistent attention with pre-recorded stimuli, all participants could perform the test. Each subject participated in more than one session to reduce performance variability due to the child's fatigue.

Verbal instructions to point to a picture and motivation were used for each child. A test assistant in the test room, seated next to the child and in visual contact with the examiner, maintained the child's interest in the task, informed the examiner about the child's response, and presented a new response slide on the screen after the child made a selection.

Each list was presented monaurally (right ear) at ten consecutive intensity levels, ranging from 0 to 45 dBHL, starting at 0 dBHL and ascending in 5 dBHL steps. Subjects responded to each item by pointing to one of six pictures of the response slide on a laptop computer screen in front of the child. None of the participants was familiarized with the test materials prior to testing. The words within each list and the order of presentation of the lists were randomized for each subject to decrease nonauditory factors such as memory effects.

2.7. Statistical Analysis

Statistical analysis was performed using the Statistical Package for the Social Sciences (SPSS) version 23.0 software (IBM Corp., Armonk, NY, USA). The normality assumption for all variables was addressed by the Kolmogorov–Smirnov test and Shapiro–Wilk test. An independent *t*-test was conducted to examine the equivalence of the two lists. To determine if there was a significant interaction between different factors (list, gender, and age) on the WRSs (dependent variable), a general linear model repeated measures analysis, consisting of WRSs at different decibel hearing levels (dBHL = 10 levels: OdBHL, 5dBHL, 1OdBHL, 15dBHL, 20dBHL, 25dBHL, 3OdBHL, 35dBHL, 4OdBHL, 45dBHL) as the within-subject variable and list (two levels: list 1 and list 2), gender (two levels: male and female) and age (three levels: 3–4 years, 4–5 years, and 5–6 years) as the between-subject variables, was used. A *p*-value of <0.05 was considered to be statistically significant.

3. Results

The limited number of suitable words that could be used with this population and the adoption of the criteria by the authors restricted the possible choices of constructing several word lists. Due to these limitations, developing more than two 50-word lists was impossible. Several 50-word lists are needed for WRS testing to minimize training and memory effects. For example, an audiologist might need to determine WRSs at different intensity levels. Therefore, a different 50-word list should be presented at each decibel hearing level. Table 2 presents the phonemic balance information for the two preschool lists. The lists were modified numerous times to satisfy better the criteria for constructing the final version of the test.

Table 2. Frequency of occurrence of phonemes in Modern Greek spoken language and frequency of preschool lists 1 and 2 (IPA: International Phonetic Alphabet).

	Phonemes IPA Symbol	Frequency Spoken Language	List 1 Frequency (%)	List 2 Frequency (%)
1	a	12.26	11.89	11.45
2	ε	10.40	8.81	8.81
3	i	14.25	13.22	13.22
4	o	9.49	8.37	8.37
5	u	2.50	1.76	2.20
6	ʎ	4.18	4.85	5.29
7	θ	1.11	0.44	0.88
8	ð	2.04	1.32	1.32
9	b	0.26	0.44	0.44
10	d	0.54	1.32	0.88
11	ts	0.11	0.44	0.44
12	dz	0.02	0.44	0.44
13	p	4.36	4.41	5.29
14	m	3.69	3.52	3.52
15	f	1.28	2.20	2.64
16	v	0.88	1.32	1.76
17	t	7.54	6.61	6.61
18	z	0.54	0.44	0.88
19	s	7.68	8.81	7.49
20	n	6.17	4.85	4.85
21	g	0.12	0.44	0.44
22	l	2.77	3.52	3.08
23	k	2.62	3.08	1.76
24	$\mathcal{X}$	0.60	1.32	1.76
25	ɣ	0.74	1.32	1.32
26	c	1.79	1.76	0.88
27	ç	0.88	1.32	2.20
28	ʝ	0.98	0.88	0.44
29	ʎ	0.11	0.44	0.88
30	ɲ	0.10	0.44	0.44
Total		100.00	100.00	100.00

Each new list consists of 50 bisyllabic nouns (Table 3). Data from the presentation of the two lists at the different intensity levels were used to construct the performance intensity (PI) functions, which represent the relationship between word recognition accuracy and the intensity (volume) level of the presented stimuli. Figure 3 presents the PI functions for the two preschool bisyllabic lists' mean scores and standard deviations.

Mean WRSs increased with an average slope per decibel of 4.79% in the first list and 4.42% in the second list in the rapidly rising portion of the function between 15 and 30 dBHL.

Table 3. The two bisyllabic word lists for use in preschool WRS testing. (IPA: International Phonetic Alphabet).

List 1				**List 2**			
Stress 1st syllable		Stress 2nd syllable		Stress 1st syllable		Stress 2nd syllable	
IPA	Modern Greek	IPA	Modern Greek	IPA	Modern Greek	IPA	Modern Greek
amos	άμμος	avγo	αβγό	ema	αίμα	arɲa	αρνιά
velos	βέλος	afti	αυτί	γadi	γάντι	avli	αυλή
je'ros	γέρος	kafes	καφές	ðiχti	δίχτυ	vuno	βουνό
ðedro	δέντρο	γati	γατί	ðodi	δόντι	vroçi	βροχή
kotes	κότες	garaz	γκαράζ	ðoro	δώρο	γries	γριές
kuɲes	κούνιες	jaʎa	γυαλιά	zugla	ζούγκλα	eʎes	ελιές
kupes	κούπες	evro	ευρώ	zones	ζώνες	pani	πανί
mati	μάτι	ceri	κερί	terma	τέρμα	karfi	καρφί
meli	μέλι	kupi	κουπί	tiγris	τίγρης	θeos	θεός
milo	μήλο	kuti	κουτί	nifes	νύφες	keftes	κεφτές
botes	μπότες	laγos	λαγός	θiki	θήκη	kubi	κουμπί
nani	νάνοι	lemos	λαιμός	mites	μύτες	krasi	κρασί
naftes	ναύτες	mali	μαλλί	niçi	νύχι	lefta	λεφτά
pana	πάνα	nona	νονά	niχta	νύχτα	majo	μαγιό
limni	λίμνη	scini	σκοινί	porta	πόρτα	moro	μωρό
roða	ρόδα	peði	παιδί	pites	πίτες	nero	νερό
teda	τέντα	kapnos	καπνός	pçano	πιάνο	palto	παλτό
dzaci	τζάκι	scepi	σκεπή	sela	σέλα	papas	παπάς
tiχos	τοίχος	spaθi	σπαθί	dzami	τζάμι	papi	παπί
topi	τόπι	tapsi	ταψί	tuvlo	τούβλο	puʎa	πουλιά
tsada	τσάντα	tiri	τυρί	treno	τρένο	nisi	νησί
çeri	χέρι	fili	φιλί	tsepi	τσέπη	taksi	ταξί
çoni	χιόνι	ftero	φτερό	fetes	φέτες	fotça	φωτιά
χtenes	χτένες	χarti	χαρτί	çines	χήνες	χali	χαλί
psari	ψάρι	çimos	χυμός	χoma	χώμα	psomi	ψωμί

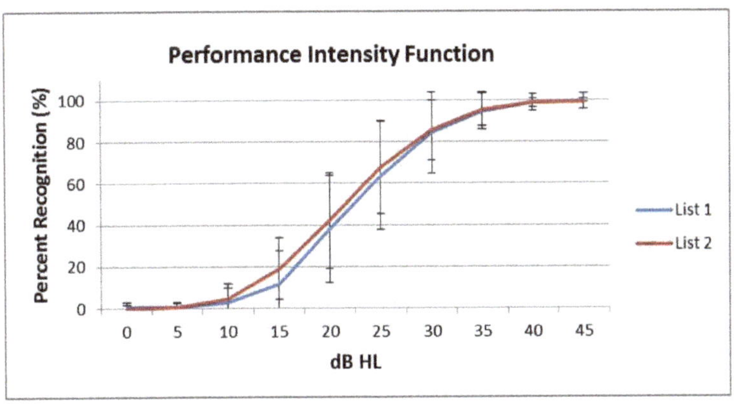

Figure 3. Monaural (right ear) PI functions of 50 normal-hearing children and standard deviations of the two preschool bisyllabic lists. The image is the property of the author.

Results of the independent t-test indicated that there are no significant differences between the two lists ($p > 0.05$) for all presentation hearing levels. Also, the repeated measures analysis results with a Greenhouse–Geisser correction showed no significant interaction on the repeated measures in terms of WRSs for all the conditions: dBHL*List ($p = 0.498$), dBHL*Age ($p = 0.538$), dBHL*Gender ($p = 0.754$), dBHL*List*Age ($p = 0.858$), dBHL*List*Gender ($p = 0.895$), dBHL*Age*Gender ($p = 0.687$), dBHL*List**Age*Gender ($p = 0.973$).

4. Discussion

The results from this study indicate that materials for a WRS test for young children are developed. The test materials have high face validity and are applicable for children as young as three years old. The test satisfies the essential components needed for a WRS test. The aforementioned means that the present test retains the most common features of the large number of WRS tests available in different languages. Such features are the type of speech material, the response format, item scored, phonemic balance, and list equivalency.

Currently, there is no available information regarding slopes of PI functions for any other Greek picture-pointing test. Therefore, a direct comparison of PI functions is not possible. However, results of an open-set WRS test for Greek-speaking school-aged children 3–6 years revealed an average slope per decibel of 5.08% in the first and 5.24% in the second list in the rising section of the function between 10 and 25 dBHL [4]. Also, PI functions from another open-set WRS test for Greek-speaking individuals over the age of 12 years presented an average slope of 3.37% for the first, 3.35% for the second, 3.31% for the third, and 3.36% for the fourth list per decibel in the rapidly rising portion of the function between 10 and 30 dBHL [18]. PI functions from analogous closed-set tests in other languages revealed comparable results. An English test by Chermak et al. (1988), the Word Intelligibility by Picture Identification Test, reported an average slope per decibel of 5.65%. Another English test, the Northwestern University Children's Perception of Speech Test, which is presented at a level of 30 dBHL over SRT, reported slopes ranging as high as 3%/dB [9]. A picture-pointing test in the Spanish language [26] presented an increase in recognition ability per decibel of 1.21% (List 1), 1.24% (List 2), and 1.02% (List 3). These values are lower than those from our PI functions because they were calculated based on SRT thresholds, not pure tone thresholds. Thus, some similarity is seen in the slopes of PI functions for both open-set and closed-set tests in Greek and other languages.

The present test consists of two phonemically balanced 50-word lists with low-redundancy bisyllabic words, with each list containing 227 phonemes. As the speech audiometry materials increase in homogeneity, the given speech audiometry test results become more precise [27].

In the present study, we preferred to test the children with full lists (50 words). However, to shorten the testing time, because young children have difficulty maintaining their attention, many clinicians prefer to test with half lists (25 words). An important next step would be to construct four 25-word lists from the above 50-word lists with an effort to maintain the construction criteria and evaluate their equivalency. However, the literature suggests that the variability of a WRS increases as the number of test items decreases and that 50 words are necessary to obtain reliable and valid results [28,29].

The test uses a closed-set response format, with a picture identification task providing a 16.67% chance score more accessible than an open-set format that allows no chance of correct responses through guessing. The closed-set task of this test requires a limited number of comparisons among the six response alternatives. In contrast, an open-set task forces the child to compare the stimulus item to all possible words in lexical memory. In the same line was the study of Clopper et al. [16] that examined the effects of open and closed-set task demands on spoken word recognition. Specifically, Clopper et al. [16] concluded that there is a significant difference in performance between closed-set and open-set word recognition tests and that task difficulty is an important variable to consider when controlling lexical competition. Also, a study [30] reinforces a closed-set domain when assessing speech perception in children. Additionally, the closed-set picture-pointing task can be administered to older children with different deficits, such as severe speech problems or lower cognitive development, impacting scoring accuracy on an open-set test. On the other hand, if the child can cooperate in an open-set format, this test could also be performed as an open-set task. Open-set testing has higher face validity since it is more representative of real-world listening performance.

The WRSs from the present study were acquired through pre-recorded stimuli. However, most audiologists use monitored live voice (MLV) presentation for WRS testing since it allows the audiologist greater flexibility during the evaluation, especially with young children [31]. Although it is easier for younger children to perform the test with MLV presentation since they cannot maintain consistent attention with pre-recorded stimuli, pre-recorded material is recommended over MLV because of the reduced variability offered by recorded test materials [27]. Moreover, a limitation that needs to be referred to is that the sample should not be considered representative of the population to generalize the results. Future studies should take that into account before their design.

5. Conclusions

In conclusion, this study has succeeded in developing a closed-set WRS test for young speakers of Modern Greek. The process included several steps in developing this test, including selecting speech material, digital recordings of speech stimuli, creating and validating digitally illustrated pictures representing the stimuli in the word lists, and establishing list equivalency. This study contributed to establishing a closed-set WRS test in the Greek language that can be valuable in assessing children's speech perception skills. This test can be of value in assessing speech perception skills in preschool children and older children or adults with limited or no Greek proficiency.

Furthermore, the test could have various applications, including diagnosis, research, and (re)habilitation. Moreover, this study can be a guideline for developing speech audiometry materials in other languages. Also, the results from Greek parts of speech development can offer valuable insights with broader implications for understanding language acquisition and development across various languages since the fundamental principles of language acquisition tend to be universal. This comparative approach can enhance our understanding of language development and support the development of more inclusive and globally applicable strategies for speech-language pathology and early intervention programs.

Regarding future direction, further research with a broad spectrum of children, those whose hearing is within the normal range, those with varying types and degrees of hearing impairment, and those with different developmental abilities is required to establish further validity and reliability of the lists. Moreover, an interesting future investigation would be to compare WRSs for significant differences between a closed and an open-set test format.

Author Contributions: Conceptualization, N.T. and K.C.; methodology, N.T.; software, K.C.; validation, N.T., K.C., V.G., F.F. and K.M.; formal analysis, K.C. and V.G.; investigation, K.C. and V.G.; resources, N.T., K.C., V.G., F.F. and K.M.; data curation, N.T., K.C. and V.G.; writing—original draft preparation, N.T., K.C., V.G., F.F., K.M., A.P. and P.P.; writing—review and editing, N.T. and A.P.; supervision, N.T. and P.P. All authors have read and agreed to the published version of the manuscript.

Funding: This research received no external funding.

Institutional Review Board Statement: The study was conducted in accordance with the Declaration of Helsinki and approved by the Departmental Committee of TEI Western Greece (currently University of Patras) with protocol 176/11/2019.

Informed Consent Statement: Informed consent was obtained from all subjects involved in the study. Written informed consent has been obtained from the patients to publish this paper.

Data Availability Statement: Available after request to the corresponding author.

Conflicts of Interest: The authors declare no conflict of interest.

References

1. Davis, J.M.; Elfenbein, J.; Schum, R.; Bentler, R.A. Effects of mild and moderate hearing impairments on language, educational, and psychosocial behavior of children. *J. Speech Hear. Disord.* **1986**, *51*, 53–62. [CrossRef]
2. Mendel, L.L. Current considerations in pediatric speech audiometry. *Int. J. Audiol.* **2008**, *47*, 546–553. [CrossRef] [PubMed]
3. American Speech-Language-Hearing Association. *Determining Threshold Level for Speech*; American Speech-Language-Hearing Association: Rockville, MD, USA, 1988.
4. Trimmis, N.; Papadeas, E.; Papadas, T.; Papathanasopoulos, P.; Gouma, P.; Goumas, P. A Modern Greek Word Recognition Score Test Designed for School Aged Children. *Mediterr. J. Otol.* **2008**, *4*, 1–8.
5. Kim, J.; Lee, J.; Lee, K.W.; Bahng, J.; Lee, J.H.; Choi, C.H.; Park, J. Test-retest reliability of word recognition score using Korean standard monosyllabic word lists for adults as a function of the number of test words. *J. Audiol. Otol.* **2015**, *19*, 68. [CrossRef] [PubMed]
6. Martin, F.N.; Clark, J.G. *Introduction to Audiology*, 8th ed.; Allyn and Bacon: Boston, MA, USA, 2003.
7. Yoon, J.; Lee, J. The test-retest reliability of Korean standard language lists for schoolchildren in speech audiometry. *Audiology* **2015**, *11*, 26–36. [CrossRef]
8. Mendel, L.; Danhauer, J. *Audiologic Evaluation and Management and Speech Perception Assessment*; Mendel, L., Danhauer, J., Eds.; Singular Publishing Company, Inc.: San Diego, CA, USA, 1997.
9. Elliott, L.L.; Katz, D.R. *Northwestern University Children's Perception of Speech:(NU-CHIPS)*; Auditec: St. Louis, MI, USA, 1980.
10. Ross, M.; Lerman, J.A.Y. A picture identification test for hearing-impaired children. *J. Speech Hear. Res.* **1970**, *13*, 44–53. [CrossRef]
11. Jerger, S.; Jerger, J. Pediatric speech intelligibility test: Performance-intensity characteristics. *Ear Hear.* **1982**, *3*, 325–334. [CrossRef]
12. Robbins, A.M.; Renshaw, J.J.; Miyamoto, R.T.; Osberger, M.J.; Pope, M.L. *Minimal Pairs Test*; Indiana University School of Medicine: Indianapolis, IN, USA, 1988.
13. Madell, J. *Behavioral Evaluation of Hearing in Infants and Young Children*; Thieme: New York, NY, USA, 1998.
14. Northern, J.; Downs, M. *Hearing in Children*, 4th ed.; Williams & Wilkins: Baltimore, MD, USA, 1991.
15. Sabo, D.L. The audiologic assessment of the young pediatric patient: The clinic. *Trends Amplif.* **1999**, *4*, 51–60. [CrossRef] [PubMed]
16. Clopper, C.G.; Pisoni, D.B.; Tierney, A.T. Effects of open-set and closed-set task demands on spoken word recognition. *J. Am. Acad. Audiol.* **2006**, *17*, 331–349. [CrossRef] [PubMed]
17. Yu, T.L.J.; Schlauch, R.S. Diagnostic Precision of Open-Set Versus Closed-Set Word Recognition Testing. *J. Speech Lang. Hear. Res.* **2019**, *62*, 2035. Available online: https://link.gale.com/apps/doc/A592138365/HRCA?u=anon~18e4f26e&sid=googleScholar&xid=02291151 (accessed on 16 November 2023). [CrossRef] [PubMed]
18. Trimmis, N.; Papadeas, E.; Papadas, T.; Naxakis, S.; Papathanasopoulos, P.; Goumas, P. Speech audiometry: The development of modern Greek word lists for suprathreshold word recognition testing. *Mediterr. J. Otol.* **2006**, *3*, 117–126.
19. Martin, M. *Speech Audiometry*, 2nd ed.; Whurr Publishers Ltd.: London, UK, 1997.
20. Epstein, A. Speech audiometry. *Otolaryngol. Clin.* **1978**, *11*, 667–676. [CrossRef]
21. Haynes, W.O.; Pindzola, R.H. *Diagnosis and Evaluation in Speech Pathology*; Pearson: London, UK; Allyn and Bacon: Boston, MA, USA, 2008.
22. Papaeliou, C.F.; Rescorla, L.A. Vocabulary development in Greek children: A cross-linguistic comparison using the Language Development Survey. *J. Child. Lang.* **2011**, *38*, 861–887. [CrossRef]
23. Koloutsou, K.; Kikidis, D.; Spanoudakis, G.; Bibas, A.; Nikolopoulos, T. Speech audiometry test with picture-related sentence lists in Modern Greek for partially hearing children. *Hear. Balance Commun.* **2017**, *15*, 187–198. [CrossRef]
24. Shipley, K.G.; McAfee, J.G. *Assessment in Speech-Language Pathology: A Resource Manual*; Plural Publishing: San Diego, CA, USA, 2019.
25. Owens, R.E., Jr. *Language Development: An Introduction*; Allyn & Bacon: Boston, MA, USA, 2001.
26. Mendel, L.L.; Pousson, M.A.; Jones, S.E.; Bass, J.K. Validation of Male Talker Recordings of the Spanish Pediatric Speech Recognition Threshold Test and the Spanish Pediatric Picture Identification Test. *Am. J. Audiol.* **2022**, *31*, 1143–1154. [CrossRef] [PubMed]
27. Wilson, R.H.; Carter, A.S. Relation between slopes of word recognition psychometric functions and homogeneity of the stimulus materials. *J. Am. Acad. Audiol.* **2001**, *12*, 7–14. [CrossRef] [PubMed]
28. Thornton, A.R.; Raffin, M.J.M. Speech-discrimination scores modeled as a binomial variable. *J. Speech Hear. Res.* **1978**, *21*, 507–518. [CrossRef] [PubMed]
29. Raffin, M.J.M.; Thornton, A.R. Confidence levels for differences between speech-discrimination scores. *J. Speech Lang. Hear. Res.* **1980**, *23*, 5–18. [CrossRef] [PubMed]
30. Kirk, K.; Diefendorf, A.O.; Pisoni, D.B.; Robbins, A.M. Assessing speech perception in children. In *Audiologic Evaluation and Managementand Speech Perception Assessment*; Singular Pub. Group: Hong Kong, China, 1997; pp. 101–132.
31. Martin, F.N.; Champlin, C.A.; Chambers, J.A. Seventh survey of audiometric practices in the United States. *J. Am. Acad. Audiol.* **1998**, *9*, 95–104. [PubMed]

brain sciences

MDPI

Review

Update on How to Approach a Patient with Locked-In Syndrome and Their Communication Ability

Kaitlyn Voity [1], Tara Lopez [2], Jessie P. Chan [2] and Brian D. Greenwald [2,*]

[1] Hackensack Meridian School of Medicine, Nutley, NJ 07110, USA; kaitlyn.voity@hmhn.org
[2] JFK Johnson Rehabilitation Institute, Edison, NJ 08820, USA; taralopz@gmail.com (T.L.); jessie.chan@hmhn.org (J.P.C.)
* Correspondence: brian.greenwald@hmhn.org

Abstract: Locked-in syndrome (LIS) is a rare and challenging condition that results in tetraplegia and cranial nerve paralysis while maintaining consciousness and variable cognitive function. Once acute management is completed, it is important to work with the patient on developing a plan to maintain and improve their quality of life (QOL). A key component towards increasing or maintaining QOL within this population involves the establishment of a functional communication system. Evaluating cognition in patients with LIS is vital for evaluating patients' communication needs along with physical rehabilitation to maximize their QOL. In the past decade or so, there has been an increase in research surrounding brain–computer interfaces to improve communication abilities for paralyzed patients. This article provides an update on the available technology and the protocol for finding the best way for patients with this condition to communicate. This article aims to increase knowledge of how to enhance and manage communication among LIS patients.

Keywords: locked-in syndrome; communication after stroke; locked-in syndrome rehabilitation; establishing communication; defects following locked-in syndrome

1. Introduction

LIS was first identified as severe tetraplegia, cranial nerve paralysis, and anarthria, but with persevering awareness and ocular movements. The prevalence and incidence of LIS are not well understood, which is likely due to a combination of rarity as well as common misdiagnosis. A survey of Dutch nursing homes in 2013 found a prevalence of LIS in 0.7 per 10,000 residents [1]. In Norway, all hospitals are required to refer LIS patients to the Norwegian National Unit for Rehabilitation of LIS, therefore creating a comprehensive database. A study utilizing this registry found that in 2021, 51 out of Norway's population of 5.5 million people had complete or incomplete LIS lasting more than 6 weeks [2].

The pathophysiology behind LIS is a vascular injury to the brain stem, specifically the ventral pons or ventral midbrain, while the cerebrum, for the most part, remains unaffected. This type of injury is traditionally caused by ischemia or a hemorrhagic infarction [2,3]. However, traumatic brain injury can also elicit this syndrome; the mechanism of injury is typically blunt or penetrating trauma, leading to an insult to the vertebrobasilar artery blood supply [2].

Classical LIS is described as quadriplegia and anarthria with preserved awareness and vertical eye movements [3]. This constellation of symptoms results from any injury to the ventral pons or the caudal ventral midbrain while the cerebrum remains unaffected [2].

LIS has three main subtypes based on impairments, which vary based on the nature of the offending lesion, but retention of consciousness is present in all subtypes. Isolated conjugate vertical eye movements are pathognomonic for classical LIS and result from lesions affecting structures, as detailed in Table 1. Incomplete LIS is characterized by the addition of voluntary movements or eye movements aside from conjugate up-gaze.

Total LIS is described as total immobility, including loss of vertical eye movements. This literature review will focus on communication methods for patients with classical LIS, but it is worth noting that impairments exist on a spectrum based on the nature of the injury and stage of recovery [3].

Table 1. Structure involved in LIS.

Structure (s)	Function	Status in Classical LIS
CN 3, 4	Vertical eye movement, pupillary reflex, eyelid control data	Intact
CN 6, paramedian pontine reticular formation	Bilateral horizontal gaze	Injured
Corticobulbar tracts (CN 5, 7, 9, 10, 11, 12)	Respiratory function	Variable
Corticospinal tracts	Limb and truncal motor functions	Injured
Medial lemniscus and spinothalamic pathways	Sensation	Intact
Reticular activating system	Arousal, consciousness, awareness	Intact

While arousal and awareness are preserved in LIS, cognitive function is variably impacted by the lesion, as shown in Figure 1. Cognitive function is most impacted immediately post-injury, and impairments in attention, memory, and cognitive endurance are commonly observed acutely [3]. Observational studies have suggested that LIS patients recover reading and oral comprehension, visual recognition, spatial orientation, right–left discrimination, short-term memory, and intellectual functioning [4]. Many of the patients, however, were found to have residual impairments in complex sentence comprehension, mental calculation, problem solving, working memory, mental flexibility, distractibility, impulse control, and executive functioning [4]. Rousseaux et al. speculated that the impairments in higher-level cognitive functioning could be secondary to the cerebellar lesion, which may not have initially been isolated to the ventral pons or resulted in impaired cerebral blood flow. Environmental causes should not be overlooked, as LIS patients often experience chronic low mental and physical stimulation as well as social isolation, which impact cognitive functioning and predispose patients to depression, which can present with attention and memory deficits [4].

LIS is a profoundly challenging condition that significantly impacts an individual's QOL. Characterized by complete paralysis of the body's voluntary muscles while cognitive functions remain intact, locked-in syndrome severely limits communication and physical movement. Patients with this condition can often feel extreme isolation from this condition as they are unable to communicate and have a lack of movement [5]. Not only can this take a toll on the patient, but also on their loved ones. LIS cannot be cured, but there are ways to improve the patient's QOL. As stated before, patients with LIS are usually able to maintain eye movement and preserve cognitive function. A major milestone in an improvement in QOL for these individuals is utilizing the eye gaze as a potential way to communicate. It is extremely important that, in an acute setting, a baseline of the patient's communication ability be explored and appropriately evaluated.

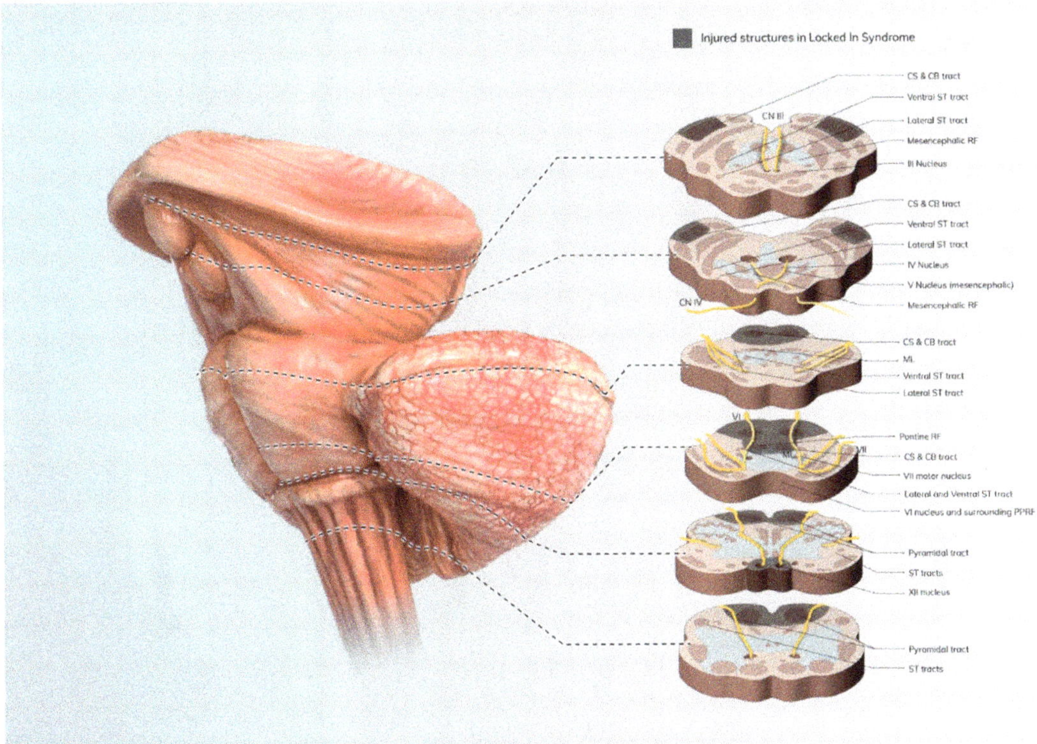

Figure 1. Pathophysiology of LIS. This image shows the structures damaged in LIS, including CS: corticospinal track; CB: corticobulbar track; ST: spinothalamic track; RF: reticular formation; ML: medical lemniscus; PPRF: paramedian pontine reticular formation.

2. Materials and Methods

A literature review was conducted using the databases PubMed and Scopus, covering the period between 2013 and 2023. The search terms used were "Locked-in Syndrome", "augmentative alternative communication", "locked in syndrome rehabilitation", "no technology augmentative alternative communication (AAC)", "LOW TECH AAC", and "HIGH TECH AAC". The Boolean operator "AND" was employed to refine the search and obtain relevant results. After the initial search, articles not written in English, published before 2013, irrelevant to the topic, and in duplicate were excluded. In total, we found 292 papers related to the research topic. A total of 203 were excluded due to the criteria listed above. We found 88 papers to use in our review, as seen in Figure 2. Additionally, three articles were not identified through the original online dataset search but from speech-language pathology articles identified through PubMed.

Data Reporting

Descriptive information was extracted from the articles, including the author, country, and year of publication. We analyzed the study type, the means of communication expressed in the articles, and the etiology of the syndrome. Information regarding the different systems of communication used was then grouped into invasive BCI vs. non-invasive BCI to highlight key features.

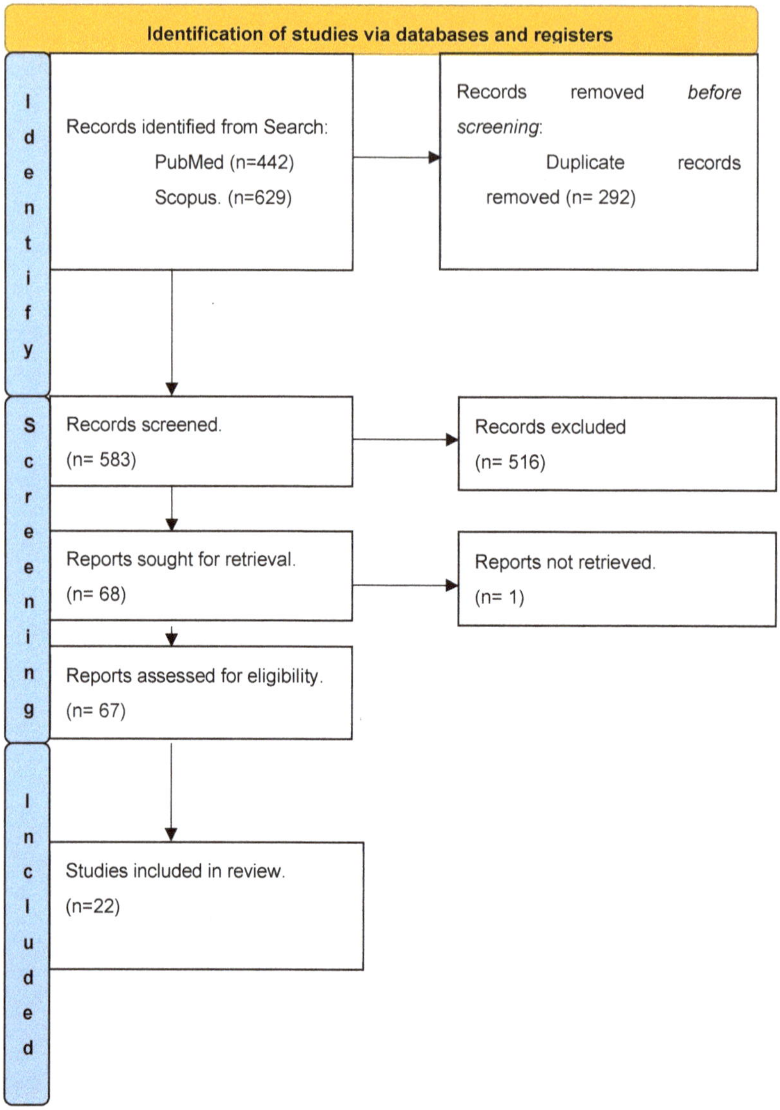

Figure 2. Process flow diagram.

3. Establishing Communication

One of the first steps in this process is confirming the level of consciousness. This involves the observation of patients' behaviors and responses to stimuli in order to determine their ability to follow commands and establish the reliability of their yes/no response. It is important to note any premorbid conditions that may affect an individual's ability to communicate, including age. Once a reliable communication means is obtained or established, the evolution of more intricate and advanced communication methods ranging from no-tech to high-tech augmentative and alternative communication (AAC) can be introduced and explored [6]. However, due to limited or minimal residual muscle activity as a result of LIS, traditional augmentative and alternative communication methods may not be sufficient. Therefore, a brain–computer interface (BCI) could provide patients with

LIS the ability to communicate with the use of solely neural signals versus requiring the use of muscle activity [7].

3.1. Overview of Solutions for Communication

Communication methods can range from no-tech, low-tech, or high-tech augmentative and alternative communication, as seen in Figure 3. No-tech does not require the use of additional materials. Communication is solely obtained through the use of bodily movements such as eye movements (i.e., blinking, looking up–down, or right–left), sign language, facial movements, or extremity movements on command, etc. [6]. When no-tech communication methods are utilized, specific movements have a specific communicative purpose. During communication exchanges, both communication partners must be aware of the individual's methods and means of communication.

Communication Methods:

Augmentative and Alternative Communication (AAC)

No Tech	Low Tech	High Tech
• Eye Movements(i.e. Blinking, vertical or horizontal eye gaze) • Facial Movements • Extremity Movement • Sign Language	• Paper-Pencil Application • Letter boards • Simple Single-Message voice output device(i.e. BigMac) • Eye Transfer(ETRAN) Board)	• Electronic Devices (i.e., ipad,/tablet/computers) • Multi Switch Systems • Brain Computer Interfaces.

Figure 3. Communication methods.

Based on the individual's eye movement and cognitive function, low-tech AAC can be explored. Low-tech AAC can involve the use of materials such as paper and pencil, letter boards, or simple single-message voice output devices (i.e., BigMac) in addition to oneself to functionally communicate. [6]. When an individual with LIS has preserved ocular movements, the use of eye gaze with letter boards (i.e., Eye Transfer (ETRAN) Board or EyeLink Board) can be an effective way to communicate. This is conducted by selecting single letters or clusters of letters via fixation or personal means of selection (i.e., blinking) [8].

Lastly, a more advanced communication method is the use of high-tech AAC. High-tech AAC involves the use of technology to aid in functional communication. This can include external communication devices such as tablets or computers [6]. For individuals with LIS, eye-gaze switches, eye tracking, or eye-gaze devices can be utilized in order to control electronic devices (i.e., tablet/computers) to communicate functionally. In addition to the use of eye-gaze devices, an additional access method includes BCI [9]. BCI allows patients with LIS to communicate by utilizing neural signals in conjunction with current high-tech AAC devices and programs [9].

3.2. Brain–Computer Interface (BCI)

Overview

BCIs involve the use of neural activity instead of muscle activity to aid in functional communication. Once raw neural signals are obtained via invasive or non-invasive means, the BCI can identify and decode signals in order to synthesize a functional message [10]. Examples of a non-invasive BCI would be communication with blinking. BCIs allow individuals with severe motor and cognitive defects to improve their QOL [11]. The goal of this review is to discuss the current research and development on the topic of BCIs and examine the risks and benefits associated with non-invasive vs. invasive methods. In Figure 4, we list the different types of Invasive vs Noninvasive. It is also important to discuss the useability, accessibility, and patient satisfaction associated with these means of communication.

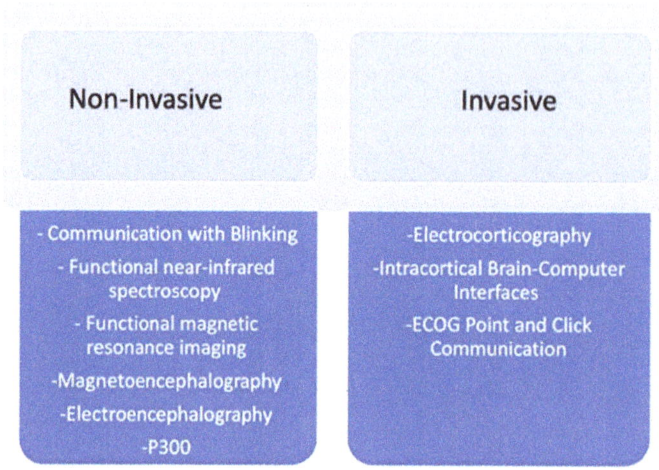

Figure 4. Invasive vs. non-invasive.

3.3. *Non-Invasive BCIs*

LIS presents a profound challenge where individuals retain cognitive function but are almost entirely paralyzed and unable to communicate using conventional means. In recent years, neuroscience and technology have advanced significantly, giving rise to innovative solutions aimed at improving the lives of those affected by LIS. One such avenue of exploration is the development of non-invasive BCIs, which hold the promise of establishing a direct link between brain activity and external devices. These BCIs enable the translation of the users' neurological signals into meaningful commands, potentially allowing them to communicate, control devices, and interact with their environment. In Table 2, we list the specific sources used and what information we took from the article.

3.3.1. Communication through Blinking

Communicating through blinking is a crucial method for individuals with LIS. Ocular movements such as blinking or specific eye movements (i.e., up/down, right/left) become a lifeline, enabling individuals to convey their thoughts, needs, and desires [12]. When communicating with ocular movements, there is a variety of AAC with different styles and capabilities ranging from blinking, the use of transparent panels with numbers and/or letters (i.e., Eye Transfer Boards), eye-gaze sensing screens, eye tracking systems, and the use of modified switches to control different applications presented on devices [8].

Table 2. Non-invasive BCI references.

Reference	Communication Involved
Kopsky et al. (2014) (Disabil Rehabil, 36(20), 1723–1727, doi:10.3109/09638288.2013.866700) [12].	Through utilizing the maintained eye movement in LIS, a specialized spelling system was tried to see if it helped make communication with blinking more user-friendly.
Park et al. (2012) (Ann Rehabil Med, 36(2), 268–272, doi:10.5535/arm.2012.36.2.268) [13].	After eye blinking therapy (AAC), there was a benefit seen in patients' communication ability.
Luo et al. (2022) (Neurother J Am Soc Exp Neurother, 19(1), 263–273, doi:10.1007/s13311-022-01190-2) [10].	With the use of EEG, brain activity was able to be tracked before and during communication to see the areas most involved.
Vansteensel et al. (2022) (Neurorehabil Neural Repair, 36(10–11), 666–677, doi:10.1177/15459683221125788) [11].	fNRI and fMRI are both emerging technologies in the area of communication, but the one downside is implanting them in daily life. They allow for the tracking and amplification of neural signals.
Butler et al. (2020) (Am J Speech Lang Pathol, 29(3), 1674–1701, doi:10.1044/2020_AJSLP-19-00050) [14].	fNRI may not only help with augmenting communication but also show promise in assisting with diagnosing communication ability.
Lugo et al. (2016) (Front Hum Neurosci, 10, 569, doi:10.3389/fnhum.2016.00569) [15]	P300 allows for better diagnosis in baseline communication ability for patients with LIS while looking at event potentials when presented with a task or stimuli.

Typically, it has been found that within the LIS population, there are preserved voluntary movements including eye blinking and vertical eye movements. When communicating, an example of the use of eye blinking is when an individual blinks once for no and twice for yes. Another common example is prolonged eye for yes [13]. Although simplistic, eye movement can be an essential method of communication [13]. While establishing an effective communication modality, it is important to keep in mind that all communication methods must be individualized to suit the strengths and reliability of the individual. Using a system of blinks and eye movements complemented by caregivers and/or devices, individuals with LIS can spell out words, select options on communication boards, and answer yes-or-no questions. This method, while slow compared to typical communication, empowers individuals with LIS to bridge the gap between their inner thoughts and the outside world, providing them with a means to express themselves and maintain vital connections with their loved ones and caregivers. It exemplifies the remarkable resilience and adaptability of the human spirit in the face of profound physical limitations.

In addition to blinking, external switches in conjunction with a switch interface can be an effective way to communicate. With this method, the switch interface receives input and directly relays it to the AAC/program/computer/toy to create the intended message or action. External switches can be attached to external wearable items, such as glasses, to be activated with the use of blinking [13].

More complex AAC involves eye-gaze technology such as screen-based eye sensors, eye tracking systems, and modified external switches to control applications on devices. Eye-gaze programs may require an individual to make a selection as the program systematically scans different potential icons and messages (i.e., when the system is scanning, items could be highlighted or bolded when toggling from icon to icon). An additional method may require the individual to blink when gazing at their target in order to make a selection [13]. Eye tracking software integrated with speech-generating devices such as Tobii Dynavox can provide individuals with an effective form of communication.

3.3.2. Electroencephalography (EEG)

Electroencephalography (EEG) is a valuable tool for assessing brain activity and communication possibilities by recording electrical activity along the scalp, providing insights into brain function. Despite its relatively lower spatial resolution compared to other imaging techniques, EEG offers high temporal resolutions, making it suitable for real-time communication applications [10]. BCIs that utilize EEG signals have been developed to enable locked-in individuals to control external devices, spell words, or answer yes-or-no questions by modulating their brainwave patterns. While the reliability and accuracy of EEG-based BCIs can be influenced by factors like signal noise and individual variability, ongoing advancements in signal processing algorithms and hardware design enhance performance. These developments hold promise for improving the QOL for individuals with severe motor disabilities, highlighting the potential of EEG-based BCIs in the fields of neuro communication and assistive technology.

3.3.3. Emerging Technologies

Functional Near-Infrared Spectroscopy

Functional near-infrared spectroscopy (fNIRS) is a non-invasive tool that uses near-infrared light to monitor brain oxygen levels, helping to assess brain activity in patients with LIS. It detects brain activity in specific regions associated with mental tasks, enabling patients to answer questions or spell words using deliberate neural activations [14]. However, its effectiveness depends on the individual's condition and responsiveness, as well as their ability to use the technology. fNIRS monitors levels of oxygenated hemoglobin, which indicate increased brain activity in the region [11]. fNIRS has emerged as a promising tool for communication; however, more research is needed prior to clinical application.

Functional Magnetic Resonance Imaging

Functional magnetic resonance imaging (fMRI) is a neuroimaging technique that has gained substantial recognition in LIS research. This technology operates by harnessing robust magnetic fields and radio waves to gauge alterations in blood flow, providing valuable insight into brain activity [16]. In the context of LIS, fMRI may have immense potential as a means of facilitating communication when used in conjunction with a BCI. It discerns unique patterns of brain activity associated with various cognitive tasks, thereby empowering individuals to express their intentions and responses effectively. Nevertheless, it is worth noting that the efficacy of fMRI with BCI as a communication tool can be influenced by the individual's capacity to consistently generate distinct brain activation patterns. Furthermore, fMRI with BCI necessitates a highly controlled environment, which may hinder its portability and accessibility for real-time communication applications [11]. As researchers continue to explore the capabilities of fMRI with BCI as a communication tool to advance QOL, it is imperative to consider these limitations.

Magnetoencephalography

Magnetoencephalography (MEG) is a neuroimaging technique that records the magnetic fields generated by neuronal activity, offering exceptional temporal resolution and precise localization of brain regions engaged in diverse cognitive processes. MEG's unique capability to capture rapid neural dynamics makes it highly valuable in the context of detecting intention and communication efforts among individuals with LIS. This is achieved through an analysis of the timing and spatial distribution of neural signals, allowing researchers to discern patterns linked to specific mental tasks or responses [11]. However, the practical application of MEG is constrained by factors such as susceptibility to movement artifact and the requirement for specialized equipment and controlled environments. Despite these limitations, MEG holds significant promise as a tool for advancing our understanding of brain function in LIS and potentially facilitating communication strategies for affected individuals.

P300

The P300 event-related potential, a distinctive neural response occurring after the presentation of a relevant stimulus, represents a promising avenue for enhancing communication among individuals with LIS. With this method, individuals are exposed to visual or auditory stimuli, and their brains generate a P300 waveform when they recognize a specific target stimulus, such as a name or designated word [15]. The P300-based BCI approach offers an efficient and non-intrusive channel for communication among locked-in individuals. By directing their attention towards the desired option, they can trigger a detectable P300 response, which can be monitored using EEG or other neuroimaging techniques. While this approach necessitates intact cognitive abilities and sustained attention, it circumvents the need for precise motor control, rendering it particularly well suited for individuals with limited physical mobility.

Summary of Risks and Benefits of Non-Invasive BCIs

Non-invasive BCIs face the challenge of detecting and interpreting weak and variable neural signals, which can result in potential inaccuracies in translating intention into action. This necessitates calibration and training periods that demand time and patience from users. Moreover, the usability and effectiveness of non-invasive BCIs exhibit substantial variability contingent upon an individual's cognitive and physical capabilities, potentially restricting their applicability to specific subsets of locked-in patients. Nevertheless, the upside of non-invasive BCIs is their immense potential to significantly enhance QOL. These interfaces empower users by allowing them to spell words, make selections, and manipulate objects solely through cognitive processes, which allow for improved communication and autonomy. This can mitigate feelings of isolation and frustration, thereby promoting emotional well-being and fostering a heightened sense of agency. Striking a balance between these potential risks and rewards necessitates a continuous cycle of innovation, close collaboration among researchers, clinicians, and patients, and consideration of each individual's unique circumstances and requirements.

3.4. *Invasive*

Invasive BCIs involve the implantation of electrodes directly into the brain and offer a remarkable avenue for establishing a direct and precise link between neural signals and devices. In Table 3, we list the specific sources used and what information we took from the article.

Table 3. Invasive BCI references.

Reference	Communication Involved
Bacher et al. (2015) (Neurorehabil Neural Repair, 29(5), 462–471, doi:10.1177/1545968314554624) [17].	In the clinical trial of the BrainGate Neural Interface System, this system, featuring the BrainGate Radial Keyboard, was compared to a standard QWERTY keyboard. The results showed a significant improvement in communication ability. The participant effectively used this interface for face-to-face communication with research staff through text-to-speech conversion and remote communication via an internet chat application.
Homer et al. (2013) (Annu Rev Biomed Eng, 15, 383–405, doi:10.1146/annurev-bioeng-071910-124640) [18].	Central components of intracortical BCIs include implanted sensors capturing neural signals and the decoding software extracting the user's intended movements from these signals. These innovations can enhance the technology's capacity, precision, and durability.

3.4.1. Electrocorticography

Electrocorticography (ECoG) emerges as a promising neuroimaging technique with substantial potential for communication. EcoG involves the strategic placement of electrodes directly on the brain's surface, affording the capability to record electrical activity with heightened spatial and temporal precision compared to non-invasive methodologies. This advantage translates into a more refined ability to pinpoint neural activity patterns associated with intention and cognition, facilitating the identification of specific brain regions and their activations. Consequently, EcoG holds the promise of substantially enhancing the accuracy of decoding an individual's thoughts and intentions. However, the utilization of EcoG in the context of LIS requires careful consideration. The procedure necessitates the surgical implantation of electrodes, which introduces potential risks and mandates meticulous monitoring and post-operative care. Additionally, the interpretation of EcoG signals is inherently complex and necessitates the application of advanced signal processing techniques to distill meaningful information. Nonetheless, there exists optimism in the realm of machine learning and computational methods, offering avenues for improving the accuracy and usability of EcoG-based BCIs.

ECOG Point and Click Communication

In this study, participants who were already enrolled in the Brain Gate program, which involves the implantation of electrodes, were examined to investigate the differences between a standard QWERTY keyboard and one developed by Brain Gate. The development known as Electrocorticography Point and Click Communication represents a noteworthy advancement within the domain of BCIs designed specifically for individuals with locked-in syndrome. This innovative approach capitalizes on the exceptional spatial and temporal resolution capabilities of EcoG to enable users to exert control over external devices through their brain signals, effectively translating their cognitive intentions into actionable commands [17]. The EcoG Point and Click Communication paradigm entails the placement of electrode arrays directly onto the surface of the brain, facilitating the detection of neural patterns linked to specific intentions, such as maneuvering a cursor to select items on a screen. The precision afforded by EcoG technology empowers users to manipulate a cursor with remarkable accuracy, enabling a point-and-click interaction akin to the traditional usage of computer mice. By concentrating their attention on various on-screen options, users can elicit distinct neural responses, which are subsequently decoded by sophisticated algorithms to initiate desired actions. This method equips individuals with the ability to engage in tasks such as word composition, message creation, or navigation of computer interfaces without necessitating physical movement, thus representing a significant leap forward in assistive technology for the LIS population.

3.4.2. Intracortical

Intracortical BCIs epitomize a cutting-edge and invasive approach with substantial potential to empower individuals with LIS. This technique involves the surgical implantation of electrodes directly into the cerebral cortex, affording the capability to meticulously and instantaneously record neural activity [17]. This technology harbors the potential to establish a direct conduit between cognition and external devices, thereby enabling communication and the manipulation of surroundings. The intracortical BCI method provides an unparalleled level of spatial and temporal resolution, facilitating the detection of specific neural activity patterns associated with distinct intentions [18]. This granular level of control opens up avenues for more sophisticated forms of communication, including intricate speech generation and the dexterous operation of robotic prosthetics closely mirroring natural limb movements. Such precision holds the promise of substantially elevating QOL for individuals with LIS, endowing them with a notable degree of agency and independence.

Summary of Risks and Benefits of Invasive BCI's

Invasive BCIs present both remarkable benefits and notable risks in the context of locked-in syndrome. With the unique combination of intact cognition and severe motor impairment, locked-in syndrome necessitates innovative solutions to bridge the gap between the mind and the body. Invasive BCIs involve the implantation of electrodes directly into the brain, offering a direct and precise link between neural signals and devices. Through the placement of electrodes on the brain's surface, EcoG provides heightened spatial and temporal precision, enhancing the accuracy of decoding an individual's thoughts and intentions. This approach can facilitate communication, interaction with technology, and the restoration of a degree of agency and independence. However, the surgical implantation of EcoG electrodes introduces potential medical risks and necessitates meticulous post-operative care.

Intracortical BCIs, on the other hand, offer unparalleled spatial and temporal resolution by implanting electrodes directly into the cerebral cortex. This technology allows for the detection of specific neural activity patterns, offering the potential for intricate forms of communication and precise control over external devices. While this precision can significantly enhance the QOL of individuals with LIS, the procedure is highly invasive, posing surgical risks and raising ethical considerations.

In conclusion, the use of invasive and intracortical BCIs in the context of LIS carries the potential for groundbreaking benefits, including improved communication and autonomy. However, the invasive nature of these procedures introduces medical risks and ethical complexities. The decision to pursue invasive BCIs should be made on a case-by-case basis, considering the individual's specific circumstances and weighing the potential benefits against the inherent risks. Continued research and technological advancements are essential to optimize the safety and effectiveness of these approaches.

4. Limitations: Affordability, Usability, Access, Grants, Insurance, Social Worker

Despite the ability of BCIs to potentially decode neural signals to synthesize functional messages, there are a variety of limitations that can negatively impact the success and reliability of the device. These limitations can include usability, efficiency, reliability, access, and cost. Clinicians must also advocate for patients to obtain the right level of assistance.

According to Elliot et al., individuals who utilized BCIs for communication exhibited slower and less accurate responses with decreased usability compared to individuals who utilized different AAC methods (i.e., eye-tracking cameras and EyeLink board). It was found that BCI required a higher cognitive workload compared to the use of an eye-tracking camera and the EyeLink board. As a result of these findings, it was found that BCI was not superior in reference to measures of speed, workload, and usability compared to other AAC means [19]. In addition, due to the emotional and cognitive involvement required for individuals and caregivers to learn and navigate BCI, individuals often resort to different communication systems in order to communicate faster [20].

Currently, within the field, there is a limited standardized clinical pathway regarding formal assessment, selection, administration, and funding for clinical use of BCI devices [20]. A lack of standardization and consistency between AAC research and BCI procedures has negatively impacted the implementation of BCI in a functional setting [20,21]. Additional limitations, such as increased setup time, slow speed of use, and the requirement of technology support for caregivers, result in limited implementation of BCI systems as primary means of communication outside of the research setting [19]. Pitt et al. found that clinicians utilizing AAC have limited interest in the use of BCIs in clinical settings and have reported poor reliability with BCIs. In addition, the increased time and knowledge for setup required to utilize the device negatively and impact useability [20].

5. Caregiver Burden

Caring for an individual with LIS can yield a variety of new challenges. Caregivers can often be impacted by a variety of stressors, such as limited or lack of communication with the patient, limited support from loved ones, a large burden of daily responsibilities, a lack of resources for support and information, financial stressors, role change, or isolation [22]. Of note, Zinkevich et al. found that caregiver burden is significantly reduced with the implementation of successful AAC interventions.

6. Conclusions

LIS is a rare and debilitating condition characterized by severe motor impairment while retaining cognitive function and awareness. The pathophysiology involves brainstem injury, often due to vascular complications or traumatic brain injury. The presentation of LIS can vary, with classical LIS being the most well-known subtype, characterized by quadriplegia, anarthria, and preserved vertical eye movements. Cognitive impairments can vary but are often observed, affecting attention, memory, and higher-level cognitive functions. Communication is a critical aspect of improving the QOL of individuals with LIS, and various invasive and non-invasive methods have been explored. Non-invasive BCIs offer communication methods with fewer risks but can present challenges related to signal detection and interpretation, usability, and standardization. Invasive BCIs, such as ECoG and intracortical interfaces, provide a more direct and precise connection between neural signals and external devices, offering the potential for advanced communication and control. While BCIs hold great potential, the use of BCIs is limited in terms of affordability, usability, access, and caregiver burden. The fields of AAC and BCI research are evolving rapidly, and ongoing efforts are needed to address these limitations and make these technologies more accessible and effective for individuals with LIS. In caring for individuals with LIS, it is crucial to consider not only the technological advancements but also the emotional and social aspects of their lives. Caregiver burden can be significant, but the implementation of AAC interventions has shown promise in reducing this burden and improving the overall QOL for both patients and caregivers. By continuing research and innovation in communication technologies, there is hope for enhancing the lives of those affected by LIS and providing them with the means to engage more fully with the world around them.

Author Contributions: Conceptualization K.V., T.L., J.P.C. and B.D.G.; methodology, K.V.; writing-original draft preparation, K.V., T.L. and J.P.C.; writing-review and editing, K.V., T.L., J.P.C. and B.D.G.; supervision, B.D.G. All authors have read and agreed to the published version of the manuscript.

Funding: This research received no external funding.

Institutional Review Board Statement: Not applicable.

Informed Consent Statement: Not applicable.

Acknowledgments: Ana Bigio, Ally Ferber.

Conflicts of Interest: The authors declare no conflicts of interest.

References

1. Kohnen, R.F.; Lavrijsen, J.C.M.; Bor, J.H.J.; Koopmans, R.T.C.M. The prevalence and characteristics of patients with classic locked-in syndrome in Dutch nursing homes. *J. Neurol.* **2013**, *260*, 1527–1534. [CrossRef] [PubMed]
2. M Das, J.; Anosike, K.; Asuncion, R.M.D. Locked-in Syndrome. In *StatPearls*; Statpearls Publishing: Treasure Island, FL, USA, 2023. Available online: http://www.ncbi.nlm.nih.gov/books/NBK559026/ (accessed on 6 November 2023).
3. Farr, E.; Altonji, K.; Harvey, R.L. Locked-In Syndrome: Practical Rehabilitation Management. *PMR* **2021**, *13*, 1418–1428. [CrossRef]
4. Rousseau, M.-C.; Baumstarck, K.; Alessandrini, M.; Blandin, V.; Billette de Villemeur, T.; Auquier, P. Quality of life in patients with locked-in syndrome: Evolution over a 6-year period. *Orphanet J. Rare Dis.* **2015**, *10*, 88. [CrossRef] [PubMed]
5. Corallo, F.; Bonanno, L.; Buono, V.L.; De Salvo, S.; Rifici, C.; Pollicino, P.; Allone, C.; Palmeri, R.; Todaro, A.; Alagna, A.; et al. Augmentative and Alternative Communication Effects on Quality of Life in Patients with Locked-in Syndrome and Their Caregivers. *J. Stroke Cerebrovasc. Dis.* **2017**, *26*, 1929–1933. [CrossRef] [PubMed]
6. Elsahar, Y.; Hu, S.; Bouazza-Marouf, K.; Kerr, D.; Mansor, A. Augmentative and Alternative Communication (AAC) Advances: A Review of Configurations for Individuals with a Speech Disability. *Sensors* **2019**, *19*, 1911. [CrossRef] [PubMed]
7. Vansteensel, M.J.; Jarosiewicz, B. Brain-computer interfaces for communication. In *Handbook of Clinical Neurology*; Elsevier: Amsterdam, The Netherlands, 2020; Volume 168, pp. 67–85. [CrossRef]
8. Ezzat, M.; Maged, M.; Gamal, Y.; Adel, M.; Alrahmawy, M.; El-Metwally, S. Blink-To-Live eye-based communication system for users with speech impairments. *Sci. Rep.* **2023**, *13*, 7961. [CrossRef] [PubMed]
9. Brumberg, J.S.; Pitt, K.M.; Mantie-Kozlowski, A.; Burnison, J.D. Brain–Computer Interfaces for Augmentative and Alternative Communication: A Tutorial. *Am. J. Speech-Lang. Pathol.* **2018**, *27*, 1–12. [CrossRef] [PubMed]
10. Luo, S.; Rabbani, Q.; Crone, N.E. Brain-Computer Interface: Applications to Speech Decoding and Synthesis to Augment Communication. *Neurotherapeutics* **2022**, *19*, 263–273. [CrossRef] [PubMed]
11. Vansteensel, M.J.; Branco, M.P.; Leinders, S.; Freudenburg, Z.F.; Schippers, A.; Geukes, S.H.; Gaytant, M.A.; Gosselaar, P.H.; Aarnoutse, E.J.; Ramsey, N.F. Methodological Recommendations for Studies on the Daily Life Implementation of Implantable Communication-Brain–Computer Interfaces for Individuals with Locked-in Syndrome. *Neurorehabilit. Neural Repair* **2022**, *36*, 666–677. [CrossRef] [PubMed]

12. Kopsky, D.J.; Winninghoff, Y.; Winninghoff, A.C.M.; Stolwijk-Swüste, J.M. A novel spelling system for locked-in syndrome patients using only eye contact. *Disabil. Rehabil.* **2014**, *36*, 1723–1727. [CrossRef] [PubMed]
13. Park, S.-W.; Yim, Y.-L.; Yi, S.-H.; Kim, H.-Y.; Jung, S.-M. Augmentative and Alternative Communication Training Using Eye Blink Switch for Locked-in Syndrome Patient. *Ann. Rehabil. Med.* **2012**, *36*, 268–272. [CrossRef] [PubMed]
14. Butler, L.K.; Kiran, S.; Tager-Flusberg, H. Functional Near-Infrared Spectroscopy in the Study of Speech and Language Impairment Across the Life Span: A Systematic Review. *Am. J. Speech-Lang. Pathol.* **2020**, *29*, 1674–1701. [CrossRef]
15. Lugo, Z.R.; Quitadamo, L.R.; Bianchi, L.; Pellas, F.; Veser, S.; Lesenfants, D.; Real, R.G.L.; Herbert, C.; Guger, C.; Kotchoubey, B.; et al. Cognitive Processing in Non-Communicative Patients: What Can Event-Related Potentials Tell Us? *Front. Hum. Neurosci.* **2016**, *10*, 569. [CrossRef] [PubMed]
16. Vansteensel, M.J.; Pels, E.G.M.; Bleichner, M.G.; Branco, M.P.; Denison, T.; Freudenburg, Z.V.; Gosselaar, P.; Leinders, S.; Ottens, T.H.; Boom, M.A.V.D.; et al. Fully Implanted Brain–Computer Interface in a Locked-In Patient with ALS. *N. Engl. J. Med.* **2016**, *375*, 2060–2066. [CrossRef] [PubMed]
17. Bacher, D.; Jarosiewicz, B.; Masse, N.Y.; Stavisky, S.D.; Simeral, J.D.; Newell, K.; Oakley, E.M.; Cash, S.S.; Friehs, G.; Hochberg, L.R. Neural Point-and-Click Communication by a Person With Incomplete Locked-In Syndrome. *Neurorehabilit. Neural Repair* **2015**, *29*, 462–471. [CrossRef]
18. Homer, M.L.; Nurmikko, A.V.; Donoghue, J.P.; Hochberg, L.R. Sensors and Decoding for Intracortical Brain Computer Interfaces. *Annu. Rev. Biomed. Eng.* **2013**, *15*, 383–405. [CrossRef] [PubMed]
19. Elliott, C.; Sutherland, D.; Gerhard, D.; Theys, C. An Evaluation of the P300 Brain–Computer Interface, EyeLink Board, and Eye-Tracking Camera as Augmentative and Alternative Communication Devices. *J. Speech Lang. Hear. Res.* **2022**, *65*, 4280–4290. [CrossRef]
20. Pitt, K.M.; Brumberg, J.S. Evaluating person-centered factors associated with brain–computer interface access to a commercial augmentative and alternative communication paradigm. *Assist. Technol.* **2022**, *34*, 468–477. [CrossRef] [PubMed]
21. Pitt, K.M.; Brumberg, J.S.; Pitt, A.R. Considering Augmentative and Alternative Communication Research for Brain-Computer Interface Practice. *Assist. Technol. Outcomes Benefits* **2019**, *13*, 1–20.
22. Zinkevich, A.; Uthoff, S.A.K.; Wirtz, M.A.; Boenisch, J.; Sachse, S.K.; Bernasconi, T.; Feldhaus, M.; Ansmann, L. Burden of informal caregivers of people without natural speech: A mixed-methods intervention study. *BMC Health Serv. Res.* **2022**, *22*, 1549. [CrossRef]

brain sciences

MDPI

Case Report

High-Frequency Language Therapy with Semantic Feature Analysis (SFA) and Transcranial Direct Current Stimulation (tDCS): A Longitudinal Single-Case Report of Semantic Variant of Primary Progressive Aphasia (svPPA)

Katharina Strunk [1,*], Sabine Weiss [1,2,3] and Horst M. Müller [1]

1 Experimental Neurolinguistics Group, Bielefeld University, Universitätsstrasse 25, 33615 Bielefeld, Germany; sabine.weiss@uni-bielefeld.de (S.W.); horst.mueller@uni-bielefeld.de (H.M.M.)
2 Center for Cognitive Interaction Technology (CITEC), Bielefeld University, Inspiration 1, 33619 Bielefeld, Germany
3 Clinical Linguistics, Bielefeld University, Universitätsstrasse 25, 33615 Bielefeld, Germany
* Correspondence: katharina.strunk@uni-bielefeld.de

Abstract: Background: The goal of this study was to investigate whether the combination of semantic feature analysis (SFA) and transcranial direct current stimulation (tDCS) is effective in treating word retrieval in the semantic variant of primary progressive aphasia (svPPA) and how long the potential effects last. Methods: A 56-year-old woman diagnosed with frontotemporal dementia (FTD) and svPPA participated in this longitudinal single-subject design. A total of four 2-week stimulation phases were conducted over a 14-month period, each of which was started depending on the participant's language performance. Follow-up testing was conducted shortly after the stimulation period, approximately 2 weeks, and approximately 4 weeks thereafter. Results: Significant improvement in word retrieval occurred after SFA and tDCS therapy. Two weeks after the end of each stimulation phase, approx. 80% of the trained words could be named correctly. For the untrained words, also significantly more words were correctly named at follow-ups compared to the baseline. Furthermore, the Boston Naming Test (BNT) demonstrated a significant increase in naming performance and showed that phonological cues facilitated word retrieval compared to semantic cues. Conclusion: The combination of SFA and tDCS was able to counteract the expected language deterioration of a participant with svPPA. This effect increased until approximately 2 weeks after each intervention. In addition, a generalization of the effect to untrained words was shown.

Keywords: semantic variant of primary progressive aphasia (svPPA); semantic feature analysis (SFA); transcranial direct current stimulation (tDCS); language

1. Introduction

Frontotemporal dementia (FTD) belongs to the neurodegenerative dementias and is characterized by progressive deficits in behavior, executive functions, and language. A distinction is made between the frontal or behavioral variant of FTD (bvFTD) and primary progressive aphasia (PPA). PPA can be further subdivided into the non-fluent agrammatic variant (nfvPPA), the logopenic variant (lvPPA), and the semantic variant (svPPA) (for reviews see [1–3]).

The semantic variant of PPA (svPPA) is characterized primarily by language impairment at the level of single-word comprehension and word retrieval, especially for low-frequency words. Surface dyslexia and dysgraphia can often be observed. Intact repetition and articulation with concomitant already pronounced semantic deficits are other features of svPPA. Speech production is usually grammatically correct and only isolated paragrammatical errors occur [1]. Semantic knowledge is increasingly impaired, as reflected in impaired recognition of objects and faces. The semantic deficits are evident for all categories of concepts (animals, fruits, etc.) and modalities. In addition, early onset behavioral changes such as loss of empathy and the development of compulsions may occur [1,3,4]. Other language domains usually remain relatively intact, at least at the early stages of the disease [1].

To counteract the degradation of semantic memory and conceptual knowledge, several promising therapeutic procedures have been developed and tested (for an overview see [5,6]). Due to the common prevalence of naming difficulties, the focus is on the study and therapy of lexical retrieval (for an overview see [7,8]). In general, therapeutic interventions to improve lexical semantics are based on either semantic (e.g., description of perceptual features, purpose) or phonological (e.g., initial phoneme or syllable) approaches. However, there are also combined or multimodal approaches [9–13]. Other therapeutic interventions aim at promoting strategies of self-control and self-directed cueing by asking the patient to activate residual semantic and phonematic but also orthographic knowledge about words that cannot be initially retrieved [5,14,15]. A recent comprehensive review found that methods for treating lexical retrieval and improving functional communication strategies are at the forefront of current practice guidelines. Moreover, lexical retrieval treatment appears to be particularly effective in the early stage of svPPA [16].

A well-known and successful approach in this context is semantic feature analysis (SFA), which is suitable for the therapeutic activation of conceptual knowledge loss. The semantic feature approach is based on the fact that parts of the distributed semantic networks still function initially [17]. The goal of SFA is to systematically activate semantic networks so that semantic features and concepts can be linked in a structured way, facilitating access to the meaning of a concept [18]. Here, "concept" is understood as an organized structure of semantic features that confer meaning. Accordingly, a concept can have different semantic features [18]. The structured procedure of SFA is also suitable for further transfer to spontaneous speech and should be used consistently to promote communicative skills [19]. Studies that used SFA or a semantic-based therapy to treat svPPA patients almost always achieved improvements in naming trained items, whereas generalization to untrained items was not guaranteed (for extensive overviews see [5,6,20]). Thus, semantic knowledge of trained items seems to last even several months whereas near transfer to untrained items and far transfer to communicative skills in everyday life occur less frequently and are, therefore, often not found [6] (for a review see [5]). However, near transfer to untrained items is more likely in milder cases of svPPA as long as semantic knowledge is available to support word retrieval [7,21]. The question of how long improvement in naming ability can be achieved by therapeutic intervention has been investigated in some studies [20]. Sustained improvement has been observed between one and twelve months [5]. However, these effects diminish over time. For example, after three to six months, only 10–65% of trained words could be named correctly [6].

Some recent studies show that the short-term and long-term effects of semantic therapy are enhanced and supported when non-invasive brain stimulation is performed simultaneously. Notably, the generalization of therapeutic effects to untrained items appears to be more frequent and stable (for a review see [20]). For instance, in the therapeutic intervention of PPA, comparatively inexpensive and safe transcranial direct current stimulation (tDCS) showed promising results (for overviews see [20,22–24]). Here, anodal tDCS increases the excitability of cortical neurons while cathodal simulation decreases it [25]. This may enhance the effect of speech and language therapies (for recent reviews see [16,20,24,26,27]).

The effect of tDCS on language processing was demonstrated in a larger group of 12 svPPA patients [28]. Subjects performed a computerized semantic matching task following a double-blind, sham-controlled, balanced crossover design. However, they performed this task only after the electrical stimulation. The results showed that offline anodal stimulation of the left anterior temporal lobe led to an improvement in semantic accuracy. These results suggest a specific functional improvement within the semantic system by tDCS compared to sham stimulation in svPPA patients, which has also been confirmed in another study [28,29]. A further study investigated picture naming abilities with anodal tDCS/sham stimulation to the left inferior parieto-temporal region in two individuals with svPPA with a double-blind crossover design [30]. They found approx. 50% improvement in naming trained items after tDCS compared with 20% improvement during therapy combined with sham stimulation immediately after 10 days of stimulation. At follow-up two weeks after the end of therapy, the two svPPA patients showed only a 7% decrease in naming trained items. In another study, three patients with svPPA completed a modified SFA training combined with anodal tDCS to the left temporo-parietal cortex [31]. Two weeks after therapy, naming accuracy increased by an average of 20% for trained items, whereas it did not change for untrained items. Naming accuracy six months after stimulation decreased by 45% for trained items and by 60% for untrained items. Thus, no generalization effects of the modified SFA therapy in combination with tDCS on untrained items could be demonstrated. In addition, no conclusions could be drawn about the respective effects of the language therapy and tDCS, since no sham condition was included [31].

Taken together, compared to patients with other forms of PPA, individuals with svPPA appear to benefit less from speech and language therapy in terms of generalization to untrained items, as they progressively lose semantic knowledge and their semantic system is disrupted. A smaller effect of transcortical electrical stimulation on word retrieval and semantic knowledge as well as smaller long-term effects than in other PPA patients or no effects at all are observed. Thus, semantic-based therapy in combination with tDCS seems to be a promising method to maintain the language abilities of svPPA patients over a longer period of time. There was evidence of improvements in naming both trained and untrained objects, presumably by learning and applying a strategy to compensate for semantic constraints [20].

In a pilot study with the svPPA participant of the present study, we investigated whether anodal tDCS over the left anterior temporal region had an effect on her speech and language performance compared with sham stimulation [32]. The participant was treated with individualized language training in combination with tDCS or sham stimulation in an ABAB design (A = sham). This double-blind study showed that the participant was able to complete 25% more language tasks correctly during the weeks with concurrent tDCS stimulation than during the training weeks in which sham stimulation was administered [32]. From this pilot study, we concluded that tDCS had a significant impact on this participant's language performance. However, the results also showed that semantic knowledge decreased slightly during the eight-month intervention, although access to the semantic system was still largely preserved. Three months elapsed between the end of this pilot study and the start of our current study.

Therefore, the primary aim of the present study was to investigate whether personalized semantic feature analysis (SFA) with concurrent tDCS at the left temporal pole would lead to improvements in naming and comprehension in this participant. To this end, we conducted a longitudinal study with several follow-up tests over 14 months. We postulated that word retrieval would remain stable or even improve over this period due to the combined therapy, although literature reporting a 20–30% decline in performance over this period if no language and speech therapy is provided [33].

The second aim was to investigate whether generalization to untrained material could be achieved and whether these items would also benefit from the present therapeutic intervention. In addition, it was to be investigated how long a possible effect lasted after the training weeks.

The third objective was to investigate the time frame in which possible effects of this intervention can be detected and how long they last. Based on the evidence from the literature, we assumed that the effect should last at least 14 days [28,31].

The fourth objective was to examine whether SFA combined with tDCS had an effect on the participant's connected speech and communicative skills. Some studies showed that the information content of language increased after training, e.g., [34], but others found no change, e.g., [35].

2. Method

2.1. Design

In a longitudinal single-subject design, the combination of SFA and tDCS was investigated over a period of 14 months.

At the beginning of the study, pre-testing with different neurolinguistic and neuropsychological test procedures was conducted. This was followed by four stimulation phases (S1–S4) in which SFA was performed simultaneously with tDCS on 10 consecutive working days. The duration of tDCS was 20 min and SFA therapy was approximately 30 min per day. Baseline testing was performed immediately before each stimulation phase, and 2–3 follow-up tests were performed immediately after, approx. two weeks after, and approx. four weeks after the end of stimulation (t_1–t_3). The intermediate tests were performed to verify how long any therapeutic effect was measurable. Short pauses took place between the stimulation phases. Some of the material used for therapy was developed in-house by an experienced clinical linguist or selected from existing databases. The picture-naming ability of trained and untrained items was tested, and the progression across stimulation phases was reviewed and compared. Two different item sets including nouns and verbs, respectively, were processed and presented randomly during therapy. At the end of the study, a post-testing was conducted, which included test procedures comparable to the pre-testing. There were 3.5 months between the pre-testing and the first stimulation phase S1. Also, there were 3.5 months between the last stimulation phase S4 and the post-testing. All testing appointments and therapy sessions were conducted at the participant's home in a quiet environment, further reducing potential distractions.

The study was approved by the Ethics Committee of Bielefeld University (2021-002) according to the German Psychological Society, the Professional Association of German Psychologists, and the Declaration of Helsinki. Written informed consent was obtained from both the participant and her husband before enrollment in the study.

2.2. Participant

The female participant was a 56-year-old right-handed native German speaker who was diagnosed with frontotemporal dementia (FTD) in 2016 by her treating neurologist. The participant had ten years of school education and completed vocational training as an industrial clerk. She is married and has a son. Everyday care tasks such as personal hygiene, household activities, grocery shopping, etc. were performed independently by the participant at the time of the study. In addition, she had an (adjusted) part-time job as an industrial clerk in a local company.

The participant stated that she had noticed progressive memory problems, word-finding difficulties, problems with language comprehension at the one-word level, and concentration difficulties for about three years prior to diagnosis. There were no other known pre-existing conditions and no positive family history related to dementia. MRI revealed left accentuated bilateral temporal lobe atrophy and a small decrease in left hippocampal volume. PET-CT evaluation showed hypoperfusion in the left inferior lateral anterior and posterior temporal cortex. Microangiopathic lesions were mild and other intracranial findings were normal for the participant's age. Other parameters such as general health and autonomic abilities were unremarkable.

There were no other therapeutic interventions and the participant did not take any medications during the 14-month study period. For transcranial direct current stimulation, contraindications such as previous neurological diseases, epilepsy, epilepsy in close relatives, severe and frequent migraine, severe traumatic brain injury, presence of metal parts in the skull, surgery on the brain, chronic skin disease, and wearing a pacemaker or taking neuroactive medications were excluded [36].

Before the start of the study, the participant was informed verbally and in writing about the procedure and content of the study. The contraindications of tDCS, possible side effects, and data storage were also explained in this way. The participant confirmed in writing that participation in the study was voluntary and that consent could be withdrawn at any time without giving reasons. Further questions could be asked at any time during the informative interview.

2.3. Neurolinguistic and Neuropsychological Assessment

Before and after the therapeutic intervention, various language and cognitive tests were administered (pre-testing/post-testing). The selected diagnostic instruments were largely based on the established diagnostic criteria for svPPA [1] (Table 1).

Table 1. Neurolinguistic and neuropsychological diagnostics.

Test	Characteristics	Pre-Testing	Post-Testing
Spontaneous speech	minutes analyzed	9.03	7.37
	number of phrases	148	124
	content words	260	230
	set phrases	2	4
	stereotypes	9	5
	Word-finding difficulties	48	24
	echolalia	2	0
	sentence breaks	1	1
	missing sentence part	1	0
	sentence part duplications	0	2
	phonematic paraphasias	0	1
	syntactic structure	simple	simple
	speech production	fluent	fluent

Test	Pre-testing	Post-testing	Norms
BNT	Score	Score	
Items correct w/o assistance	15 */60	10 */60	55.73 (SD 4.42)
No. of semantic cues given	11/45	13/50	
No. correct following sem. cue	2/11	0/13	
No. of phonemic cues given	8/43	44/50	
No. correct following phon. cue	0/8	33/44	
Items correct with assistance	2/19	33/57	
BOSU	Raw value	Raw value	Cut-off-value
Assigning objects to situations	0	2 *	≥2
Sorting objects by main semantic features	0	0	≥2
Sorting objects according to semantic secondary features	0	2	≥3
Semantic sorting of written words	1 *	2 *	≥1
Sorting objects by color	0	1 *	≥1
TROG-D	Score	Score	
	18/21	19/21	
Grammatical errors	• double object constructions • topicalization • relative clause	• topicalization • relative clause	

Table 1. *Cont.*

Test	Characteristics	Pre-Testing	Post-Testing
RWT	Percentile rank	Percentile rank	Percentile rank
Semantic fluency	<1 *	<1 *	≥10
Semantic category change	<2 *	<1 *	≥10
Lexical fluency	<3 *	<1 *	≥10
Lexical category change	<1 *	<1 *	≥10
AAT (Repeat)	Score		Score
	146		≥144
	Percentile rank		
	94		
VLMT	T-value	T-value	T-value
Learning	52	47	50 (SD 10)
Delayed recall	45	39 *	50 (SD 10)
Recognition	46	42	50 (SD 10)
DemTect	10 */18	9 */18	>13 points
MoCA	22 */30	20 */30	>26 points
GDS	3/15	7.5 */15	≤5 points

Neurolinguistic and neuropsychological tests administered before (pre-testing) and after the therapeutic intervention (post-testing). There were 14 months between the results of pre- and post-testing. Comparison of the results with the respective norm values. * Deviations from the (age-appropriate) norm data.

The test procedures for assessing the participant's cognitive performance and speech and language production included an analysis of the spontaneous speech, which is based on the criteria of the Aachen Aphasia Test (AAT) [37]. The topics selected by the therapist were close to the participant's everyday life in order to stimulate guided spontaneous speech. In the pre-testing, the topic of traveling was discussed, as this was a hobby of the participant. The topics of the spontaneous speech in the post-testing were food, travel, and both general and personal changes due to the COVID-19 pandemic. The spontaneous speech analysis was performed by the investigator at the beginning of the diagnostic session to give the participant a casual start. The conversation was recorded with a recording device (Olympus, Tokyo, Japan, VN-731PC) and later transcribed and analyzed manually. Further, the Boston Naming Test for confrontation naming (BNT; [38]); the Bogenhausen Semantics Examination, a predominantly nonverbal semantic test for object knowledge (BOSU; [39]); the TROG-D [40], a test of grammar function and single-word comprehension; the Regensburg Word Fluency Test (RWT; [41]); and the repetition subtest of the Aachen Aphasia Test (AAT; [37]) were performed. In addition, verbal memory and general cognitive abilities were assessed with a verbal learning and memory test (VLMT; [42]), the DemTect (version A) [43], and the Montreal Cognitive Assessment (MoCA; [44]). Further, the Geriatric Depression Scale (GDS; [45]) was applied (results see Table 1).

2.4. Therapy Material

Various worksheets for nouns and verbs, respectively, were created for the therapy (Figure 1). These were based on a template used in a single case study [46]. A separate worksheet was used for each item. The goal of the therapy was to fill in all semantic fields of the worksheets in writing in order to be able to name the item during processing or at the end of processing. The participant could decide which fields to fill in first. In case of difficulties with the semantic fields, the therapist provided assistance. Regarding the nouns, first, the category to which an object belongs was asked, then the properties of an object, the place where a certain object can be used or purchased, the usage of an object, and a spontaneous association to an object similar to the object depicted (Figure 1, top).

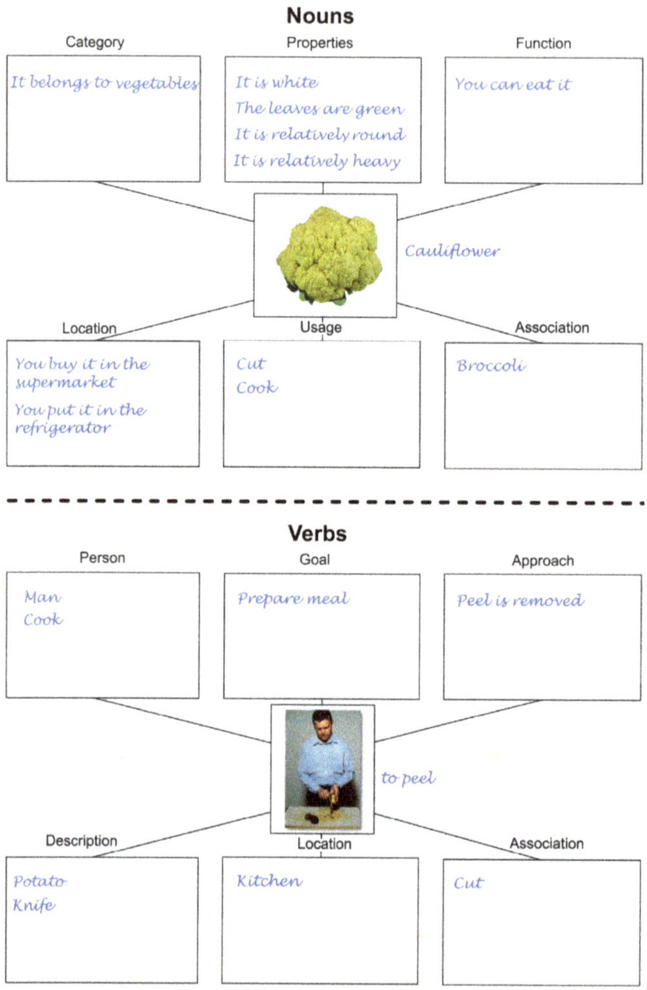

Figure 1. SFA sheets for nouns and verbs. (**top**): SFA sheet for nouns for the word "Blumenkohl" (*cauliflower*) as an example. (**bottom**): SFA sheet for verbs for the word "schälen" (*to peel*) as an example. The participant's original responses were translated into English.

The worksheet used for processing the verbs was similar in structure to that used for the nouns. The semantic fields were based on a template [47]. First, the participant had to identify the person performing the action, and then the goal pursued by the action (Figure 1, bottom). It was also possible to mention previous or further actions that were directly related to the action presented. Then, the location where the action takes place and an association had to be described. Finally, the target word for nouns and verbs should be written down at the lower right margin of the sheet.

The visual material for processing nouns was taken from the Bank of Standardized Stimuli (BOSS) which consists of colored photographs that can be divided into 23 semantic categories [48,49]. The nouns used can be assigned to the semantic categories of animals, food, household items, office supplies, furniture, electrical items, clothing, and tools. The nouns used were divided into two lists, each used during stimulation phases S1 and S3 and during stimulation phases S2 and S4, respectively. Each list consisted of 40 words, which were tested at the beginning of the stimulation phases during the baseline survey. The average word frequency of the nouns of set 1 (M = 1152.54, SD = 4473.37) and set 2 (M = 601.74, SD = 1212.99) was statistically equivalent ($t = 1.47$, $p = 0.14$). The following psycholinguistic parameters were defined using five-point Likert scales [49]. In list 1, the selected items had a high familiarity score (M = 4.36, SD = 0.56), as did those in list 2 (M = 4.24, SD = 0.68). No significant differences were found ($t = 1.10$, $p = 0.27$). In terms of visual complexity average values for list 1 (M = 2.45, SD = 0.56) and list 2 (M = 2.47, SD =0.65) were obtained, but no significant differences were found ($t = 0.96$, $p = 0.34$). Moderately high values were obtained for object agreement in both set 1 (M = 3.85, SD = 1.31) and set 2 (M = 3.91, SD = 1.81) with no significant differences between the two sets ($t = 0.62$, $p = 0.53$). For viewpoint agreement, moderately high values were achieved for set 1 (M = 3.65, SD = 1.24) and set 2 (M = 3.79, SD = 1.21), with no significant differences found ($t = 1.26$, $p = 0.21$). Furthermore, the manipulability was examined for set 1 (M = 2.8, SD = 0.79) and set 2 (M = 2.65, SD = 0.88). Again, no significant difference was found between sets 1 and 2 for this parameter ($t = 0.29$, $p = 0.77$). The last parameter studied was the number of syllables in the words of set 1 (M = 2.08, SD = 0.88) and set 2 (M = 2.56, SD = 0.89). A significant difference was found for the word sets ($t = -2.81$, $p = 0.005$).

The visual material for verbs was taken from the Everyday Life Activities (ELA) item set [50]. In the present study, ELA set 1 was used. This set shows the actions of a family's everyday life. These everyday activities are divided into different semantic fields which include gestures and facial expressions, movements and postures, daily physical needs, dressing, food and drink preparations, eating and drinking, household activities, shopping, necessary daily actions, social interactions, leisure, hobbies, and sports [51]. The verbs used were all full verbs and concretes. The verbs represented activities in the kitchen and in the household, with tools, sports and exercise, leisure activities, music, and emotions.

Verbs were also divided into two lists. The first list was used during stimulation phases S1 and S3, and the second list was used during S2 and S4. As with nouns, each list consisted of 40 words that were tested at the beginning of a stimulation phase in the baseline survey. For verbs, two-tailed Student's t-tests for independent measures were performed, too. For the average word frequency in set 1 (M = 3025.16, SD = 6755.79) and set 2 (M = 619.79, SD = 2044.44), no significant difference was found ($t = 1.97$, $p = 0.051$). The average syllable number in set 1 (M = 2.2, SD = 0.57) and set 2 (M = 2.4, SD = 0.75), was also not significantly different ($t = -1.01$, $p = 0.31$).

2.5. tDCS Equipment

Transcranial current stimulation was performed using a battery-powered direct current stimulator (neuroConn, Ilmenau, Germany). Stimulation was performed using two conductive rubber electrodes (5 × 7 cm^2) placed in sponges soaked in 0.9% NaCl solution. Before placing the electrodes, the corresponding skin sites were gently rubbed with a cotton swab soaked in 70% alcohol to improve impedance. The average value of impedance was 3.9 kOhm at the beginning of stimulation, 3.1 kOhm after 10 min of stimulation, and 3.0 kOhm toward the end of stimulation. Placement of the electrodes was performed according to the international 10-10 EEG system. The anode was placed on the intersection between FT7 and FT9, with the long side of the sponge electrode oriented horizontally and slightly inclined next to the left ear (Figure 2). This electrode position is located approximately over the left anterior temporal lobe and has already been used successfully in a previous study [28]. The cathode was placed on the contralateral side on the supraorbital region (Fp2), horizontally aligned with the long side of the sponge electrode. During the experiment, anodal tDCS was applied for 20 min with a current intensity of 1.5 mA, and fade-in and fade-out phases of 10 s each. Based on the results of the pilot study and the participant's personal request, no sham stimulation was performed in this study to provide the participant with the optimal intervention over the 14 months. tDCS was performed concurrently with the training phases with SFA. The subject cooperated well and her attention and concentration remained stable during the sessions. During the intermediate tests, when no tDCS was performed, the participant needed more motivation toward the end of the study. This was always achieved by positive verbal reinforcement by the therapist.

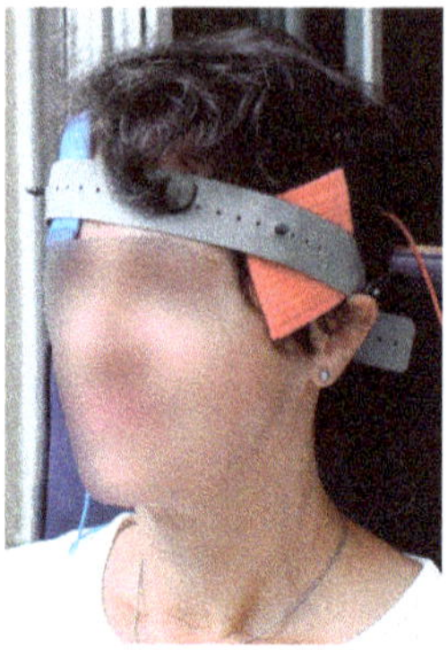

Figure 2. tDCS application. Placement of the tDCS electrodes: anode (red) over the left temporal pole, cathode (blue) at the right supraorbital region.

The participant's performance in all stimulation phases and test sessions was recorded with an audio recorder (Olympus, Tokyo, Japan, VN-731PC).

2.6. Experimental Procedure

A diagnostic pre-testing was performed prior to the therapeutic intervention (Figure 3). Further, at the beginning of each stimulation phase (S1–S4), a baseline survey took place in which the participant was asked to name 40 nouns and 40 verbs shown in the form of picture cards. Afterward, SFA training and concurrent tDCS started.

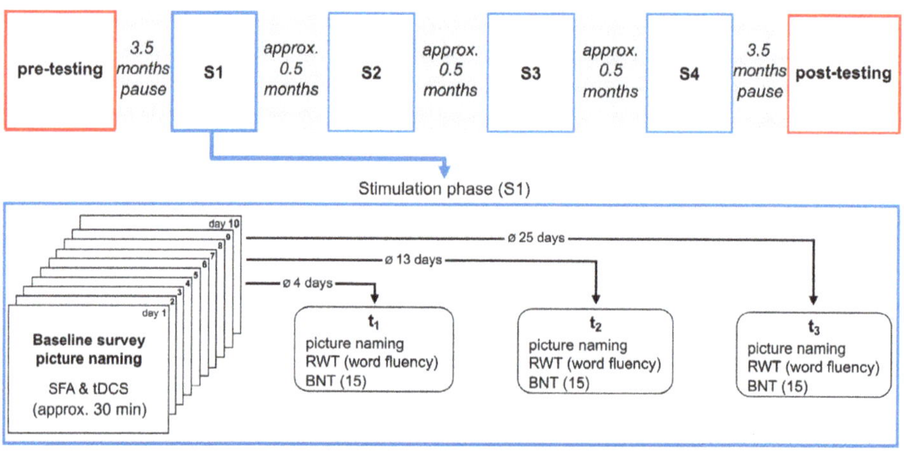

Figure 3. Experimental procedure. Flow chart of the stimulation phases S1–S4. For S1, the detailed sequence of the intervention is shown as an example.

The items that were not named correctly in the baseline survey were trained with the combination of SFA and tDCS for ten consecutive working days in each stimulation phase. On average, 24 items were practiced per stimulation phase. The stimulation lasted 20 min and therapy in total of 30 min per day. After ten days of therapy, intermediate tests

(t_1, t_2, t_3) were performed. In each stimulation phase, the intermediate tests took place approximately 4, 13, and 25 days after the last stimulation session. During the intermediate tests, a picture naming task, four subtests of the RWT, and the 15-item version of the BNT were administered. If a 10% drop in performance was detected on at least one of the tests, the next stimulation phase was initiated. In the stimulation phase S3, only two intermediate tests were performed because a drop in performance of more than 10% in picture naming was found in the interim test t_2 when the trained items were examined. Therefore, t_3 was omitted in this stimulation phase. In addition, there were pauses of about 0.5 months between all stimulation phases due to organizational reasons.

The intermediate tests took place without tDCS. The picture naming test consisted of the noun and verb material which was used in the baseline survey (untrained words) and in the therapy (trained words). During this test, the participant did not receive any feedback as to whether she had named an item correctly or not. In addition to the words from the baseline survey, the untrained words consisted of words that the participant had not seen before. The words in the baseline surveys and the previously 30 unseen untrained words were identical in S1 and S3 and S2 and S4, respectively. The naming performance of the trained and untrained items was analyzed separately. The four subtests of the RWT each consisted of a semantic fluency task, a semantic category-switching task, a formal lexical fluency task, and a formal lexical category-switching task. The shortened version of the BNT was used as a standardized procedure for naming performance. At the end of the entire intervention, a post-testing was performed (Figure 3).

Pre-testing was performed 3.5 months before the start of the first stimulation phase S1 on two consecutive working days. The three and a half-month break between diagnostic pre-testing and S1 was due to the participants' work and leisure activities, which had to be taken into account when planning the study. The tests on both pre-testing days lasted about two hours and were adapted to the participant's ability to concentrate and pay attention. It is not to be expected that the result of the diagnostics with regard to the clinical picture would have changed during this period. However, before each stimulation phase, a baseline for picture naming was obtained, which served as a reference for a possible improvement due to the stimulation unit.

The post-testing also took place 3.5 months after the end of the last stimulation phase S4 and was also performed on two consecutive working days. The delay between the last stimulation phase and the post-testing was due to the COVID-19 pandemic and contact restrictions.

Post-testing also took approximately two hours per day. The participant had the opportunity to request a break at any time during the tests. The whole period between pre- and post-testing was 14 months.

2.7. Data Analysis

The data collected were statistically analyzed using Jamovi (vers. 2.3; [52]). The statistical significance level was set at 5% ($\alpha = 0.05$) for all analyses. The *p*-values were Bonferroni corrected where necessary. Binominal logistic regression was used to investigate whether the participant's picture naming performance improved after the intervention. First, we investigated whether word type (nouns vs. verbs) or the word lists used predicted whether an item was named correctly or not. Second, we analyzed whether the training status of the words in each set (trained vs. untrained) predicted naming performance. No feedback was given to the patient during the testing to avoid learning via the tests. Furthermore, we tested whether the individual test time points t_1 to t_3 of the intermediate results predicted better naming performance of the untrained items to investigate generalization effects. This was done for each single phase (S1–S4) and further consolidated for all four intervention blocks.

Paired *t*-tests were used to compare the item sets concerning their linguistic parameters. The pre- and post-testing conducted at the beginning and the end of the study were analyzed descriptively. In addition, linear regression analyses were used to compare the data from the BNT as well as the RWT administered during pre-testing, post-testing, and interim-testing.

3. Results

3.1. Behavioral Results of Pre- and Post-Testing

The participant was examined and diagnosed by an experienced clinical linguist before and after the therapeutic intervention; the time between the pre- and post-tests was 14 months. A series of neurolinguistic and neuropsychological tests were performed and the results, norm scores, and criteria for interpretation of the test results are listed in Table 1.

When analyzing spontaneous speech, the ratio of content words to phrases was 1.8 in both the pre-test and post-test, with an average of 6.2 words produced per sentence in the pre-test and 6.0 words per sentence in the post-test. This indicates that speech continued to be fluent. Speech rate, averaging 108 words per minute at pre-testing and 107 words at post-testing, could also be considered normal. The phrases were characterized by a short and simple syntactic structure with a stringing together of main clauses and few subordinate clauses. Strikingly, the analyses of the spontaneous speech of the pre- and post-testing revealed clear differences in the number of word-finding difficulties (Table 1). While the ratio of word-finding difficulties per phrase in the pre-testing was 0.32, the value in the post-testing was 0.19, reflecting an improvement of more than 40%. This indicates a possible effect of the therapeutic intervention on the single-word level in spontaneous speech and thus on the participant's communicative abilities. All in all, at the time of pre- and post-testing, the language abnormalities identified during standardized tests could still be compensated well by the participant in spontaneous speech.

Naming performance in the BNT showed significant performance deficits both before and after the intervention (Table 1). During the pre-testing, the patient could spontaneously name 25% of the items correctly. After semantic cues, a total of approx. 28% of the items were named correctly. After phonematic cues, no item could be named correctly. During post-testing, 17% of the items were spontaneously named correctly. After the semantic cues, the participant could not name a further item correctly. However, she was able to correctly name an additional 33 items after the phonematic cues, so she was able to correctly name a total of 75% of the items in the post-testing.

The BOSU examined functional and associative semantic memory performance with five subtests. In four of the five subtests, the participant showed a significant decrease in the ability to assess semantic relations at the post-testing. In particular, situational semantic relations or the hierarchy of semantic knowledge was increasingly affected over the 14-month study period.

The TROG-D tested grammar comprehension using 21 subtests. It was found that the subject's grammatical skills were comparatively unaffected and had even improved minimally over the course of the therapeutic intervention. The errors that occurred were all lexical in nature, which could indicate deficits in single-word and sentence comprehension.

The RWT was used to examine semantic and lexical word fluency as well as word fluency during a semantic and lexical category change. A testing time of two minutes per subtest was chosen both for pre-testing and post-testing. The results were well below the average word fluency, corresponding to a percentile rank of $\geq$10. At post-testing, the impairment was even more pronounced.

The ability to repeat was tested using the "Repeat" subtest of the AAT. Sounds, monosyllables, loan- and foreign words, compound words, and sentences were tested. The participant achieved a score of 146 (out of 150), which corresponds to a percentile rank of 94. This test was administered only during the pre-testing and the participant did not show any deficits in repeating.

All in all, the language test procedures make it clear that the subject's difficulties were primarily in word retrieval with visual stimuli and in word fluency.

The VLMT showed slightly decreased memory performance in verbal learning, recall, and recognition which became more pronounced in the post-testing. Both the DemTect and the MoCA showed a mild cognitive impairment of the participant already during the pre-testing, with scores worsening during the post-testing. In the performance of the MoCA, the conspicuities were particularly in the repetition of sentences (main and subordinate clauses), in the word fluency task, and in the delayed word recall of a word list. During the post-testing, short-term memory (repeating number sequences backward) and semantic abstraction (naming the commonality of two concepts) additionally deteriorated. In the DemTect, the word fluency task was especially conspicuous during pre-testing. During post-testing, a deterioration in the ability to convert numbers was also observed. The GDS indicated mild to moderate depression only at the post-testing.

Overall, neurolinguistic and neuropsychological testing revealed semantic deficits in this participant in terms of impaired single-word comprehension of low-frequency words and impaired object-naming skills. Thus, the two mandatory criteria for the diagnosis of svPPA were met [1]. Furthermore, spontaneous speech was characterized by well-preserved surface structure and fluency. The BOSU showed difficulty with object knowledge. In contrast, language comprehension at the sentence level and repetition of verbal items were virtually unimpaired. From these findings, it can be inferred that at the beginning of the therapeutic intervention, the participant had svPPA in an early stage since word-finding difficulties and difficulties in single-word comprehension were the most prominent impairments, while other linguistic domains were relatively intact. Also, no changes in behavior were noted that would indicate a later stage of svPPA [53]. In addition, decreases in other semantic abilities were noted in functional and associative semantic knowledge, semantic and formal lexical word fluency, and word retrieval. The cognitive screening indicated a mild cognitive impairment which became more severe over the course of the study. All in all, the participant showed no obvious impairments in the areas of phonology, syntax, grammar comprehension, and repetition at the time of therapeutic intervention. This finding indicates intact speech and language skills outside of (lexical-) semantic processing.

3.2. Results of SFA in Combination with tDCS

A total of 25 items (12 nouns, 13 verbs) were processed in the stimulation phase S1, 24 items (12 nouns, 12 verbs) in S2, 23 items (11 nouns, 12 verbs) in S3, and a total of 24 items (11 nouns, 13 verbs) in S4.

First, we investigated whether word type (nouns vs. verbs) predicted whether an item was named correctly or not. Binominal logistic regression showed no effect of word type ($B = 0.15$, $SE = 0.12$, $Z = 1.2$, $p = 0.22$). Therefore, the words were considered as one group in the following analyses. A further question concerned the two word lists used in the different stimulation phases. The word list did not predict whether an item was named correctly or not ($B = 0.04$, $SE = 0.12$, $Z = 0.34$, $p = 0.73$). However, there was a significant effect of training on the correctness of naming. More trained items were named correctly than untrained items ($B = 0.82$, $SE = 0.13$, $Z = 6.14$, $p < 0.001$).

Another aspect addresses the question of which of the respective test time points within a stimulation phase were predictors for the correct or incorrect naming of items in the other test time points. For this purpose, binomial logistic regression was performed for each test time point separately for trained and untrained words. The results for trained items indicate that the ability to retrieve trained words actually increased until two weeks after the end of stimulation (Figure 4). In S4, this was significant ($B = 1.705$, $SE = 0.856$, $Z = 1.99$, $p = 0.05$). In S3, naming performance already decreased significantly between t_1 and t_2 ($B = -1.455$, $SE = 0.752$, $Z = -1.94$, $p = 0.05$). In S4, it decreased significantly only between t_2 and t_3 (approximately 2–4 weeks after the intervention) ($B = -2.398$, $SE = 0.844$, $Z = -2.84$, $p = 0.004$). Figure 4 shows the percentage of correctly named trained words for all stimulation phases and intermediate tests. In the baseline (BL), none of the subsequently trained words could be retrieved. The percentage of correctly named words increased by approximately 20% from t_1 to t_2 in all stimulation phases except S3.

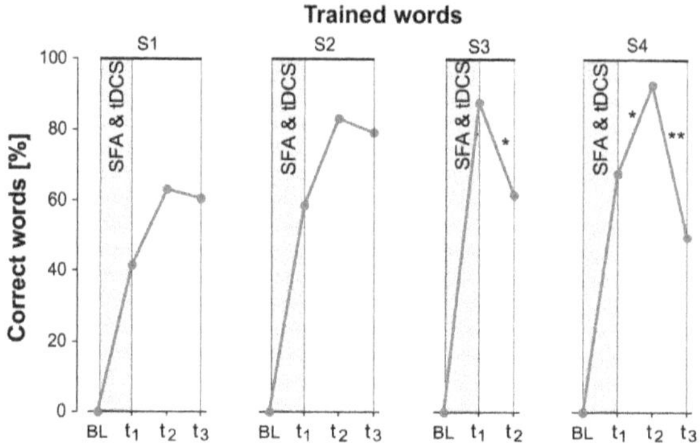

Figure 4. Results for trained words. Percentage of correctly named trained words in the baseline (BL), and the intermediate testing time points t_1, t_2, and t_3 for all four stimulation phases. t_1 took place on average 4 days after the simultaneous SFA and tDCS therapeutic intervention (SFA & tDCS), t_2 after about 2 weeks, and t_3 after about another 4 weeks. p-values are Bonferroni corrected (* = $p < 0.05$, ** = $p < 0.01$).

Binominal linear regression for untrained words showed that the participant's word retrieval ability changed significantly in S2, S3, and S4 (Table 2). Compared to the baseline, S2 showed significantly higher naming performance in all three intermediate tests. In S3, naming performance increased significantly from BL to t_1. In S4, naming performance increased significantly and decreased significantly again from t_2 to t_3 (Table 2).

Table 2. Comparison of the intermediate tests for untrained words.

Stimulation Phase	Intermediate Tests	B	SE	Z	p
S2	BL-t_1	0.825	0.393	2.10	0.036
	BL-t_2	1.434	0.415	3.45	<0.001
	BL-t_3	1.311	0.411	3.19	<0.001
S3	BL-t_1	1.219	0.427	2.86	0.004
S4	BL-t_1	1.096	0.406	2.70	0.007
	BL-t_2	2.107	0.507	4.16	<0.001
	t_1-t_2	1.011	0.528	1.91	0.05
	t_2-t_3	−1.434	0.513	−2.80	0.005

Stimulation phases (S2, S3, S4) and test time points (BL, t_1, t_2, t_3) predicting significant naming differences for untrained items. p-values are Bonferroni corrected.

For the untrained items, naming performance increased by about 20% on average in all simulation phases, from baseline to t_1 (Figure 5). In S2 and S4, performance increased again by about 20% in t_2. In t_2 of S4, the participant was even able to retrieve 80% of the untrained items correctly, which corresponds to an increase of about 40% compared to the baseline.

It is also interesting to examine how the participant's naming performance changed at each intermediate test, averaged over all four stimulation phases, to obtain an overview of what the intervention did overall and how long the performance improvement lasted on average for this subject.

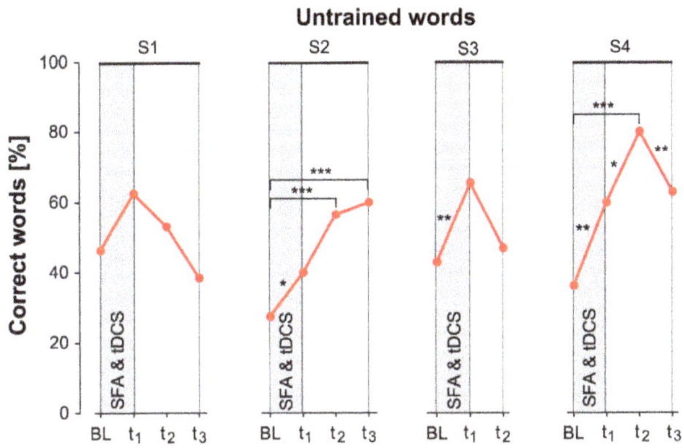

Figure 5. Results for untrained words. Percentage of correctly named untrained words in the baseline (BL), and the intermediate testing time points t_1, t_2, and t_3 for all four stimulation phases. t_1 took place on average 4 days after the simultaneous SFA and tDCS therapeutic intervention (SFA & tDCS), t_2 after about 2 weeks, and t_3 after about another 4 weeks. *p*-values are Bonferroni corrected (* = $p < 0.05$, ** = $p < 0.01$, *** = $p < 0.001$).

For the trained words, binomial logistic regression revealed no significant changes between each averaged test time point. On average, naming improved from the baseline by 60% in t_1 and by an average of 75% in t_2 after therapy compared to baseline values (Figure 6, left).

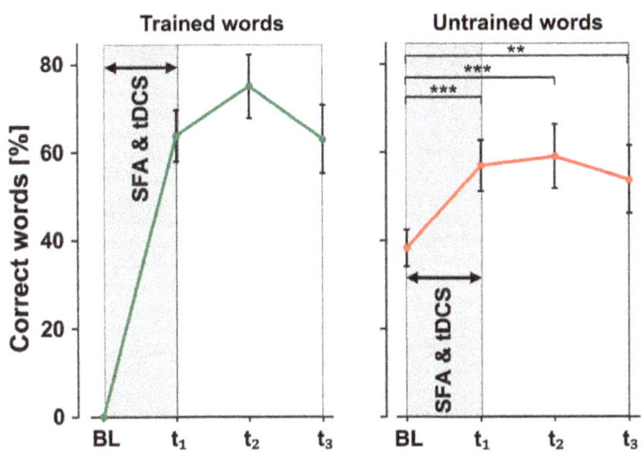

Figure 6. Averaged results for trained and untrained words. Mean percentage of correctly named trained (**left**) and untrained words (**right**) at the four test times (baseline, t_1, t_2, t_3) averaged across all stimulation phases. SFA & tDCS means simultaneous therapeutic intervention. *p*-values are Bonferroni corrected (** = $p < 0.01$, *** = $p < 0.001$).

For the untrained words, binomial logistic regression showed that significantly more words could be correctly named at all three test time points compared to the baseline (t_1: B = 0.800, SE = 0.200, $Z = 3.99$, $p < 0.001$; t_2: B = 1.021, SE = 0.206, $Z = 4.95$, $p < 0.001$; t_3: B = 0.648, SE = 0.218, $Z = 2.98$, $p = 0.003$). Again, there were no significant changes between test time points (Figure 6, right). These results indicate that the participant was able to transfer the learned strategies of SFA to the untrained words. These changes persisted for at least up to 4 weeks, with a peak two weeks after the end of the intervention.

3.3. Results of the BNT

A standardized test procedure for naming at all test time points was the BNT. In the intermediate tests, 15 of the 60 items were tested. The versions of the shortened BNT were identical in stimulation phases S1 and S3 and in stimulation phases S2 and S4. BNT results showed an increase from pre-testing to post-testing. In pre-testing, about 30% of the items were correctly named; in post-testing, about 70% (Figure 7).

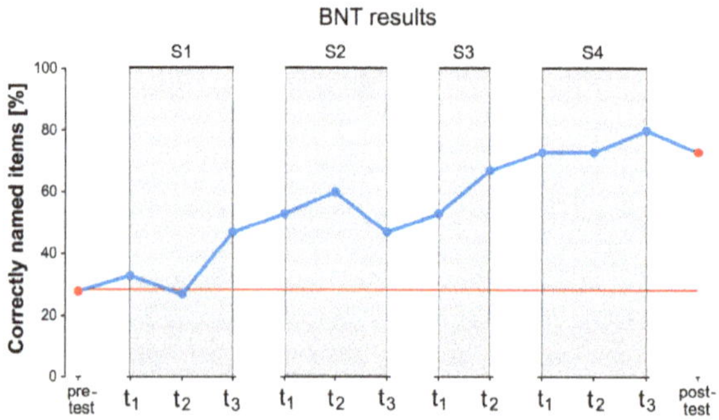

Figure 7. Results of the Boston Naming Test. Percentage of all correctly named items in the BNT in pre-testing, post-testing, and all intermediate tests for all stimulation phases, regardless of whether correct spontaneously or after given cues (blue line). The red line represents the initial percentage of correctly named items in the pre-test over the entire period of the intervention and illustrates in particular the difference between the results of the pre- and post-test.

Linear regression showed that the stimulation phase predicted the number of correctly retrieved items independent of cues ($R^2 = 0.93$). There were significant results for the second stimulation phase (S2) compared to S1 (B = 3.33, SE = 0.713, $t = 4.68$, $p = 0.014$), for S3 compared to S1 (B = 4.33, SE = 0.797, $t = 5.44$, $p = 0.006$), and for S4 compared to S1 (B = 6.67, SE = 0.713, $t = 9.35$, $p < 0.001$). Further, S4 predicted a higher number of correctly named items than S2 (B = 3.33, SE = 0.713, $t = 4.68$, $p = 0.014$). All *p*-values were Bonferroni corrected. Strikingly, after a semantic cue, the participant was unable to correctly name an item at any of the intervening test time points. In contrast, the number of correctly retrieved items after a phonological cue increased with the number of phonological cues given (Pearson's r = 0.66, $p = 0.014$).

3.4. Results of the RWT

The RWT, which tested semantic and lexical word fluency, switching between semantic categories, and switching between letters, was performed in the pre-testing, post-testing, and interim tests t_1 to t_3 of all stimulation phases. Compared to the pre-testing and post-testing, the probability of increased semantic word fluency averaged across all stimulation phases, only tended to be higher in t_1, t_2, and t_3, but values were not significant after Bonferroni correction (Figure 8). For lexical fluency, only t_1 showed a trend toward higher probability. Also, no significant predictions were found for semantic category switching and letter category switching (Figure 8). This means that the testing time points of the intermediate tests predicted performance in semantic fluency that improved slightly but not significantly in this participant compared to pre-testing but also post-testing. She improved two and five weeks after the intervention, although she still remained below the age-appropriate norm and showed decreased performance in the post-testing.

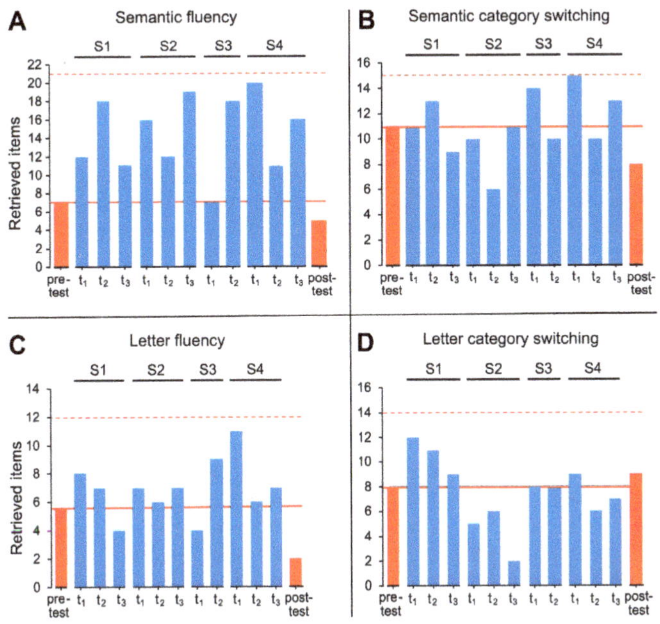

Figure 8. Results for word fluency. The absolute number of words retrieved in the semantic fluency, letter fluency tasks, semantic category switching, and letter switching tasks are shown separately for the pre-and post-testing (red columns) and the intermediate tests at t_1, t_2, and t_3 of all stimulation phases (blue columns). The dashed red line represents age-appropriate norms, the solid red line represents the value collected in pre-testing.

3.5. Results of the Comparison between First and Second Half of Intervention

In PPA there is often a difference in language in the first phase of a therapeutic intervention and the second phase [54]. We, therefore, compared the data on picture naming from the first half of the therapy (S1, S2) with those from the second half (S3, S4). We found no significant prediction of the intervention phase for the trained items (B = 0.175, SE = 0.223, Z = 0.786, $p_{bonferroni}$ = 0.432). However, for the untrained items, we found a significant difference between the first and second half of the intervention (B = 0.362, SE = 0.15, Z = 2.44, $p_{bonferroni}$ = 0.015; *exp* (*B*) = 0.695). The second half of the therapy resulted in a higher number of correctly named items. The BNT also showed significant results. In the second half of the therapy, significantly more items could be named (B = 4.07, SE = 1.078, t = 3.77, $p_{bonferroni}$ = 0.004; η^2 = 0.613). No difference was found in the RWT between S1, S2, and S3, S4.

4. Discussion

In this longitudinal study over 14 months, the word retrieval performance of a participant diagnosed with svPPA was examined at several time points and in several tests.

Prior to the study, the participant's speech, language, and cognitive performance was assessed in a pre-testing using different neurolinguistic and neuropsychological tests. This was followed by four periods of therapy with SFA and simultaneous tDCS over 10 days. A baseline survey of picture naming was carried out before each therapy period, and 2–3 follow-up tests were carried out between the therapy phases. The start of a new therapy phase depended on whether the participant's performance in word fluency and/or picture naming dropped by 10%. After the four therapy phases and all follow-up tests, a post-test using similar tests was carried out at the end.

The combined SFA/tDCS intervention resulted in improved word retrieval performance for the participant. The stimulation and therapy effects persisted for approx. two weeks after the end of each stimulation period. This was true for both trained and untrained words. Thus, a generalization effect could be measured. In addition, the number of word-finding difficulties decreased during guided spontaneous speech. The participant showed a significant effect of therapeutic intervention with the combination of SFA and tDCS. Compared to the baseline before the therapeutic intervention, she was able to retrieve about 60% more trained words correctly in the averaged first interim test. After about two weeks (t_2), even 75% of the items could be named correctly. For the untrained items, an increase of about 20% was observed shortly after the stimulation phase (t_1) compared to the baseline, and a further increase of about 5% was observed from t_1 to t_2. For both trained and untrained items, naming performance decreased again by an average of 10% from t_2 to t_3 (approx. 2–4 weeks after the end of the intervention).

4.1. Benefit of SFA

Previously, other authors demonstrated that SFA therapy per se is well suited to achieve an improvement in word retrieval in svPPA [5,7,21]. Through SFA, retrieval of concrete words in picture naming is facilitated by combining specific features and establishing connections between them. The elaboration of semantic concepts can also favor the generalization effect [15]. It has also been shown that therapy methods that have an additional conversational focus strengthen general communication skills [55]. The combined SFA and tDCS intervention may have an impact not only on single-word retrieval, but also on general communication skills, especially in spontaneous speech.

4.2. Benefit of tDCS

Since it has been shown that word retrieval performance in svPPA in general decreases between 20 and 30% within a year the increase in performance of almost 15% in our study, especially in the untrained items, is remarkable [33]. In our study, SFA was combined with tDCS, because its effect on improved word retrieval for trained and untrained words was further enhanced by tDCS in several studies [20,28,29]. The ability to name pictures of trained items improved by 30% in svPPA patients after anodal tDCS of the left inferior parieto-temporal region compared to sham stimulation immediately after 10 days of stimulation [30]. tDCS is an effective addition to conventional speech and language therapy because the neural networks involved are enlarged by anodal electrical stimulation, making training effects easier to achieve [56]. Since in our study tDCS was administered on ten consecutive working days for 20 min each, we believe that this enhanced and prolonged the effects of SFA. The increase in BNT from the pre-test to the post-test (14 months) by about 40% also supports this assumption. In addition, some studies have shown that patients with PPA often show a difference in language performance between the first half of a therapeutic intervention and the second half [54]. In the present study, the participant was able to name significantly more untrained items correctly in the second phase of therapy and was also able to name more words correctly in the BNT. The significant increase in performance from the first half to the second half of the therapy could possibly have been achieved through the combined use of SFA and tDCS.

4.3. Untrained Items

The results with respect to the untrained items are particularly interesting. There was a significant increase in the naming of the untrained items compared to the baseline, up to two weeks after the end of the therapeutic intervention. Semantic-based therapy methods were found to produce significant learning effects for trained items, but improvements for untrained items were not demonstrated at significant levels [7,21]. To date, there is also limited evidence of generalization and maintenance effects on untrained items in patients with svPPA [12,15,16]. One explanation for the generalization of naming ability to untrained items could be that the participant learned the principles of SFA and applied them strategically when processing a new concept. Therapy methods that use self-cueing strategies favor generalization effects [14]. Another study showed that the elaboration of semantic concepts is more effective for word retrieval and repeated naming of items proved to be ineffective [57]. Another factor that influences generalization effects is item selection. The selection of meaningful words is important for retention and generalizability [5,15]. In our study, the trained and untrained words were chosen to correspond to a large extent to the participant's personal life. Another aspect that may have positively influenced generalization is the fact that the participant was treated in a relatively early stage of the disease. Brain atrophy that is not yet far advanced and a relatively preserved semantic memory favors generalization effects [15]. In our view, a major factor that supported the increase in performance on the untrained words might also have been the combination of SFA and tDCS. Naming of untrained words has been shown to often improve after a combined intervention of speech and language therapy and transcranial brain stimulation [20]. tDCS can be considered an intervention that enhances therapy and generalization effects [58]. The increase in naming performance persisted for both trained and untrained words up to two weeks after the end of each stimulation phase. From this course, it could be concluded that stimulation periods at two-week intervals with short pauses were optimal for increasing the participant's language performance. Similar to this study, the effect of improved language performance two weeks after the end of stimulation [29,31] and increased duration of treatment effects [58] have been demonstrated.

The BNT also showed that the number of untrained items that could be correctly named increased significantly over the course of the study. Interestingly, the number of words named spontaneously and without cues remained relatively constant during the intervention, but the number of words named correctly after phonological cues increased significantly. The number of words retrieved after phonemic cues correlated significantly with the number of phonemic cues. In addition, fewer semantic cues were required during the intervention. The participant was able to retrieve the target words by independently and adequately rewriting semantic connections in combination with phonological support. By combining semantic and phonological cues, the concept of a represented object could be more easily linked to the correct name [5,59].

4.4. Word Fluency

A slight but not significant improvement in word fluency, as measured by the RWT was observed in the course of the 14 months. Semantic word fluency increased on average especially at t_2 and t_3 compared to pre- and post-testing. Phonological word fluency also tended to increase while letter switching and semantic category switching were comparatively stable. The slight improvement in word fluency during a stimulation phase could be due to the fact that the participant used self-directed cueing strategies, such as considering the semantic features of an object or activity. However, the age-appropriate norm values could not be reached at any test time point, so normalization of word fluency was not possible. The below-average performance in word fluency indicates that limitations in executive functions were already present at the time of the testing since word fluency tasks require cognitive flexibility.

4.5. Spontaneous Speech

A very promising trend emerged in the spontaneous speech analyses. While phonological, grammatical, and morphosyntactic features were relatively stable from pre- to post-testing, the number of word-finding difficulties decreased in the post-test 14 months later. A direct comparison of spontaneous speech showed that the number of word-finding difficulties decreased by 40%. This indicates that the participant was able to apply the compensation strategies of the SFA in spontaneous speech and thus in everyday communication. SFA can be considered a cueing strategy that simultaneously facilitates transfer to a specific context because it is universally transferable to the context used in everyday communication. Although studies have shown that word learning can generalize across trained items and that this can lead to improvement in everyday language, little research has investigated generalization across linguistic levels or tasks [7]. Some studies have reported that the information content of patients' speech increases and that they show more correct information units in discourse after access to lexical items is facilitated [34]. However, others found no change in communication [35]. There is a possibility that the participant with svPPA in this study experienced improvements in connected speech because the effect of a conventional language therapy approach (SFA) was augmented with transcranial brain stimulation (tDCS). It has been shown that therapy methods with a conversational focus strengthen general communication skills. This effect might be enhanced by tDCS. Thus, an overarching goal of such therapy is not only to restore or improve language per se, but to prompt productive strategies to enhance existing communication skills [55].

Despite the significant improvements in naming ability for single words and spontaneous speech, a comparison of pre- and post-tests showed that participants' cognitive abilities deteriorated over the course of the study. Thus, number processing ability, word formation, attention, and verbal memory decreased. Furthermore, there was a significant decline in semantic nonverbal abilities from the pre-test to post-test. Grammar comprehension and understanding of morphological syntactic structures remained largely intact. However, it is possible that the progressive course of svPPA would have developed more rapidly in the absence of language therapy combined with transcranial stimulation [33].

4.6. Possible Limitations

Although the results of our study are promising, certain limitations must be discussed. It cannot be differentiated whether the described findings would have occurred to this extent even without tDCS. In the present study, as already mentioned, a sham condition was deliberately omitted, because of its positive effect on the participant as a motivational factor and because she reported positive effects on her sleep, mood, general performance, and concentration. In addition, the pilot study showed significant improvements in language abilities with tDCS compared to sham [32]. Also, several studies demonstrated that the naming ability of trained, but especially of untrained words in patients with PPA can be improved in the long term by combining speech and language therapy with tDCS (for an overview see [20]). For these reasons, sham stimulation was ethically unacceptable to us.

Electrode positioning can also be debated. In most studies that investigated the effectiveness of tDCS in patients with PPA, stimulation was applied over left frontal areas [60–62]. In our study, we applied tDCS over the left anterior temporal areas, and thus over the brain areas affected by atrophy. However, significant effects were obtained in patients with svPPA with stimulation of the anterior temporal lobe and other studies also suggest a possible importance of stimulating the atrophied areas [23,28]. It has been found that for naming pictures of untrained items, only parieto-temporal tDCS yielded significantly higher scores after two months [29]. For the trained items, stimulation at both left inferior parieto-temporal sites and left dorsolateral prefrontal sites showed significantly greater improvement with real tDCS immediately at the end of the stimulation sessions. The parieto-temporal montage resulted in significantly better picture naming even two weeks after stimulation. This underlines the stimulation of left anterior temporal sites in our study.

5. Conclusions

In summary, although the current study is a single case study of a monolingual German native speaker, the results show a very promising possibility for the treatment of speech and language disorders in svPPA. Daily application of SFA combined with tDCS for ten consecutive working days in four stimulation phases improved naming ability for trained but also for untrained items in a participant with svPPA. We adapted the start of each new intervention phase to the patient's performance in verbal fluency and picture naming and only started a new therapy phase when there was a drop of around 10% in one of the tests performed. Through this adaptive design, we found that stimulation intervals of two weeks were effective in counteracting the expected language and cognitive decline in this participant. The combined application of SFA and tDCS is a time-consuming and strenuous procedure for the patient. An interruption of the therapy of approx. 2–3 weeks showed no change in performance. The combined therapy may, therefore, not have to be continuous, but can take place at intervals. Most importantly, the linguistic compensation strategies of the SFA could be transferred to the participant's everyday communication.

Author Contributions: K.S. and S.W. designed the experiment. K.S. conducted the experiment. K.S. and S.W. analyzed the data and wrote the manuscript. K.S. and H.M.M. prepared the figures and tables. S.W. and H.M.M. reviewed and edited the manuscript. All authors have read and agreed to the published version of the manuscript.

Funding: We acknowledge support for the publication costs by the Open Access Publication Fund of Bielefeld University and the Deutsche Forschungsgemeinschaft (DFG).

Institutional Review Board Statement: This study was conducted in accordance with the Ethics Committee of Bielefeld University stated by the German Psychological Society, the Professional Association of German Psychologists, and the Declaration of Helsinki with protocol 2021-002.

Informed Consent Statement: Written informed consent was obtained from the participant involved in this study.

Data Availability Statement: The data presented in this study are available on request from the corresponding author due to permission restriction.

Acknowledgments: We would like to thank our participant and her family for participating in this study over a long period of time.

Conflicts of Interest: The authors declare no conflicts of interest.

References

1. Gorno-Tempini, M.L.; Hillis, A.E.; Weintraub, S.; Kertsez, A.; Mendez, M.; Cappa, S.F.; Ogar, J.M.; Rohrer, J.D.; Black, S.; Boeve, B.F.; et al. Classification of primary progressive aphasia and its variants. *Neurology* **2011**, *76*, 1006–1014. [CrossRef]
2. Hodges, J.R.; Piguet, O. Progress and challenges in frontotemporal dementia research: A 20-year review. *J. Alzheimer's Dis* **2018**, *62*, 1467–1480. [CrossRef]
3. Mesulam, M.M.; Coventry, C.; Bigio, E.H.; Geula, C.; Thompson, C.K.; Bonakdarpour, B.; Gefen, T.; Rogalski, E.; Weintraub, S. Nosology of primary progressive aphasia and the neuropathology of language. In *Frontotemporal Dementias*; Ghetti, B., Buratti, E., Boeve, B., Rademakers, R., Eds.; Springer: Cham, Switzerland, 2021; Volume 1281, pp. 33–49.
4. Kertesz, A.; Jesso, S.; Harciarek, M.; Blair, M.; McMonagle, P. What is semantic dementia? A cohort study of diagnostic features and clinical boundaries. *Arch. Neurol.* **2010**, *67*, 483–489. [CrossRef] [PubMed]
5. Pagnoni, I.; Gobbi, E.; Premi, E.; Borroni, B.; Binetti, G.; Cotelli, M.; Manenti, R. Language training for oral and written naming impairment in primary progressive aphasia: A review. *Transl. Neurodegener.* **2021**, *10*, 24. [CrossRef] [PubMed]
6. Suárez-González, A.; Savage, S.A.; Bier, N.; Henry, M.L.; Jokel, R.; Nickels, L.; Taylor-Rubin, C. Semantic Variant Primary Progressive Aphasia: Practical Recommendations for Treatment from 20 Years of Behavioural Research. *Brain Sci.* **2021**, *11*, 1552. [CrossRef]
7. Jokel, R.; Graham, N.L.; Rochon, E.; Leonard, C. Word retrieval therapies in primary progressive aphasia. *Aphasiology* **2014**, *28*, 1038–1068. [CrossRef]
8. Robinaugh, G.; Henry, M.L. Behavioral interventions for primary progressive aphasia. In *Handbook of Clinical Neurology*; Hillis, A.E., Fridriksson, J., Eds.; Elsevier: Amsterdam, The Netherlands, 2022; Volume 185, pp. 221–240.
9. Newhart, M.; Davis, C.; Kannan, V.; Heidler-Gary, J.; Cloutman, L.; Hillis, A.E. Therapy for naming deficits in two variants of primary progressive aphasia. *Aphasiology* **2009**, *23*, 823–834. [CrossRef]
10. Jokel, R.; Anderson, N. Effectiveness of semantic versus phonological cues in semantic dementia. *Procedia Soc. Behav. Sci.* **2010**, *6*, 191–192. [CrossRef]
11. Jokel, R.; Kielar, A.; Anderson, N.D.; Black, S.E.; Rochon, E.; Graham, S.; Freedman, M.; Tang-Wai, D.F. Behavioural and neuroimaging changes after naming therapy for semantic variant primary progressive aphasia. *Neuropsychologia* **2016**, *89*, 191–216. [CrossRef]
12. Jafari, S.; Khatoonabadi, A.R.; Noroozian, M.; Mehrim, A.; Ashayeri, H.; Nickels, L. The effect of word retrieval therapying primary progressive aphasia: A single case study. *Arch. Neurosci.* **2018**, *5*, e67577.
13. Haentjens, K.; Auclair-Ouellet, N. Naming gains and within-intervention progression following semantic feature analysis (SFA) and phonological components analysis (PCA) in adults with chronic post-stroke aphasia. *Aphasiology* **2021**, *35*, 1024–1047. [CrossRef]

14. Henry, M.; Rising, K.; Demarco, A.; Miller, B.; Gorno-Tempini, M.; Beeson, P. Examining the value of lexical retrieval treatment in primary progressive aphasia: Two positive cases. *Brain Lang.* **2013**, *127*, 145–156. [CrossRef] [PubMed]
15. Cadório, I.; Lousada, M.; Martins, P.; Figueiredo, D. Generalization and maintenance of treatment gains in primary progressive aphasia (PPA): A systematic review. *Int. J. Lang. Commun. Disord.* **2017**, *52*, 543–560. [CrossRef] [PubMed]
16. Wauters, L.D.; Croot, K.; Dial, H.R.; Duffy, J.R.; Grasso, S.M.; Kim, E.; Schaffer Mendez, K.; Ballard, K.J.; Clark, H.M.; Kohley, L.; et al. Behavioral Treatment for Speech and Language in Primary Progressive Aphasia and Primary Progressive Apraxia of Speech: A Systematic Review. *Neuropsychol. Rev.* 2023; *ahead of print*. [CrossRef]
17. Martin, A.; Chao, L.L. Semantic memory and the brain: Structure and processes. *Curr. Opin. Neurobiol.* **2001**, *11*, 194–201. [CrossRef]
18. Boyle, M. Semantic Feature Analysis treatment for aphasic word retrieval impairments: What's in a name? *Top. Stroke Rehabil.* **2010**, *17*, 411–422. [CrossRef]
19. Efstratiadou, E.A.; Papathanasiou, I.; Holland, R.; Archonti, A.; Hilari, K. A systematic review of Semantic Feature Analysis therapy studies for aphasia. *J. Speech Lang. Hear. Res.* **2018**, *61*, 1261–1278. [CrossRef]
20. Cotelli, M.; Manenti, R.; Ferrari, C.; Gobbi, E.; Macis, A.; Cappa, S.F. Effectiveness of language training and non-invasive brain stimulation on oral and written naming performance in Primary Progressive Aphasia: A meta-analysis and systematic review. *Neurosci. Biobehav. Rev.* **2020**, *108*, 498–525. [CrossRef]
21. Henry, M.L.; Hubbard, H.I.; Grasso, S.M.; Dial, H.R.; Beeson, P.M.; Miller, B.L.; Gorno-Tempini, M.L. Treatment for word retrieval in semantic and logopenic variants of primary progressive aphasia: Immediate and long-term outcomes. *J. Speech Lang. Hear. Res.* **2019**, *62*, 2723–2749. [CrossRef]
22. Norise, C.; Hamilton, R.H. Non-invasive brain stimulation in the treatment of post-stroke and neurodegenerative aphasia: Parallels, differences, and lessons learned. *Front. Hum. Neurosci.* **2017**, *10*, 675. [CrossRef] [PubMed]
23. Coemans, S.; Struys, E.; Vandenborre, D.; Wilssens, I.; Engelborghs, S.; Paquier, P.; Tsapkini, K.; Keulen, S. A Systematic Review of Transcranial Direct Current Stimulation in Primary Progressive Aphasia: Methodological Considerations. *Front. Aging Neurosci.* **2021**, *13*, 710818. [CrossRef]
24. Sheppard, S.M. Noninvasive brain stimulation to augment language therapy for primary progressive aphasia. In *Handbook of Clinical Neurology*; Hillis, A.E., Fridriksson, J., Eds.; Elsevier: Amsterdam, The Netherlands, 2022; Volume 185, pp. 251–260.
25. Nitsche, M.A.; Paulus, W. Excitability changes induced in the human motor cortex by weak transcranial direct current stimulation. *J. Physiol.* **2000**, *527*, 633–639. [CrossRef]
26. Marangolo, P. The potential effects of transcranial direct current stimulation (tDCS) on language functioning: Combining neuromodulation and behavioral intervention in aphasia. *Neurosci. Lett.* **2020**, *719*, 133329. [CrossRef] [PubMed]
27. Nissim, N.R.; Moberg, P.J.; Hamilton, R.H. Efficacy of noninvasive brain stimulation (tDCS or TMS) paired with language therapy in the treatment of primary progressive aphasia: An exploratory meta-analysis. *Brain Sci.* **2020**, *10*, 597. [CrossRef]
28. Teichmann, M.; Lesoil, C.; Godard, J.; Vernet, M.; Bertrand, A.; Levy, R.; Dubois, B.; Lemoinem, L.; Truong, D.Q.; Bikson, M.; et al. Direct current stimulation over the anterior temporal areas boosts semantic processing in primary progressive aphasia. *Ann. Neurol.* **2016**, *80*, 693–707. [CrossRef]
29. Roncero, C.; Service, E.; De Caro, M.; Popov, A.; Thiel, A.; Probst, S.; Chertkow, H. Maximizing the treatment benefit of tDCS in neurodegenerative anomia. *Front. Neurosci.* **2019**, *13*, 1231. [CrossRef] [PubMed]
30. Roncero, C.; Kniefel, H.; Service, E.; Thiel, A.; Probst, S.; Chertkow, H. Inferior parietal transcranial direct current stimulation with training improves cognition in anomic Alzheimer's disease and frontotemporal dementia. *Alzheimer's Dement. Transl. Res. Clin. Interv.* **2017**, *3*, 247–253. [CrossRef]
31. Hung, J.; Bauer, A.; Grossman, M.; Hamilton, R.H.; Coslett, H.B.; Reilly, J. Semantic feature training in combination with transcranial direct current stimulation (tDCS) for progressive anomia. *Front. Hum. Neurosci.* **2017**, *11*, 253. [CrossRef]
32. Balduin-Philipps, L.S.; Strunk, K.; Weiss, S.; Müller, H.M. Transcranial direct current stimulation (tDCS) and semantic feature analysis (SFA) in semantic dementia: A single case study. *Clin. Neurophysiol.* **2020**, *131*, e194. [CrossRef]
33. Bright, P.; Moss, H.E.; Stamatakis, E.A.; Tyler, L.K. Longitudinal studies of semantic dementia: The relationship between structural and functional changes over time. *Neuropsychologia* **2008**, *46*, 2177–2188. [CrossRef]
34. Beales, A.; Cartwright, J.; Whitworth, A.; Panegyres, P.K. Exploring generalisation processes following lexical retrieval intervention in primary progressive aphasia. *Int. J. Speech Lang. Pathol.* **2016**, *18*, 299–314. [CrossRef]
35. Croot, K.; Taylor, C.; Abel, S.; Jones, K.; Krein, L.; Hameister, I.; Ruggero, L.; Nickels, L. Measuring gains in connected speech following treatment for word retrieval: A study with two participants with primary progressive aphasia. *Aphasiology* **2015**, *29*, 1265–1288. [CrossRef]
36. Antal, A.; Alekseichuk, I.; Bikson, M.; Brockmüller, J.; Brunoni, A.R.; Chen, R.; Cohen, L.G.; Dowthwaite, G.; Ellrich, J.; Flöel, A.; et al. Low intensity transcranial electric stimulation: Safety, ethical, legal regulatory and application guidelines. *Clin. Neurophysiol.* **2017**, *128*, 1774–1809. [CrossRef]
37. Huber, W.; Poeck, K.; Weniger, D.; Willmes, K. *Aachener Aphasie Test (AAT)*; Hogrefe: Göttingen, Germany, 1983.
38. Kaplan, E.; Goodglass, H.; Weintraub, S.; Segal, O. *Boston Naming Test*; Lea & Febiger: Philadelphia, PA, USA, 1983.
39. Glindemann, R.; Klintwort, D.; Ziegler, W.; Goldenberg, G. *Bogenhausener Semantik-Untersuchung BOSU*; Urban & Fischer: Munich, Germany, 2002.
40. Fox, A.V. *TROG-D Test zur Überprüfung des Grammatikverständnisses*; Beltz Test GmbH: Göttingen, Germany, 2009.

41. Aschenbrenner, S.; Tuchs, O.; Lange, K.W. *RWT Regensburger Wortflüssigkeits-Test*; Hogrefe: Göttingen, Germany, 2001.
42. Helmstaedter, C.; Lendt, M.; Lux, S. *VLMT Verbaler Lern-und Merkfähigkeitstest*; Beltz Test GmbH: Göttingen, Germany, 2001.
43. Kalbe, E.; Kessler, J.; Calabrese, P.; Smith, R.; Passmore, A.P.; Brand, M.; Bullock, R. DemTect: A new, sensitive cognitive screening test to support the diagnosis of mild cognitive impairment and early dementia. *Int. J. Geriatr. Psychiatry* **2004**, *19*, 136–143. [CrossRef]
44. Nasreddine, Z.S.; Phillips, N.A.; Bédirian, V.; Charbonneau, S.; Whitehead, V.; Collin, I.; Cummings, J.L.; Chertkow, H. The Montreal Cognitive Assessment, MoCA: A brief screening tool for mild cognitive impairment. *J. Am. Geriatr. Soc.* **2005**, *53*, 695–699. [CrossRef]
45. Sheikh, J.I.; Yesavage, J.A. Geriatric Depression Scale (GDS): Recent evidence and development of a shorter version. *Clin. Gerontol.* **1986**, *5*, 165–173.
46. Boyle, M.; Coelho, C.A. Application of semantic feature analysis as a treatment for aphasic dysnomia. *Am. J. Speech Lang. Pathol.* **1995**, *4*, 94–98. [CrossRef]
47. Wambaugh, J.L.; Mauszycki, S.; Wright, S. Semantic feature analysis: Application to confrontation naming of actions in aphasia. *Aphasiology* **2014**, *28*, 1–24. [CrossRef]
48. BOSS. Available online: http://sites.google.com/site/mathieubrodeur/Home/boss (accessed on 29 November 2023).
49. Moreno-Martínez, F.J.; Montoro, P.R. An ecological alternative to Snodgrass & Vanderwart: 360 high quality colour images with norms for seven psycholinguistic variables. *PLoS ONE* **2012**, *7*, e37527.
50. Stark, J.A. *Everyday Life Activities Photo Series*; Psychology Press Limited: London, UK, 1994.
51. Stark, J.A. ELA "Everyday Life Acticities"-Photoserie im kreativen Einsatz bei der Diagnose und Therapie von Sprachstörungen. In *Studien zur Klinischen Linguistik*; Rickheit, G., Ed.; Westdeutscher Verlag: Opladen, Germany, 1997; pp. 379–409.
52. The jamovi project. *Jamovi*, (version 2.3.28); Jamovi: Sydney, Australia, 2023.
53. Karageorgiou, E.; Miller, B.L. Frontotemporal lobar degeneration: A clinical approach. *Semin. Neurol.* **2014**, *34*, 189–201. [CrossRef] [PubMed]
54. Krajenbrink, T.; Croot, K.; Taylor-Rubin, C.; Nickels, L. Treatment for spoken and written word retrieval in the semantic variant of primary progressive aphasia. *Neuropsychol. Rehabil.* **2018**, *30*, 915–947. [CrossRef]
55. Marangolo, P.; Pisano, F. Conversational Therapy in Aphasia: From Behavioral Intervention to Neuromodulation. *Semin. Speech Lang.* **2020**, *41*, 61–70. [CrossRef] [PubMed]
56. Nitsche, M.A.; Paulus, W. Transkranielle Gleichstromstimulation. In *Das TMS-Buch*; Siebner, H., Ziemann, U., Eds.; Springer: Berlin/Heidelberg, Germany, 2007; pp. 533–542.
57. Reilly, J. How to constrain and maintain a lexicon for the treatment of progressive semantic naming deficits: Principles of item selection for formal semantic therapy. *Neuropsychol. Rehabil.* **2016**, *26*, 126–156. [CrossRef] [PubMed]
58. de Aguiar, V.; Zhao, Y.; Faria, A.; Ficek, B.; Webster, K.T.; Wendt, H.; Wang, Z.; Hillis, A.E.; Onyike, C.U.; Frangakis, C.; et al. Brain volumes as predictors of tDCS effects in primary progressive aphasia. *Brain Lang.* **2020**, *200*, 104707. [CrossRef]
59. Strunk, K.; Wollenberg, M.; Weiss, S.; Müller, H.M. Semantic feature analysis (SFA) and phonological components analysis (PCA) combined with transcranial direct current stimulation (tDCS) in semantic variant of primary progressive aphasia (svPPA): A longitudinal single case study. In Proceedings of the 12th World Congress for Neurorehabilitation, Vienna, Austria, 14–17 December 2022; p. eP271.
60. Ficek, B.N.; Wang, Z.; Zhao, Y.; Webster, K.T.; Desmond, J.E.; Hillis, A.E.; Frangakis, C.; Faria, A.V.; Caffo, B.; Tsapkini, K. The effect of tDCS on functional connectivity in primary progressive aphasia. *NeuroImage Clin.* **2018**, *19*, 703–715. [CrossRef] [PubMed]
61. Tsapkini, K.; Webster, K.T.; Ficek, B.N.; Desmond, J.E.; Onyike, C.U.; Rapp, B.; Frangakis, C.E.; Hillis, A.E. Electrical brain stimulation in different variants of primary progressive aphasia: A randomized clinical trial. *Alzheimer's Dement.* **2018**, *4*, 461–472. [CrossRef]
62. Tao, Y.; Ficek, B.; Rapp, B.; Tsapkini, K. Selective Functional Network Changes Following tDCS-Augmented Language Treatment in Primary Progressive Aphasia. *Front. Aging Neurosci.* **2021**, *13*, 681043. [CrossRef]

brain sciences

MDPI

Systematic Review

Evidence of Language Development Using Brief Animated Stimuli: A Systematic Review

Triantafyllia I. Vlachou [1], Maria Kambanaros [2], Panagiotis Plotas [1,3] and Voula Chris Georgopoulos [1,3,*]

[1] Department of Speech and Language Therapy, University of Patras, 26504 Patras, Greece
[2] The Brain and Neurorehabilitation Lab, Department of Rehabilitation Sciences, Cyprus University of Technology, Limassol 3041, Cyprus
[3] Primary Health Care Laboratory, School of Health Rehabilitation Sciences, University of Patras, 26504 Patras, Greece
[*] Correspondence: vgeorgop@upatras.gr

Abstract: There is limited evidence regarding the effect of animation compared to static pictures on children's language development. The aim was to systematically review the available literature for evidence concerning the effect of brief animation on spoken language responses (receptive—listening or expressive—speaking) in typically developing (TD) children aged 3 to 9 years. Five databases were searched, resulting in seven included studies. The characteristics of animated stimuli, the manner of presentation, and the language-related tasks were recorded, and questions were posed about the effect of brief animation on children's receptive and expressive language abilities. The evidence suggests that animation may have a positive effect on expressive language abilities of children compared to static pictures. As far as the effect of animation on receptive language performance is concerned, the evidence is less concrete. Future directions regarding the potential of animation on language development are discussed.

Keywords: animation; language development; typically developing children; speech language pathology

1. Introduction

Limited research has been conducted on the impact of animations compared to static pictures on children's language development within the field of speech language pathology (SLP). This systematic review aimed to identify studies employing brief animations to explore receptive and expressive verbal language responses in typically developing children (TD) aged 3 to 9 years. The objective was to investigate any positive or negative effects of brief animation on language development and related non-verbal behaviors (e.g., motivation, distraction, and concentration).

1.1. What Is Animation?

Modern technology offers diverse animation forms (e.g., 2D and 3D; [1]) integrated into daily life through computers and mobile devices (e.g., applications, users' interfaces, etc.) [2,3]. However, the scope of this study is not the technological advancements of animations, but rather their potential effect on language development in TD children.

The main theoretical frameworks concerning human perception of animation are grounded in the principles of gestalt psychology [4–6], visual communication [6–8], the cognitive theory of multimedia learning [9], the cognitive load theory [10–12], and the animation processing model [13]. In this study, the terminology employed aligns with that of Bétrancourt and Tversky [7], who defined animation as "... any application that generates a sequence of frames, wherein each frame represents a modification of the preceding one, and where the sequencing of frames is determined by either the designer or the user" [7] (p. 313).

The existing literature has extensively explored the efficacy of animations in educational contexts, comparing them to static images for conveying complex subjects (e.g., biology and chemistry), with a focus on adolescent and young adult learners [6,7,14–17].

The use of animation in instructional materials initially gained interest due to its dynamic nature, conveying additional information regarding temporal changes. Also, animation was deemed attractive, amusing, and motivating. However, concerns have arisen regarding its efficacy due to its transient nature, distracting attributes, and the cognitive overload it imposes on working memory [6,8,14].

Animations have been indirectly studied in electronic book formats for instructing young children [18–21]. Many studies associate animations with children's vocabulary and narration, but they primarily rely on multimedia approaches [9]. Consequently, they are unable to offer clear insights specifically on animation due to the interference of other variables. Beyond the positive or negative results reported in these studies, they also reflect children's familiarity with technology from an early age [22].

1.2. Animation in Speech and Language Pathology Research

In the field of SLP, static pictures are commonly used as visual stimuli for language-related tasks, while the potential utility of animation has received limited research attention. To the best of our knowledge, only a few studies have explored the effect of animation versus static pictures in young TD children, and this is primarily within the field of augmentative and alternative communication (AAC).

AAC employs various symbols to represent client needs and communication concepts and facilitates children with developmental disabilities "to understand, produce, and learn language" [23] (p. 1). Iconicity, the visual relationship of symbols to their referents (visual resemblance), is categorized by the ease of guessing their meanings: transparent (easily guessed), translucent (obvious with context), and opaque (not readily perceived) [24]. The aim of AAC is to enhance users' communicative competence, encompassing linguistic, operational, social, and strategic abilities, as well as psychosocial and environmental factors [25,26].

Jagaroo and Wilkinson [27] discussed the role and function of motion perception from the perspective of cognitive neuroscience, arguing for the use of animated symbols in AAC and presenting a theoretical framework to guide speech–language pathologists' clinical decisions. Frick et al. [2] reviewed previous research on animation in AAC and suggested potential applications of animation to enhance linguistic and operational competence for individuals with complex communication needs. In the field of the brain–computer interface in AAC (BCI-AAC), Pitt et al. [28] investigated the use of functional animated symbols in children aged 9 to 13 years old, reporting preliminary positive results. Finally, Schlosser et al. [29] conducted a scoping review that describes the roles of animation in AAC for TD individuals and those with developmental disabilities (children and adults).

In the field of AAC, studies comparing animations to static pictures as symbols in TD children were conducted by Mineo et al. [30], Schlosser et al. [31,32], Gunduz [33], and Brock et al. [23]. Additionally, Harmon et al. [34] investigated animated symbols alongside environmental sounds.

Also, Frizelle et al. [35] explored the effect of animation compared to static pictures on TD children's language development using animations to assess complex syntax. Finally, Diehm et al. [36] investigated children's narrative retelling skills in response to an animated video story versus a static pictures book, employing a multimedia features approach.

Considering all of the above, there is a significant gap in the literature regarding the effect of animation on young children's typical language development.

1.3. Children's Language Development

Language is considered a behavior that involves comprehension and use. It is typically described as spoken and written (or other communication symbol systems). These two forms include receptive (listening and reading) and expressive (speaking and writing) components. Language also comprises of five domains: phonology, morphology, syntax, semantics, and pragmatics [37].

Speech language pathologists (SLPs) do not apply instructional approaches when working on language development with young children for several reasons: (a) Language development cannot be taught as a set of rules [38]. (b) Assessment and/or intervention is individualized. (c) The primary goal of intervention is to enhance language skills and communication [39]. (d) Formal literacy skills are not fully developed or may still be emerging [40]. (e) Various language aspects or progress are evaluated using specific standardized tests [41]. Consequently, SLPs cannot adapt and apply instructional practices designed for adult learners to children. Finally, it should be noted that both language and cognitive abilities are under development throughout this stage of life [42,43], emphasizing the need for specific research tailored to this population.

Given these considerations, it is necessary to investigate whether animation serves as an effective tool for fostering language development in children.

1.4. The Present Study

Twenty years ago, Tversky et al. described animation as an "attractive graphic device" [8] (p. 250). However, the authors raised doubts about the efficacy of animation in instruction and its ability to convey concepts described with many words or to externalize internal knowledge. This prompts us to question whether animation presentations have positive or negative effects on the language development of young children, and if so, how do these effects manifest.

The primary objective of this systematic review is to gather evidence of spoken language responses (receptive—listening or expressive—speaking) in TD children exposed to a brief animated stimulus. A secondary aim is to document any other reported non-verbal behaviors (e.g., motivation, concentration, and distraction).

The following research questions were asked:

1. Can animation better support children's receptive language abilities compared to a static picture?
2. Does animation enhance children's expressive language more than a static picture?

This systematic review is necessary because there are a lack of data on how animation impacts children's language development in SLP literature. The findings are expected to contribute to future research questions and suggestions for clinical practice.

2. Methods

The present systematic review is based on the preferred reporting items for systematic reviews and meta-analyses protocol (PRISMA) and the relevant flow diagram [44,45]. This review was not pre-registered.

2.1. The Present Study

The selection criteria for the included studies were established following the PICO model (population, intervention, comparison, and outcome) as described in Table 1 [46].

The focus centered on studies involving TD children within the specific age range to establish a basis for this population. Papers should include brief animations lasting less than one minute. The studies should either compare animations to static pictures or compare other animation characteristics. The sentence level was defined as the maximum linguistic response.

In terms of outcomes, the goal was to report data on language responses examined in each study, as well as language responses reported as secondary to the main study objectives. Furthermore, it was meaningful to record any other reported behaviors, whether positive or negative, potentially triggered by the presence of animation.

A brief animation was defined as the minimum duration from which evidence would be extracted. The main research purpose of the included papers should be animation involving some aspect of TD children's language (receptive or expressive). Studies from the instructional field, and those related to e-books, applications, and television viewing were excluded.

Table 1. Inclusion and exclusion criteria.

Selection Criterion	Description
Population/problem	TD children 3 to 9 years old
Intervention	Presentation of brief animations (under one minute duration) individually provided by SLPs or psychologists regardless of the frequency of administration.
Comparisons	Animations versus static pictures or comparisons of different animation features.
Outcomes	1. Language response: (a) targeted within the research; (b) reported but not targeted. 2. Other behaviors: (a) wanted (e.g., attention, concentration, participation, and motivation) or (b) undesired (e.g., distraction, withdrawal, frustration, and overuse).
Settings	Research clinic, schools, nurseries, and in the home.
Limits	Animation is the main research purpose which involves some language aspect. Studies published in English language without local restrictions.
Exclusion criteria	Multimedia applications and e-books. Research that examines animations in adults and non-typically developing children. Animated stories aimed at narration. Studies using animation as feedback and/or reward. Papers regarding instruction and television animation viewing.

2.2. Information Sources

A literature search from October to December 2022 was conducted by the first author, considering selection criteria covering the past 20 years (2002–2022). The databases Scopus, Web of Science, PubMed, ScienceDirect, and the Educational Resources Information Center (ERIC) were searched.

Due to the limited availability of relevant studies, further research was conducted by examining reference lists and the journal *"Computer Animation and Virtual Worlds"*, and consulting Google Scholar up to February 2023. In both search avenues, articles, conference papers, and theses that employed quantitative research methods were included.

2.3. Search Strategy and Eligible Studies

Different combinations of the following search terms were used: "animation", "language", "preschool children", "language development", "vocabulary", "verbs", "learning", "animated books", and "animated pictures".

The search process involved standard steps: identification, screening (title, abstract, and full text), and inclusion of the final studies. Initially, 5165 studies were identified across all databases. After applying filters based on time and scientific areas, and removing duplicates and ineligible records, 51 studies remained. During the title screening, 8 studies were excluded (other reviews and meta-analyses), leaving 43. Subsequently, 22 more studies were excluded after reviewing the abstracts (qualitative studies or ineligibility based on inclusion criteria), leaving 21. In the full-text reading stage, 17 studies were excluded because of their indirect relevance to the research questions. Ultimately, only 4 studies were included.

Additional searches, including reference lists and Google Scholar, uncovered 31 studies. Four were excluded based on title content (reviews and meta-analyses), and twelve more were excluded due to differing research approaches. After reviewing the remaining 15 studies, 12 were excluded for indirect animation investigation or different approaches. The second search added 3 studies to the systematic review, resulting in a total of 7 included studies.

The complete process is illustrated in Figure 1, using the PRISMA 2020 flow diagram [44,45] for new systematic reviews, encompassing searches across databases, registers, and other sources.

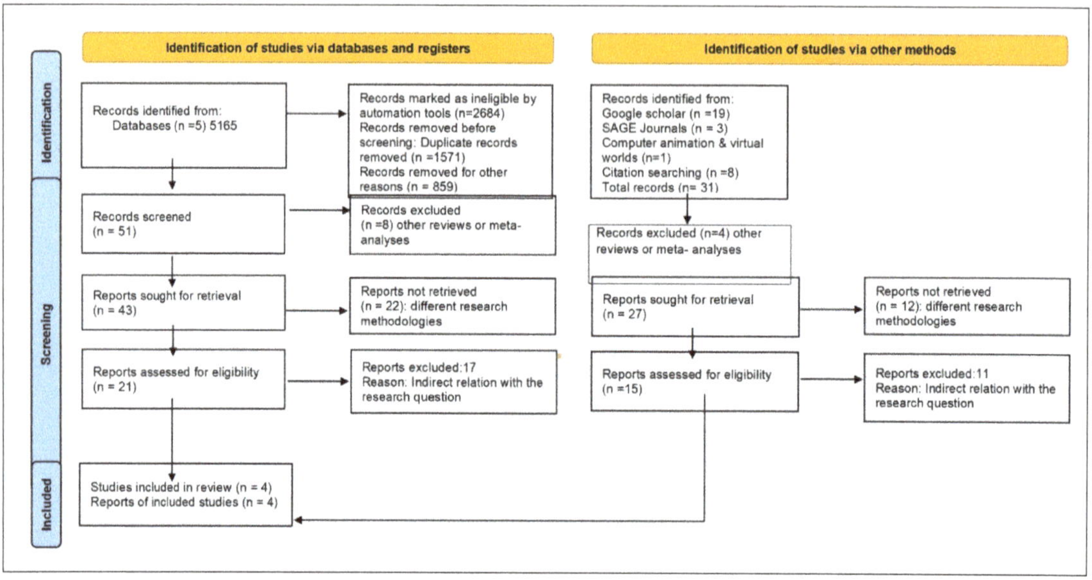

Figure 1. PRISMA flow diagram.

The studies meeting the inclusion criteria were as follows: Mineo et al. [30], Schlosser et al. [31], Schlosser et al. [32], Harmon et al. [34], Gunduz [33], Frizelle et al. [35], and Brock et al. [23].

3. Results

3.1. Aims of the Eligible Studies

The key attributes of the included studies, encompassing research design, research purpose, participants, animated items, comparisons, and animation-related outcomes (positive, no difference, and negative) are presented in Table 2.

The first study investigating the effect of animation in the field of AAC was conducted by Mineo et al. in 2008 [30]. They employed a within-group design to examine the identification of 24 verbs across four conditions (video, animation, static line drawings with disequilibrium cues, and static line drawings with movement cues) in a sample of 93 TD children aged 3, 4, and 5 years. The results indicated that children performed more effectively in dynamic conditions (i.e., video and animation) as opposed to static conditions. Additionally, a developmental effect was observed, with symbol recognition improving as children grew older.

Schlosser et al. [31] used a mixed-group design to examine transparency, name agreement, and identification of 24 animated verbs and 8 animated prepositions compared to their corresponding static symbols with 53 TD children aged 3, 4, and 5 years. The results showed a positive effect on transparency and name agreement in verbs, with no significant difference found for prepositions. Additionally, no significant difference was observed in the identification task of verbs and prepositions. A similar developmental effect to the study by Mineo et al. [30] was evident in all measures.

Table 2. Studies' characteristics.

Study	Research Design	Research Purpose/Animation Function	Participants	Animated Items	Comparisons	Outcomes Regarding Animation
Mineo et al. [30]	Within-group design	Iconicity of symbols—identification	93 TD children (3-, 4-, and 5-years old)	24 verbs	Four conditions: video—animation—static/dis—static/move * symbols	Positive for identification of verbs.
Schlosser et al. [31]	Mixed-group design	Iconicity of symbols—transparency, name agreement, and identification	53 TD children (3-, 4-, and 5-years old)	24 verbs and 8 prepositions	Animated vs. static symbols	Positive for transparency and name agreement in verbs, no difference in prepositions, and no difference for identification in verbs and prepositions.
Schlosser et al. [32]	Randomized 2 × 2 × 2 × 3 factorial mixed design	Iconicity of symbols—naming and identification	220 TD children (3-, 4-, and 5-years old)	24 verbs and 8 prepositions	Animated vs. static/ two sets of symbols	Positive for naming in verbs and prepositions/no difference for identification.
Harmon et al. [34]	Between-group design	Iconicity of symbols—naming	46 TD children (3; 1–3; 11 years old)	18 verbs	Animated symbols with environmental sounds vs. without sounds	Positive for naming mostly in environmental sounds condition.
Gunduz [33]	2 × 2 × 2 × 3 randomized factorial design	Iconicity of symbols—naming and identification	97 TD children (3-, 4-, and 5-years old)	24 verbs and 8 prepositions	Animated vs. static/two sets of symbols	Positive for naming mainly in verbs/no difference for identification.
Frizelle et al. [35]	Within-group design	Assessment of syntax understanding—language comprehension ability	103 TD children (3;6–4;11 years old)	50 sentences (five types of relative clauses)	Sentence verification animation task vs multiple-choice picture selection task	Positive for syntax understanding.
Brock et al. [23]	Counterbalanced 2 × 2 × 2 mixed design	Iconicity of symbols—identification and labeling	24 children (7;0–8;11 years old)	21 sentences	Animated vs. static symbols and psycholinguistic effects	Positive for identification and labeling accuracy mainly in verbs and prepositions/animation reduced the psycholinguistic effects (word frequency and imageability),

* Static/dis: static line drawings with disequilibrium cues; static/move: line drawings with movement cues.

Schlosser et al. [32] employed a randomized 2 × 2 × 2 × 3 factorial mixed design to investigate the naming and identification of 24 animated verbs and 8 animated prepositions in comparison to their static versions. Additionally, this study compared two symbol sets: the Autism Language Program [ALP] developed by Boston Children's Hospital (https://www.childrenshospital.org/programs/autism-language-program (accessed on 27 August 2023)) [47] and the Picture Communication System [PCS] [48]. This study featured the largest sample, consisting of 220 TD children aged 3, 4, and 5 years. The results showed positive outcomes for naming verbs and prepositions, with no significant difference in identification. Furthermore, children exhibited better performance with ALP symbols over PCS symbols, and the developmental effect was reaffirmed.

In 2014, Harmon and colleagues [34] examined whether the addition of environmental sounds (e.g., breaking and door opening) improves the comprehension of animated symbols. They employed a between-group design, comparing 18 animated verbs with and without environmental sounds in a naming task. They recruited a sample of 46 TD children aged 3; 1–3; 11 years, who were randomly assigned to two groups. The results revealed a significant increase in naming accuracy when symbols were presented with environmental sounds.

Gunduz [33] replicated Schlosser et al.'s [32] study using a sample of 97 TD Turkish-speaking children aged 3, 4, and 5 years in Turkey. A 2 × 2 × 2 × 3 randomized factorial design was employed to investigate the naming and identification of 24 animated verbs and 8 animated prepositions using the same two sets of symbols (ALP and PCS). The key findings were replicated as follows: (a) naming accuracy was higher for animated verbs than for prepositions, (b) naming accuracy was superior for ALP animated symbols, (c) the developmental effect of animation was once again observed, and (d) no significant difference was found in the identification task.

In 2019, Frizelle et al. [35] compared two assessment methods for the understanding of complex syntax. They examined a sentence verification animation task and a multiple-choice picture selection task within a sample of 103 TD children aged 3; 6 to 4; 11 years. The study included five types of relative clauses, and the findings indicated that children exhibited superior performance in the sentence verification animation task.

Brock et al. [23] utilized a counterbalanced 2 × 2 × 2 mixed design to investigate the effect of animated symbols on forming sentences. They conducted identification and labeling tasks for 21 sentences with a sample of 24 TD children (aged 7; 0–8; 11 years). Additionally, they introduced a psycholinguistic analysis considering word frequency, imageability, and concreteness. The findings revealed a positive effect of animation on identification and labeling accuracy, primarily for verbs and prepositions, along with a reduction in the psycholinguistic effects of word frequency and imageability.

Most AAC studies have explored the iconicity of single animated symbols (e.g., [30–33]. These experiments compared animated symbols to static symbols, while only Harmon et al. [34] compared animated symbols with and without environmental sounds. Additionally, Brock et al. [23] and Frizelle et al. [35] focused on the sentence level (see Table 2).

In summary, all of the AAC studies mentioned follow a chronological progression in animation research employing consistent research methods, enabling meaningful comparisons. It is important to note that these studies were carried out at an experimental level, not a clinical one, and all report positive results for animation effect.

3.2. Quality Appraisal of Studies

The eligible studies are reported as experimental by the authors and included typically developing (TD) children as participants. The researchers aimed to investigate new variables [23] to establish a research base for children's developmental capacities in response to animation before applying it in clinical populations [30,32–35]. Consequently, the group research designs used did not involve clinical and control groups. Randomization was employed in presentation order or to assign participants to different groups or conditions, and children were informed by the researchers that they were playing a computer-based game. Given these circumstances, two research quality appraisal methods were employed to ensure completeness and fairness.

In the appraisal process, the first and fourth authors independently assessed each study, resolving any discrepancies in discussion with the second author.

Initially, the PEDro-P scale for randomized (RCTs) and non-randomized control trials (NRCTs) was applied (https://speechbite.com/group-comparison-studies (accessed on 18 July 2023)). The PEDro-P scale derives from the original PEDro scale [49], with its reliability investigated by [50]. Both scales consist of 11 criteria evaluating various bias risks, resulting in a final score ranging from 0 to 10. PEDro-P differs in the definition and application of criterion 4. Although the total score of the original PEDro scale lacks validation, it is widely accepted as 0–3 for 'poor', 4–5 for 'fair', 6–8 for 'good', and 9–10 for 'excellent' evidence quality (https://pedro.org.au (accessed on 20 July 2023)). The authors sought guidance from the speechBITE training program. The PEDro-P ratings for the studies are presented in Table 3.

Table 3. The PEDro-P ratings out of 10.

Study	1	2	3	4	5	6	7	8	9	10	11	Total Score
Mineo et al. [30]	1	1	0	1	1	0	NA *	1	1	1	1	7/10
Schlosser et al. [31]	1	1	0	1	1	0	NA	1	1	1	1	7/10
Schlosser et al. [32]	1	1	0	1	1	0	NA	1	1	1	1	7/10
Harmon et al. [34]	1	1	0	1	1	0	NA	1	1	1	1	7/10
Gunduz [33]	1	1	0	1	1	0	NA	1	1	1	1	7/10
Frizelle et al. [35]	1	1	0	1	1	0	NA	1	1	1	1	7/10
Brock et al. [23]	1	1	0	1	1	0	NA	1	1	1	1	7/10

* NA: not applicable.

Also, the quality assessment tool for quantitative studies (QAT) was employed (https://www.ephpp.ca/quality-assessment-tool-for-quantitative-studies (accessed on 18 July 2023)). QAT is a result of the Effective Public Health Practice Project (EPHPP) initiated by the Ontario Ministry of Health [51]. This tool evaluates the following factors: (a) selection bias, (b) study design, (c) confounders, (d) blinding, (e) data collection methods, (f) withdrawals and dropouts, (g) intervention integrity, and (h) analyses. The quality of evidence is categorized as 1 for "strong", 2 for "moderate", and 3 for "weak". The final rating is "strong" if no "weak" ratings are present, "moderate" if one "weak" rating is identified, and "weak" if two or more "weak" ratings are found. The authors referenced the QA dictionary, and the QAT ratings are presented in Table 4.

Table 4. The QAT Ratings.

Study	Selection Bias	Study Design	Confounders	Blinding	Data Collection Method	Withdrawals and Dropouts	Final Rating
Mineo et al. [30]	Strong	Moderate	Strong	Moderate	Strong	Strong	Strong
Schlosser et al. [31]	Strong	Moderate	Strong	Moderate	Strong	Strong	Strong
Schlosser et al. [32]	Strong	Strong	Strong	Moderate	Strong	Strong	Strong
Harmon et al. [34]	Strong	Strong	Strong	Moderate	Strong	Strong	Strong
Gunduz [33]	Strong	Strong	Strong	Moderate	Strong	Strong	Strong
Frizelle et al. [35]	Strong	Moderate	Strong	Moderate	Strong	Strong	Strong
Brock et al. [23]	Moderate	Moderate	Strong	Moderate	Weak	Strong	Moderate

As shown in Tables 3 and 4, all studies received a score of 7 on the PEDro-P scale and were rated as "strong" in quality on the QAT, except for the study by Brock et al. [23], which was rated as "moderate". Consequently, both tools suggest a good to moderate-to-strong level of evidence. The research quality of the included studies is further discussed below.

Regarding the eligibility criteria and selection bias, the authors targeted specific samples with specific characteristics from the typical population for research purposes. Only Frizelle et al. [35] applied an opt-out protocol to minimize volunteer bias. Sample sizes ranged from 46 to 220 subjects, except for the study by Brock et al. [23], which included 24 children (see Table 2). Additionally, some studies performed power calculations [31,32,35]. All studies employed both standardized and informal tests to assess children before their participation in research projects. Withdrawals and dropouts were documented in [34,35].

Randomization was applied in [32–34]. Mineo et al. [30], Schlosser et al. [31], Frizelle et al. [35], and Brock et al. [23] randomized the presentation order of items and/or the order of conditions. None of the studies reported concealed allocation, likely because they did not involve clinical populations or therapy methods.

Furthermore, all of the potential nuisance variables were controlled, and the research limitations were reported by the authors. The study by Brock et al. [23] faced a unique limitation due to the COVID-19 pandemic, necessitating the use of the Zoom video conferencing tool. This is reflected in the "moderate" rating of QAT.

Most studies conducted complete statistical analyses. However, some studies provided additional insights through error analysis (e.g., [30,34]), cross-linguistic analysis [33], or psycholinguistic analysis [23], in line with their research objectives. Frizelle et al. [35], in their comparison of two syntax comprehension tasks for assessment purposes, employed a standardized test to validate their findings.

In all studies, blinding of researchers was not applied, likely due to the introduction of research tasks on a computer or tablet, which enhanced the validity and fidelity of administration and scoring. As a result, criterion 7 on the PEDro-P scale was marked as "Not Applicable" (NA). Also, prior to the main research phase, children had the chance to acquaint themselves with the software and computer or tablet demands.

In the study of Frizelle et al. [35], it was reported that the research was explained to the children without providing additional details by the authors. In the AAC studies, subjects were informed that they would participate in a guessing game. Since both cases involved young children who were not clinical populations, these methods were deemed blinding.

In conclusion, the eligible studies offer a sufficient level of evidence of the positive effect of animation on their respective research objectives.

3.3. Animations and Research Tasks

In consideration of the research questions, the characteristics of the animations and the relative tasks used in the included studies were recorded: (a) duration and number of animated items presented in each task, (b) realism level as classified by [14] (level 1—L1: schematic, level 2—L2: rather simple, level 3—L3: rather realistic, and level 4—L4: photo-realistic = video), (c) participants' prior knowledge of the represented animated content (lexical items), (d) involvement of verbal linguistic components (receptive/expressive language) and/or domain (syntax, semantics, etc.), (e) target responses and participants' responses (according to authors' reports), coded as follows: no response (for receptive language), one or two words, or a phrase/sentence (for expressive language), (f) the language used in the research project, including potential reports for bilingual participants, and (g) reports on other observed behaviors of children (see Table 1). All of these characteristics are summarized in Table 5.

The durations of animations in the included studies varied from 6 to 29 s. Brock et al.'s [23] study lacks this specific information. Nevertheless, given their partial use of the same symbols as Schlosser et al. [31,32], it can be inferred that the duration falls within a similar range.

Regarding the number of animated items in each task, only the studies by Harmon et al. [34] and Frizelle et al. [35] presented a single animated item. Mineo et al. [30], Schlosser et al. [31,32], and Gunduz [33] introduced four and six animations concurrently (identification task), while Brock et al. [23] gradually introduced five animations within a sentence sequence.

Table 5. Characteristics of animations and research tasks.

	Mineo et al. [30]	Schlosser et al. [31]	Schlosser et al. [32]	Harmon et al. [34]	Gunduz [33]	Frizelle et al. [35]	Brock et al. [23]
Duration	10 s/30 frames per second	14 s transparency; 20 s identification	14 s naming; 20 s identification	13–29 s	14 s naming; 20 s identification	6 s average	No report
Animations presented per task	4	1 or 4	1 or 4	1	1 or 6	1	5
Realism level	L1 *	L2	L1 and L2	L2	L1 and L2	L2	L2
Previous knowledge of the items	YES	YES	YES	YES	YES	Under assessment	YES
Verbal linguistic component(s) involved and/or domain	Receptive language (verbs)	Receptive/expressive language (verbs and prepositions)	Receptive/expressive language (verbs and prepositions)	Expressive language (verbs)	Receptive/expressive language (verbs and prepositions)	Receptive language/syntax (relative clauses)	Receptive/expressive language (sentences) and psycholinguistic characteristics
Target responses/ number of words	No word	One word for verbs and 1–2 words for prepositions	One word for verbs and 1–2 words for prepositions	One word	One word for verbs and 1–2 words for prepositions	No word	A sentence (five or more words)
Responses of the participants	No response was demanded	One word, an alternative world, a phrase, or sentence	One word, an alternative word, a phrase, or sentence	One word, an alternative word, a phrase, or sentence	One word, an alternative word, a phrase, or sentence	No response was demanded	Complete grammatical sentences or symbol-by-symbol labeling
Language used	English	English	English	English	Turkish	English	English
Bilingualism reported	NO	NO	NO	NO	NO	Bilinguals excluded	NO
Other behaviors	Viewing response to simultaneous animated stimuli	No report	No report	Visual attention to the screen	No report	Rarely requested animation repetition	No report

* L: level.

All of the animations were created with colorful graphics designed for children in a two-dimensional (2D) format and their realism level varied from L1 to L2 [14]. Only in the study of Mineo et al. [30] did the authors create an experimental set of animated black and white line drawings (stick figures).

Most AAC studies utilized the same sets of animated symbols (ALP and PCS symbols) for different research objectives [31–34] that were not originally created for these studies. Schlosser et al. [32] and the replication study of Gunduz [33] compared these two sets of animated symbols (PCS and ALP) across two different languages. In contrast, Frizelle et al.'s [35] animations were designed specifically for their study. Lastly, Brock et al. [23] combined animated symbols from these two sets to construct sentences and introduced non-animated photographs in an identification task.

All studies in the field of AAC ensured that children possessed prior receptive or expressive knowledge of the lexical items (content of representation) being investigated, as they were exploring the level of iconicity of the symbols. On the other hand, Frizelle et al. [35] used animations as an assessment method to determine the presence or absence of the knowledge under examination (relative clauses).

Regarding language features, the included studies involved verbal forms (listening and speaking). All studies involved animated tasks that required receptive language demonstration, with five of them necessitating expressive language from children (see Table 5). Harmon et al. [34] introduced only animated expressive language tasks, adding environmental sounds in the control condition. Five studies [30–34] focused on the word level (animated verbs and/or prepositions), while two studies [23,35] addressed the sentence level (syntax). In these instances, one animation represented a complex sentence, or five animated symbols represented the sentence components, respectively.

The authors introduced animated symbols aiming for specific responses. In identification tasks (receptive language), no answer was required. The children simply had to select one animation from the others. In the case of Frizelle et al. [35], children had to verify if the animation was represented by the sentence that they heard. In naming or labeling tasks (expressive language), children had to produce one word for verbs, one or more words for prepositions (e.g., next to), or a sentence with five or more words (e.g., [23]) in response to a prompt question.

On the other hand, children's expressive responses for verbs and prepositions varied from one word to a phrase or sentence containing the target word. Additionally, the authors reported that children sometimes used alternative words, such as synonyms and free morphemes [31–34]. Finally, in Brock et al.'s study [23], children demonstrated the ability to produce complete grammatical sentences or symbol-by-symbol labeling.

The English language predominated in the eligible studies, with only Gunduz [33] offering information on animated symbols in the Turkish language. Furthermore, Frizelle et al. [35] stands out as the sole study that explicitly mentions the exclusion of bilingual children. All other studies indicate that English is the primary language spoken at home.

Limited information was reported in the studies about other observed behaviors in children during the presentation of animations. Mineo et al. [30] (p. 169) noted that "... some children seemed somewhat agitated by the need to attend to multiple moving stimuli at the same time" and explained how children reacted and performed during the specific task. Harmon et al. [34] argue that even with the addition of environmental sounds, children seemed to rely more on the visual animated stimuli and reported that all participants visually attended to the screen. Finally, Frizelle et al. [35] reported that participants rarely requested a repetition of the animation for which they had to make a judgment.

In conclusion, apart from the environmental sounds in Harmon et al.'s [34] study, none of the animations featured simultaneous linguistic sounds (e.g., words). Essentially, most animations presented only visual stimuli, while Harmon et al.'s [34] study specifically focused on environmental, not linguistic, stimuli. The only instance of animation and a spoken sentence occurring concurrently was found in Frizelle et al.'s [35] study.

Clearly, the choices made in each study were contingent upon their respective objectives and methodologies.

3.4. Focus on Aspects of Language

While studies in the field of AAC primarily focus on symbol research, the authors have addressed some language-related aspects in connection with their methodologies and related tasks. Thus, according to Mineo et al. [30], a language comprehension task (identification) was employed to examine receptive language while comparing four forms of picture-based language representations. Similar tasks were also employed by Schlosser et al. [31,32] and Gunduz [33]. Furthermore, Schlosser et al. [31,32] argue that the identification task requires children to associate language with graphic symbols, and this adds a novel dimension to the study of animation.

On the other hand, tasks such as naming demanded expressive language from the participants. Schlosser et al. [32] discussed how this task triggered numerous alternative responses among the participants. Nevertheless, Harmon et al. [34] argued that the type of prompt used in these tasks might have influenced the children's responses. Specifically, the instruction "What is this?" could lead to responses containing nouns or sentences consisting of both nouns and verbs, rather than the desired verb. Eventually, Harmon et al. [34] suggested that the training tasks aided the participants in grasping the experimental task objectives. However, it remains unclear whether this phenomenon (i.e., the production of sentences) was exclusive to animated stimuli or also occurred with the static items (e.g., [31–33]). This question is essentially addressed in the Harmon et al. [34] study, which exclusively involved animated stimuli and arrived at similar observations.

Harmon et al.'s [34] analysis of semantic errors of children sheds light on a distinct aspect of language. The researchers observed semantic type superordinate errors (among others) across conditions (i.e., animations without sound vs. animations with environmental sounds) where the trend was to name the animated symbol with a more general verb instead of the target verb. The authors purported that this finding is consistent with the children's early vocabulary development.

Gunduz [33] replicated the study of Schlosser et al. [32] by using the same animated symbols due to the unavailability of related symbols for the Turkish language. The author conducted a cross-linguistic analysis of Turkish by examining specific linguistic characteristics (e.g., word class, syntax, etc.) and suggested that the acquisition of verbs and prepositions may be language specific. The results (Table 2) generally aligned with those of Schlosser et al. [32], indicating that the animated symbols, without any adaptation, allowed for the representation of the same meanings in a different language, at least to some extent.

The study of Brock et al. [23] focused on syntax, utilizing both identification and labeling tasks involving animated symbols. Experimental sentences followed the subject–verb–object–prepositional phrase structure and were constructed by five animated symbols. As depicted in Table 5, children demonstrated the ability to follow this structure, generating grammatical sentences or symbol-by-symbol labeling, all without prior training. Furthermore, the authors introduced a novel dimension to AAC, conducting a psycholinguistic analysis that considered word frequency, imageability, and concreteness. They argued that animation reduced the effects of these factors.

Frizelle et al. [35] explored language assessment by testing children's comprehension of complex sentences. They discovered that the sentence verification animation task enhanced children's sentence comprehension abilities, offering a novel assessment approach. Additionally, they observed that each method revealed a different hierarchy of constructions and the impact of assessment method on the participants' scores was greater for some constructions than others. Furthermore, the authors discussed the influence of pragmatically appropriate context on the children's scores, as well as cognitive factors like attention and memory, which could affect scores in multiple-choice assessments using static pictures.

Finally, the evidence on bilingualism is severely limited. While Frizelle et al. [35] excluded bilingual children, the remaining included studies did not yield any clear indication of whether bilingual children were participants in the research.

4. Discussion

In this systematic review, we conducted a search for evidence regarding the effect of brief animations on the language development of TD children aged 3 to 9 years. This research topic was chosen due to the limited information available regarding the effectiveness of animations compared to static pictures in children's language development. After a search and evaluation process, we identified seven relevant studies. One of these studies [35] focused on the use of animation to assess the understanding of complex syntactic structures. The remaining six studies [23,30–34] investigated the iconicity of animated symbols in the context of AAC. The studies included brief 2D animations (maximum duration of 29 s) with realism levels 1 and/or 2 [14] and involved a total of 636 TD children. The data collected from these studies are used to address our research questions.

For the first research question, "Can animation better support children's receptive language abilities than a static picture?", the primary source of evidence was the AAC studies. Although the core research objectives differed, five of the included studies [23,30–33] involved an identification task related to aspects of receptive language (see Table 5). That is, children were required to demonstrate their existing knowledge of the lexical item. Additionally, Frizelle et al. [35] directly addressed receptive language for sentences with complex syntax. However, the results vary, with Mineo et al. [30], Brock et al. [23], and Frizelle et al. [35] presenting positive findings, while Schlosser et al. [31,32] and Gunduz [33] found no significant difference between animated and static symbols for these tasks. Consequently, the evidence collected does not provide a conclusive answer to the first research question.

In addressing the second research question—"Does animation enhance children's expressive language more than a static picture?"—the evidence rested upon the five studies from the field of AAC [23,31–34]. For diverse research objectives, these authors employed naming or labeling tasks where children were required to verbally express themselves following a verbal prompt while attempting to guess the meaning of the animated symbol. As detailed in Section 3.4, the participants' responses regarding the length of utterance (number of words), alternative words (e.g., synonyms), semantic errors (vocabulary development), cross-linguistic replication (English to Turkish language), and the ability to produce complete grammatical sentences or perform symbol-by-symbol labeling without training [23] yielded valuable insights into expressive language abilities. This evidence suggests that animation may (a) enhance children's expressive language skills, (b) indicate some aspects of language development, (c) exhibit cross-linguistic effects on word meaning, and (d) contribute to sentence production. These pieces of evidence provide a promising response to the second research question concerning the effect of animation on the expressive language of children.

4.1. Converging Results

Although Mineo et al. [30] and Frizelle et al. [35] conducted studies in different fields of SLP practice, both concur on the influence of assessment type and manner on the children's performance. These studies employed animations for distinct research objectives, (AAC symbol identification and language assessment), with a shared focus on uncovering children's receptive linguistic knowledge, addressing word knowledge and syntactical complexity, respectively. Mineo et al. [30] aimed to determine which form of visual stimulus (among four options) best facilitated children in demonstrating their lexical knowledge, while Frizelle et al. [35] explored whether two distinct assessment types (sentence verification animation task vs. multiple-choice picture selection task) yielded similar results when assessing comprehension of complex syntax. Both research teams designed their animations to suit their research objectives. In the study of Mineo et al. [30], participants already possessed lexical knowledge, while in Frizelle et al. [35], participants were undergoing assessment for the target knowledge. Furthermore, both studies compared their methodologies to those of standardized language test constructions (e.g., Peabody Picture Vocabulary Test—PPVT; Clinical Evaluation of Language Fundamentals—CELF4). Ultimately, despite their different research scopes, both studies reached a comparable conclusion regarding the positive impact of animation. This convergence of views and consistent positive results allow for some optimism for the potential future role of animation in SLP research and clinical practice.

4.2. Research and Clinical Implications

The current systematic review presents evidence of the effect of brief animations on the language development of TD children compared to static pictures. The only available data have been drawn from experimental studies within the field of SLP, rather than clinical investigations. Research concerning the application of animations in clinical populations fell outside the scope of this systematic review.

In this study, the minimum characteristics of brief animations concerning duration, design, and manner of presentation are outlined. These can serve as a valuable foundation for developing brief animations for clinical practice, particularly for children with language disorders. It is essential to highlight that apart from the positive outcomes revealing the effectiveness of animations, no adverse effects on the behavior of TD children were documented, which also is important as a basis for clinical studies.

4.3. Limitations and Future Research

The primary limitation of this systematic review is related to the scarcity of children's language-related studies concerning brief animations compared to static pictures. The search process revealed only one study [35] that directly investigated the correlation between animation and language development of TD children. The AAC field provided important, albeit indirect, evidence concerning some aspects of receptive and expressive language knowledge through identification tasks and naming using animation. Overall, within the field of SLP, the AAC domain remains the most informed area of animation research over the past 15 years.

Speech–language pathologists can explore the various research pathways regarding the effects of animation on children's language development (e.g., semantics, syntax, etc.), particularly for those facing language difficulties. Animation may facilitate the receptive or expressive language of children, but more research is needed (e.g., on the design of animation). Examining data from diverse language, cultural, and bilingual populations would be interesting. Nevertheless, the priority of this paper was to determine from the research evidence the most effective brief animation protocols for children acquiring their first language. This systematic review serves as a foundational step for future research.

The relationship between animation and cognitive load is a crucial topic that links duration, presentation speed, and the number of simultaneously moving stimuli to cognitive processing requirements and working memory limitations [6–8]. This study focused on brief animations, framing the minimum duration (under 30 s), speed, and realism level [8,14]. The findings revealed no adverse effects for a single animated item with these characteristics. However, the introduction of four, five, or six animations simultaneously on a screen may warrant further investigation. In the study of Mineo et al. [30], the viewing time for simultaneously introduced animations was the shortest compared to other studies included in Table 2. The authors argued that this format is like other standardized language tests (e.g., PPVT). Conversely, Frizelle et al. [35] discussed how the same style of standardized language tests with static pictures may impose on cognitive load.

Furthermore, Schlosser et al. [32] and Gunduz [33] actually compared two animated symbol sets with different realism levels. Also, Schlosser et al. [32] argued that the identification task (forcing a choice between four animations) was easier compared to the naming task (naming a single animated symbol). Finally, Brock et al. [23] discussed the practical implications of simultaneously looped animations and suggested modifications related to user control. Therefore, future research on methods for introducing animations should consider the cognitive abilities under development and the target populations for assessment or intervention.

5. Conclusions

The present systematic review investigated the effect of brief animations on language development of typically developing (TD) children compared to static pictures. The characteristics of the animations, the tasks used, and the children's responses were recorded. The evidence collected on verbal language responses suggested that animations may positively affect expressive language. However, the evidence regarding receptive language is less conclusive. While our findings permit some optimism, further research specifically focused on the effect of animations on children's language development is necessary.

Author Contributions: Conceptualization, T.I.V. and V.C.G.; formal analysis, T.I.V. and V.C.G.; resources, T.I.V., V.C.G., M.K. and P.P.; data curation, T.I.V. and V.C.G.; writing—original draft preparation, T.I.V., M.K. and V.C.G.; writing—review and editing, V.C.G., M.K. and P.P.; visualization, T.I.V. and V.C.G.; supervision, V.C.G. and M.K. All authors have read and agreed to the published version of the manuscript.

Funding: This research received no external funding.

Institutional Review Board Statement: Not applicable.

Informed Consent Statement: Not applicable.

Data Availability Statement: No new data were created.

Conflicts of Interest: The authors declare no conflicts of interest.

References

1. Kainz, O.; Jakab, F.; Kardoš, S. The computer animation in education. In Proceedings of the 11th International Conference on Emerging eLearning Technologies and Applications, Stara Lesna, Slovakia, 24–25 October 2013; Institute of Electrical and Electronics Engineers: Piscataway, NJ, USA, 2013; pp. 201–206. [CrossRef]
2. Frick, B.; Boster, J.B.; Thompson, S. Animation in AAC: Previous research, a sample of current availability in the United States, and future research potential. *Assist. Technol.* **2022**, *35*, 302–311. [CrossRef] [PubMed]
3. Chevalier, F.; Riche, N.H.; Plaisant, C.; Chalbi, A.; Hurter, C. Animations 25 years later: New roles and opportunities. In Proceedings of the International Working Conference on Advanced Visual Interfaces, (AVI '16), Bari, Italy, 7–10 June 2016; Association for Computing Machinery: New York, NY, USA, 2016; pp. 280–287. [CrossRef]
4. Graham, L. Gestalt theory in interactive media design. *J. Humanit. Soc. Sci.* **2008**, *2*, 1–12.
5. Pierson, R. Gestalt, Animation, and the Culture of Design. Cinéma & Cie. *Film Media Stud. J.* **2022**, *22*, 69–81. [CrossRef]
6. Schnotz, W.; Lowe, R.K. A unified view of learning from animated and static graphics. In *Learning with Animation: Research Implications for Design*; Lowe, R.K., Schnotz, W., Eds.; Cambridge University Press: Cambridge, UK, 2008; pp. 304–356. Available online: https://www.researchgate.net/publication/303206444_A_Unified_View_of_Learning_from_Animated_and_Static_Graphics (accessed on 6 October 2022).
7. Betrancourt, M.; Tversky, B. Effect of computer animation on users' performance: A review/(Effet de l'animation sur les performances des utilisateurs: Une sythèse). *Le Trav. Hum.* **2000**, *63*, 311–329. Available online: https://www.proquest.com/scholarly-journals/effect-computer-animation-on-users-performance/docview/1311633511/se-2 (accessed on 20 March 2023).
8. Tverky, B.; Morrison, J.B.; Betrancourt, M. Animation: Can it facilitate? *Int. J. Hum.-Comput. Stud.* **2002**, *57*, 247–262. [CrossRef]
9. Mayer, R.E. *Multimedia Learning*; Cambridge University Press: Cambridge, UK, 2001.
10. Mayer, R.E.; Moreno, R. Animation as an aid to multimedia learning. *Educ. Psychol. Rev.* **2002**, *14*, 87–99. [CrossRef]
11. Mayer, R.E.; Moreno, R. Nine ways to reduce cognitive load in multimedia learning. *Educ. Psychol.* **2003**, *38*, 43–52. [CrossRef]
12. Sweller, J. Chapter two-Cognitive load theory. *Psychol. Learn. Motiv.* **2011**, *55*, 37–76. [CrossRef]
13. Lowe, R.; Boucheix, J.-M. A composition approach to design of educational animations. In *Learning from Dynamic Visualization: Innovations in Research and Application*; Lowe, R., Ploetzner, R., Eds.; Springer: Berlin/Heidelberg, Germany, 2017; pp. 5–30. Available online: https://link.springer.com/book/10.1007/978-3-319-56204-9 (accessed on 12 October 2022).
14. Höffler, T.N.; Leutner, D. Instructional animation versus static pictures: A meta-analysis. *Learn. Instr.* **2007**, *17*, 722–738. [CrossRef]
15. Berney, S.; Bétrancourt, M. Does animation enhance learning? A meta-analysis. *Comput. Educ.* **2016**, *101*, 150–167. [CrossRef]
16. Ploetzner, R.; Berney, S.; Bétrancourt, M. A review of learning demands in instructional animations: The educational effectiveness of animations unfolds if the features of change need to be learned. *J. Comput. Assist. Learn.* **2020**, *36*, 838–860. [CrossRef]

17. Lowe, R.; Ploetzner, R. (Eds.) *Learning from Dynamic Visualization. Innovations in Research and Application*; Springer: Berlin/Heidelberg, Germany, 2017. Available online: https://link.springer.com/content/pdf/10.1007/978-3-319-56204-9.pdf (accessed on 12 October 2022).
18. Smeets, D.J.; Bus, A.G. The interactive animated e-book as a word learning device for kindergartners. *Appl. Psycholinguist.* **2015**, *36*, 899–920. [CrossRef]
19. Richter, A.; Courage, M.L. Comparing electronic and paper storybooks for preschoolers: Attention, engagement, and recall. *J. Appl. Dev. Psychol.* **2017**, *48*, 92–102. [CrossRef]
20. Takacs, Z.K.; Bus, A.G. Benefits of motion in animated storybooks for children's visual attention and story comprehension. An eye-tracking study. *Front. Psychol.* **2016**, *7*, 1591. [CrossRef] [PubMed]
21. Sarı, B.; Başal, H.A.; Takacs, Z.K.; Bus, A.G. A randomized controlled trial to test efficacy of digital enhancements of storybooks in support of narrative comprehension and word learning. *J. Exp. Child Psychol.* **2019**, *179*, 212–226. [CrossRef]
22. Broemmel, A.D.; Moran, M.J.; Wooten, D.A. The Impact of Animated Books on the Vocabulary and Language Development of Preschool-Aged Children in Two School Settings. *Early Child. Res. Pract.* **2015**, *17*, n1.
23. Brock, K.L.; Zolkoske, J.; Cummings, A.; Ogiela, D.A. The Effects of Symbol Format and Psycholinguistic Features on Receptive Syntax Outcomes of Children Without Disability. *J. Speech Lang. Hear. Res.* **2022**, *65*, 4741–4760. [CrossRef] [PubMed]
24. Fuller, D.; Lloyd, L. Toward a common usage of iconicity terminology. *Augment. Altern. Commun.* **1991**, *7*, 215–220. [CrossRef]
25. Light, J. Toward a definition of communicative competence for individuals using augmentative and alternative communication systems. *Augment. Altern. Commun.* **1989**, *5*, 137–144. [CrossRef]
26. Light, J.; McNaughton, D. Communicative competence for individuals who require augmentative and alternative communication: A new definition for a new era of communication. *Augment. Altern. Commun.* **2014**, *30*, 1–18. [CrossRef]
27. Jagaroo, V.; Wilkinson, K. Further considerations of visual cognitive neuroscience in aided AAC: The potential role of motion perception systems in maximizing design display. *Augment. Altern. Commun.* **2008**, *24*, 29–42. [CrossRef] [PubMed]
28. Pitt, K.M.; Cole, Z.J.; Zosky, J. Applying Functional Animation to Pictorial Symbols for Supporting P300–Brain–Computer Interface Access to Augmentative and Alternative Communication Devices by Children. *Int. J. Hum.-Comput. Interact.* **2022**, 1–13. [CrossRef]
29. Schlosser, R.W.; Choe, N.; Koul, R.; Shane, H.C.; Yu, C.; Wu, M. Roles of Animation in Augmentative and Alternative Communication: A Scoping Review. *Curr. Dev. Disord. Rep.* **2022**, *9*, 187–203. [CrossRef]
30. Mineo, B.A.; Peischl, D.; Pennington, C. Moving targets: The effect of animation on identification of action word representations. *Augment. Altern. Commun.* **2008**, *24*, 162–173. [CrossRef] [PubMed]
31. Schlosser, R.W.; Shane, H.; Sorce, J.; Koul, R.; Bloomfield, E.; Debrowski, L.; DeLuca, T.; Miller, S.; Schneider, D.; Neff, A. Animation of graphic symbols representing verbs and prepositions: Effects on transparency, name agreement, and identification. *J. Speech Lang. Hear. Res.* **2012**, *55*, 342–358. [CrossRef] [PubMed]
32. Schlosser, R.W.; Koul, R.; Shane, H.; Sorce, J.; Brock, K.; Harmon, A.; Moerlein, D.; Hearn, E. Effects of animation on naming and identification across two graphic symbol sets representing verbs and prepositions. *J. Speech Lang. Hear. Res.* **2014**, *57*, 1779–1791. [CrossRef]
33. Gunduz, S. Effect of Animation on Naming and Identification of Graphic Symbols Representing Verbs and Prepositions in Turkish Preschoolers. Doctoral Dissertation, Texas Tech University, Lubbock, TX, USA, 2017. Available online: https://hdl.handle.net/2346/86958 (accessed on 27 March 2023).
34. Harmon, A.C.; Schlosser, R.W.; Gygi, B.; Shane, H.C.; Kong, Y.Y.; Book, L.; Macduff, K.; Hearn, E. Effects of environmental sounds on the guessability of animated graphic symbols. *Augment. Altern. Commun.* **2014**, *30*, 298–313. [CrossRef] [PubMed]
35. Frizelle, P.; Thompson, P.; Duta, M.; Bishop, D.V. Assessing children's understanding of complex syntax: A comparison of two methods. *Lang. Learn.* **2019**, *69*, 255–291. [CrossRef]
36. Diehm, A.E.; Wood, C.; Puhlman, J.; Callendar, M. Young children's narrative retell in response to static and animated stories. *Int. J. Lang. Commun. Disord.* **2020**, *55*, 359–372. [CrossRef]
37. American Speech-Language-Hearing Association. (n.d.). Language in Brief. Available online: https://www.asha.org/practice-portal/clinical-topics/spoken-language-disorders/language-in-brief/ (accessed on 25 July 2023).
38. Bloom, L.; Lahey, M. *Language Development and Language Disorders*; John Wiley and Sons: Somerset, NJ, USA, 1978.
39. Law, J.; Garrett, Z.; Nye, C. Speech and language therapy interventions for children with primary speech and language delay or disorder. *Campbell Syst. Rev.* **2005**, *1*, 1–85. [CrossRef]
40. Cabell, S.Q.; Justice, L.M.; Konold, T.R.; McGinty, A.S. Profiles of emergent literacy skills among preschool children who are at risk for academic difficulties. *Early Child. Res. Q.* **2011**, *26*, 1–14. [CrossRef]
41. Hegde, N.M. *Hegde's PocketGuide to Assessment in Speech-Language Pathology*, 3rd ed.; Thomson Delmar Learning: Clifton Park, NY, USA, 2008.
42. Huitt, W.; Hummel, J. Piaget's theory of cognitive development. *Educ. Psychol. Interact.* **2003**, *3*, 1–5.
43. Ahmad, S.; Ch, A.H.; Batool, A.; Sittar, K.; Malik, M. Play and Cognitive Development: Formal Operational Perspective of Piaget's Theory. *J. Educ. Pract.* **2016**, *7*, 72–79.
44. Moher, D.; Shamseer, L.; Clarke, M.; Ghersi, D.; Liberati, A.; Petticrew, M.; Shekelle, P.; Stewart, L.A.; PRISMA-P Group. Preferred reporting items for systematic review and meta-analysis protocols (PRISMA-P) 2015 statement. *Syst. Rev.* **2015**, *4*, 1. [CrossRef]

45. Page, M.J.; McKenzie, J.E.; Bossuyt, P.M.; Boutron, I.; Hoffmann, T.C.; Mulrow, C.D.; Shamseer, L.; Tetzlaff, J.M.; Akl, E.A.; Brennan, S.E.; et al. The PRISMA 2020 statement: An updated guideline for reporting systematic reviews. *Int. J. Surg.* **2021**, *88*, 105906. [CrossRef] [PubMed]
46. Straus, S.E.; Sackett, D.L. Using research findings in clinical practice. *BMJ* **1998**, *317*, 339–342. [CrossRef]
47. Boston Children's Hospital. Autism Language Program. Available online: https://www.childrenshospital.org/programs/autism-language-program (accessed on 27 August 2023).
48. Mayer-Johnson, T. *Boardmaker Plus*; Mayer-Johnson, Inc.: Solana Beach, CA, USA, 2008.
49. Sherrington, C.; Herbert, R.D.; Maher, C.G.; Moseley, A.M. PEDro. A database of randomized trials and systematic reviews in physiotherapy. *Man. Ther.* **2000**, *5*, 223–226. [CrossRef]
50. Murray, E.; Power, E.; Togher, L.; McCabe, P.; Munro, N.; Smith, K. The reliability of methodological ratings for speechBITE using the PEDro-P scale. *Int. J. Lang. Commun. Disord.* **2013**, *48*, 297–306. [CrossRef]
51. Thomas, B.H.; Ciliska, D.; Dobbins, M.; Micucci, S. A process for systematically reviewing the literature: Providing the research evidence for public health nursing interventions. *Worldviews Evid.-Based Nurs.* **2004**, *1*, 176–184. [CrossRef] [PubMed]

brain sciences

MDPI

Brief Report

On the Dynamics of Inferential Behavior while Reading Expository and Narrative Texts

Yongseok Yoo

School of Computer Science and Engineering, Soongsil University, Seoul 06978, Republic of Korea; yyoo@ssu.ac.kr; Tel.: +82-2-820-0678

Abstract: Inference plays a key role in reading comprehension. This study examines changes in inferential behavior while reading different genres. The inferential behavior of 28 students with reading disabilities (RDs) and 44 students without RDs was quantified while they read expository and narrative texts. First, the average rates of inference attempts and correct inferences were measured during reading. Then, the same rates were measured separately during early and late reading to see if there was a change in inferential behavior. The results show that the change in inferential behavior depends on the genre. While reading the expository text, both groups showed no significant change in their inference making. In contrast, while reading the narrative text, both groups showed higher rates of inference attempts, and only the students without RD showed a significant increase in correct inferences. The implications of these findings for the design of more engaging and effective reading programs are discussed.

Keywords: reading; reading disabilities; inference; think aloud; dynamics

1. Introduction

1.1. Inference Making during Reading

Reading involves complex interactions between multiple cognitive processes, such as letter–sound correspondence, phonological memory, word recognition, sentence processing, and comprehension. A deficiency in any of the low-level processes can lead to reading disabilities (RDs) [1,2]. For example, deficits in lexicography and phonics can lead to difficulties with word identification and dyslexia [1,2]. Higher-level cognitive processes, such as reading comprehension, are less well understood [2].

Inference has been shown to play a key role in reading comprehension, serving as a critical component in the construction of meaning from text [3–5]. Inference allows readers to fill in gaps in explicit textual information, facilitating deeper understanding and integration of knowledge. Previous research has emphasized its importance not only for comprehending literal content but also for engaging with the text at a deeper level [4], allowing for the application of prior knowledge and the anticipation of subsequent narrative developments [5]. Carlson et al. [6] emphasized the importance of assessing reading comprehension by measuring the process of making inferences while reading. They emphasized the importance of examining the cognitive processes that struggling readers use while reading, such as literal comprehension and thinking about the implications of the text, and suggested think-aloud tasks as a useful assessment method.

To assess readers' comprehension processes, think-aloud tasks are used. The think-aloud task was originally used as a tool to demonstrate thinking while performing a given task by asking for the subjects to say out loud what they are thinking while performing the task [7]. A common way to administer the think-aloud task in reading is to ask students to read one sentence at a time and say anything that comes to mind. Research using the think-aloud task has found that readers make a variety of inferences while reading [8,9]. For example, Laing and Kamhi [10] divided third graders by reading level to see if they differed on a think-aloud task. They found that average students made more explanatory inferences than struggling readers. Carlson et al. [6] administered a think-aloud task to struggling readers in Grades 3–5 and found that the struggling readers were more likely to simply repeat what they had just read, while the control group attempted to elaborate on what they had read.

Recent studies show that the genre of the text can influence the inferences made while reading. The difference in inferences between reading narrative and expository texts was investigated [11]. The results showed that students used inferential comprehension more often when reading narrative texts than expository texts, and the researchers argued that the comprehension of expository texts was more difficult. Students may have difficulty with expository texts because they require more background knowledge and skills to make connections to what they are reading [12,13]. In [14], the relationship between the characteristics of undergraduate students and their ability to comprehend expository texts was investigated. The results showed that readers' background knowledge plays an important role in making inferences during reading. In [15], the effects of genre on the reading comprehension of elementary school students were investigated. The results showed that individual differences in world knowledge were more related to the comprehension of expository texts than narrative texts. However, there is still limited research on how elementary students with and without RDs differ in making inferences for their comprehension of different types of texts, such as narrative and expository texts.

1.2. Contributions of the Study

The results of previous studies are based on average inference behavior during reading, under the assumption that inference is a stationary process. Given the complex nature of inference, we suggest that inference behavior may change dynamically during reading. In this study, we investigate changes in the inferential behavior of students with and without RD during reading, focusing on how readers adapt their reading strategies in response to different genres. This investigation is based on the hypothesis that inferential behavior is not static but rather evolves dynamically with the reading context, reflecting a sophisticated interplay between the text and the reader's adaptation. These interactions of students with and without RDs were quantified as they read texts of different genres (expository and narrative).

This study strategically targeted specific age groups with and without RDs. The third and fourth grades play a crucial role in a student's education, representing a pivotal point in the development of essential comprehension skills that are critical for later learning [13]. The ability to understand the meaning of a text and make inferences is expected to be mastered before fourth grade, and comprehension difficulties during this period have been linked to later academic underachievement in all subjects [16]. Thus, understanding challenges in these grades is vital to establish a strong foundation for future academic success. Therefore, this study focused on the inferential behavior of students with and without RDs during reading at this stage.

Specifically, we used the standard think-aloud protocol used in previous studies of reading [6–10]. The novel contributions of this study include the focus on inference making and its change over time. The research questions for this study were as follows.

First, when reading expository and narrative texts, do students with and without RDs differ in the rate at which they make inferences and the accuracy of their inferences?

Second, when reading expository and narrative texts, do students with and without RDs show a change in their inference making during reading?

2. Materials and Methods

2.1. Participants Information

Seventy-two third- and fourth-grade students in 14 public elementary schools in South Korea participated in this study. Each school had approximately 500 students enrolled.

Within each grade, classes were divided into four or five sections, with 20–24 students in each class. None of the participants received any financial or social assistance.

Table 1 shows participants information. Among the participants, 28 students were identified as having RD by screening followed by the standardized Reading Achievement and Reading Cognitive Process (RA-RCP) tests [17] (scoring below the 16th percentile). The median scores of the students with RDs and without RDs were the 10th and 73rd percentiles, respectively. These students with RDs had intact word recognition and communication skills. The remaining 44 students without RDs served as the control group.

Table 1. Participants information.

Grade	With RDs		Without RDs	
	Male	Female	Male	Female
3	8	5	8	7
4	10	5	13	16
Total	18	10	21	23

2.2. Data Collection Using the Think-Aloud Protocol

Three stimulus texts were taken from teachers' guides for reading courses. The first text consisted of three short sentences and was used for the training session to familiarize the participants with the experiment.

The other two texts for the main session were similar in readability and came from different genres (expository and narrative). The texts for the main session were developed as follows. First, the researchers reviewed the textbooks, recommended books, and reading comprehension instruments for Grades 3–4 and extracted six candidate passages. Next, the content validity was checked by a group of experts; two elementary school teachers, a professor of language education, and a professor of Korean language and literature were asked to check whether the difficulty level and length of the passages were appropriate and met the purpose of the test. Then, two of the six passages were selected according to the experts' opinions. Changes were made to adjust the difficulty level of the sentences and to shorten the length. The experts reviewed the revised texts again, and an agreement was reached among all the experts. Consequently, the expository text contained 10 sentences with 92 words and the average, minimum, and maximum numbers of words per sentence were 9.2, 5, and 15, respectively. The number of words in each sentence did not differ significantly between the first five sentences and the remaining sentences (t-test, $p = 0.68$). The narrative text contained 11 sentences with 92 words, and the average, minimum, and maximum numbers of words per sentence were 8.4, 4, and 13, respectively. The number of words in each sentence did not differ significantly between the first five sentences and the remaining sentences (t-test, $p = 0.20$). Each text was printed on A4 paper, with one sentence per line.

The tests were administered individually by a trained instructor in a quiet classroom. One sentence at a time was presented by the instructor while the other sentences were covered. The participants were asked to verbalize their thoughts right after reading each sentence, and those responses were transcribed.

The instructor training was implemented over the span of one week prior to the beginning of the study. This training encompassed the objectives of the assessment, the administration procedures, and considerations to observe during the administration of the assessment.

2.3. Coding

The reader's response to each sentence was coded as one of correct inference, incorrect inference, and no inference by three trained evaluators (professors who specialize in learning disabilities). If their classifications for a response differed, they discussed the matter until agreement was reached.

The definitions of the three inference types are as follows. First, correct inferences are inferences that are made accurately based on the reading. Correct inferences may be explanatory, predictive, or associative inferences [9]. Explanatory inferences involve elaboration of the target sentence by relating it to previous sentences or background. Predictive inference is when the reader guesses what would come next based on what came before. Associative inference is making new inferences based on background knowledge or experiences that come to mind as the reader reads a sentence. Second, incorrect inferences are defined as inference attempts made by readers that are out of context or do not make logical sense. Third, no inferences are responses where no inference was made. They can be simple repetitions of what has been read or nonsensical answers.

2.4. Data Analysis

The frequency of the three inference types (correct inference, incorrect inference, and no inference) made by each reader were calculated for each text.

To measure any change in the inferential behavior, the frequencies during early and late reading were calculated separately. For the expository text, the frequencies corresponding to the first five sentences were measured for the early behavior, and those corresponding to the remaining five sentences were measured for the late behavior. For the narrative text, the frequencies corresponding to the first five sentences were measured for the early behavior, and those corresponding to the remaining six sentences were measured for the late behavior.

Statistical significance was measured as follows. Some of the measured frequencies did not pass the normality test using the Kolmogorov–Smirnov test. This is probably due to the small sample size and the limited range of frequencies. Therefore, nonparametric models were used for hypothesis testing. To compare the difference in inference frequency between students with and without RDs, the Mann-Whiney U test was performed. To test the change between early and late reading time, the Wilcoxon signed rank sum test was performed. All statistical analyses ware performed using SciPy version 1.9.1 [18].

3. Results

3.1. Comparison of the Frequency of Inferences between Students with and without RDs

Table 2 shows the frequency of inferences between students with and without RDs for expository and narrative texts. For the expository text, participants with RDs made significantly fewer correct inferences ($p < 0.001$, t-test) and significantly more incorrect inferences ($p < 0.05$, t-test) than those without RDs. The frequency of no inferences was also significantly higher for students with RDs than those without RDs ($p < 0.001$, t-test). For the narrative text, students with RDs made significantly fewer correct inferences ($p < 0.001$, t-test) than those without RDs. The frequency of no inferences was also significantly higher for students with RDs than for those without RDs ($p < 0.001$, t-test). Thus, the students with RDs attempted to make inferences less often, and their inferences were not correct more often than those of the students without RDs.

Table 2. Inference frequencies during reading.

Genre	Inference Type	With RDs		Without RDs		Statistic (p-Value)
		Mean	SE	Mean	SE	
Expository	Correct	0.23	0.05	0.66	0.05	183.5 ($p < 0.001$)
	Incorrect	0.17	0.04	0.08	0.02	753.0 ($p > 0.05$)
	No	0.60	0.08	0.26	0.05	915.0 ($p < 0.001$)
Narrative	Correct	0.29	0.05	0.70	0.04	157.5 ($p < 0.001$)
	Incorrect	0.12	0.03	0.06	0.01	714.0 ($p > 0.05$)
	No	0.59	0.07	0.24	0.04	949.5 ($p < 0.001$)

3.2. The Change in Inferential Behavior during Reading

Table 3 shows the inference frequencies during early and late reading of the expository text. The inference frequencies did not change between early and late reading for the both groups ($p > 0.05$, t-test).

Table 3. Inference frequencies during early and late reading of the expository text.

Group	Inference Type	Early		Late		Statistic (p-Value)
		Mean	SE	Mean	SE	
RD	Correct	0.24	0.04	0.22	0.04	29.0 ($p > 0.05$)
	Incorrect	0.15	0.04	0.20	0.04	17.0 ($p > 0.05$)
	No	0.61	0.07	0.58	0.08	23.0 ($p > 0.05$)
Without RD	Correct	0.65	0.05	0.67	0.06	159.0 ($p > 0.05$)
	Incorrect	0.09	0.03	0.07	0.03	58.0 ($p > 0.05$)
	No	0.26	0.05	0.26	0.05	99.0 ($p > 0.05$)

Table 4 shows the inference frequencies during early and late reading of the narrative text. The frequency of correct inference increased during the late reading only for the students without RDs ($p < 0.001$, t-test), not for the students with RDs ($p > 0.05$, t-test). The frequencies of no inferences decreased during the late reading for both students without RDs ($p < 0.001$, t-test) and students with RDs ($p < 0.01$, t-test).

Table 4. Inference frequencies during early and late reading of the narrative text.

Group	Inference Type	Early		Late		Statistic (p-Value)
		Mean	SE	Mean	SE	
RD	Correct	0.24	0.06	0.33	0.06	74.5 ($p > 0.05$)
	Incorrect	0.07	0.03	0.15	0.04	13.0 ($p > 0.05$)
	No	0.69	0.07	0.52	0.07	24.0 ($p < 0.01$)
Without RD	Correct inference	0.60	0.05	0. 79	0.04	144.0 ($p < 0.001$)
	Incorrect inference	0.05	0.02	0.06	0.02	68.0 ($p > 0.05$)
	No inference	0.35	0.05	0.16	0.04	55.0 ($p < 0.001$)

4. Discussion

The inferential behavior of the readers for different genres of the text was studied. First, the frequencies of inference types were quantified for students with and without RDs for the entire reading time. For both expository and narrative texts, the students with RDs attempted to make inferences less often, and their inferences were not correct more often than those of the students without RD. Second, the change in the inferences between early and late reading time was measured. For the expository text, the frequency of making inferences did not change between early and late reading. In contrast, for the narrative text, both groups attempted to make inferences more frequently during late reading, resulting in more correct inferences only for the students without RDs. Implications of this genre-dependent inferential behavior are discussed below.

The frequency of inferences of both groups remained constant throughout the reading of expository texts. This could be explained on the basis of previous findings that background knowledge plays a crucial role in reading expository texts [12–15]. The level of background knowledge related to the given text would determine the overall inference attempts and correctness but would not change significantly during reading.

In contrast, both groups increased the frequency of inferences in the later stages of narrative reading, whereas the frequency of inferences remained constant during the reading of expository texts. This change during reading has more to do with cognitive processes during reading than with previously acquired knowledge. It has been reported that students with RDs lack higher-order cognitive skills, such as focusing on essential content, extending sentence content, and predicting future events [19]. Our findings suggest that the dynamic engagement of such higher-order cognitive skills is more encouraged when reading narrative texts than expository texts. Thus, initiating reading instruction with story books may be an effective strategy for fostering comprehension skills. This approach leverages the natural inclination toward increased engagement with stories as they progress, potentially facilitating the development of inferential reasoning.

A clinical implication of our findings is the potential to improve diagnostic accuracy. Current standardized diagnostic tests measure reading ability over a relatively long period of time, which may not capture more subtle differences and changes in reading behavior during reading. Our findings suggest that monitoring changes in inferential behavior may provide additional information about reading behavior. This could improve diagnostic criteria and help differentiate between different types of reading disorders, allowing for more targeted interventions.

However, there remain several limitations that need to be addressed in future studies. In this study, representative expository and narrative texts were carefully selected by a group of experts. Investigating the generalizability of our findings to diverse texts is an important direction for future studies. In addition, another direction would be to confirm the dynamic change in inference behavior using other methods, such as neuroimaging technologies, and to understand the neural underpinnings of this change.

5. Conclusions

In this study, we investigated changes in inferential behavior while they read expository and narrative texts. Students with RDs participants showed lower rate of inference attempts, and these inferences were not correct more often than those of students without RDs. While reading the expository text, students both with and without RDs showed no significant change in their inference making. In contrast, while reading the narrative text, both groups showed higher rates of inference attempts, and only the students without RDs showed a significant increase in correct inferences. These findings suggest that consideration of genre is important in designing more engaging and effective reading programs.

Funding: This research received no external funding.

Institutional Review Board Statement: This study was conducted in accordance with the Declaration of Helsinki and approved by the Institutional Review Board of Chonnam National University (1040198-170920-HR-074-02) on 13 October 2017.

Informed Consent Statement: A comprehensive written consent form was provided to all participants and their parents. These forms outlined the objectives of the research, the extent of participation, potential benefits and risks, assurances of confidentiality, and the right of the participants to withdraw from the study at their discretion.

Data Availability Statement: Available upon request. The data are not publicly available due to Personal Information Protection Act.

Conflicts of Interest: The author declares no conflicts of interest.

References

1. American Psychiatric Association. *Diagnostic and Statistical Manual of Mental Disorders: DSM-5*, 5th ed.; American Psychiatric Publishing: Washington, DC, USA, 2013.
2. Hulme, C.; Snowling, M.J. *Developmental Disorders of Language Learning and Cognition*; John Wiley & Sons: Hoboken, NJ, USA, 2013.
3. Yuill, N.; Oakhill, J. *Children's Problems in Text Comprehension: An Experimental Investigation*; Cambridge University Press: New York, NY, USA, 2010.
4. Cain, K.; Oakhill, J.V. Inference making and its relation to comprehension failure in young children. *Read. Writ. Interdiscip. J.* **1999**, *11*, 489–503. [CrossRef]
5. Bowyer-Crane, C.; Snowling, M.J. Assessing children's inference generation: What do tests of reading comprehension measure? *Br. J. Educ. Psychol.* **2005**, *75*, 189–201. [CrossRef] [PubMed]
6. Carlson, S.E.; Seipel, B.; McMaster, K. Development of a new reading comprehension assessment: Identifying comprehension differences among readers. *Learn. Individ. Differ.* **2014**, *32*, 40–53. [CrossRef]
7. Ericsson, K.A.; Simon, H.A. *Protocol Analysis Verbal Reports as Data*; The MIT Press: London, UK, 1993.
8. Rapp, D.N.; Broek, P.V.D.; McMaster, K.L.; Kendeou, P.; Espin, C.A. Higher-order comprehension processes in struggling readers: A perspective for research and intervention. *Sci. Stud. Read.* **2007**, *11*, 289–312. [CrossRef]
9. McMaster, K.L.; van Den Broek, P.; Espin, C.A.; White, M.J.; Rapp, D.N.; Kendeou, P.; Bohn-Gettler, C.M.; Carlson, S. Making the right connections: Differential effects of reading intervention for subgroups of comprehenders. *Learn. Individ. Differ.* **2012**, *22*, 100–111. [CrossRef]
10. Laing, S.P.; Kamhi, A.G. The use of think-aloud protocols to compare inferencing abilities in average and below-average readers. *J. Learn. Disabil.* **2002**, *35*, 437–448. [CrossRef] [PubMed]
11. Clinton, V.; Taylor, T.; Bajpayee, S.; Davison, M.L.; Carlson, S.E.; Seipel, B. Inferential comprehension differences between narrative and expository texts: A systematic review and meta-analysis. *Read. Writ.* **2020**, *33*, 2223–2248. [CrossRef]
12. Graesser, A.C.; McNamara, D.S.; Louwerse, M.M.; Cai, Z. Coh-Metrix: Analysis of text on cohesion and language. *Behav. Res. Methods Instrum. Comput.* **2004**, *36*, 193–202. [CrossRef] [PubMed]
13. Snowling, M.J.; Hulme, C. *The Science of Reading: A Handbook*; Blackwell Publishing: Oxford, UK, 2005.
14. Kendeou, P.; van den Broek, P. The effects of prior knowledge and text structure on comprehension processes during reading of scientific texts. *Mem. Cogn.* **2007**, *35*, 1567–1577. [CrossRef] [PubMed]
15. Best, R.M.; Floyd, R.G.; McNamara, D.S. Differential competencies contributing to children's comprehension of narrative and expository texts. *Read. Psychol.* **2008**, *29*, 137–164. [CrossRef]
16. Shanahan, T.; Shanahan, C. Teaching disciplinary literacy to adolescents: Rethinking content-area literacy. *Harv. Educ. Rev.* **2008**, *78*, 40–59. [CrossRef]
17. Kim, A.; Kim, U.; Hwang, M.; Yoo, H. *Test of Reading Achievement and Reading Cognitive Processes Ability (RA-RCP)*; Hakjisa: Seoul, Republic of Korea, 2014.
18. SciPy—Fundamental Algorithms for Scientific Computing in Python. Available online: https://scipy.org/ (accessed on 15 March 2024).
19. Seipel, B.; Carlson, S.E.; Clinton, V.E. When do comprehender groups differ? A moment-by-moment analysis of think-aloud protocols of good and poor comprehenders. *Read. Psychol.* **2017**, *38*, 39–70. [CrossRef]

brain sciences

MDPI

Article

Cognitive vs. Linguistic Training in Children with Developmental Language Disorder: Exploring Their Effectiveness on Verbal Short-Term Memory and Verbal Working Memory

Theodora Bachourou [1,2], Stavroula Stavrakaki [2,*], Vasiliki Koukoulioti [3] and Ioanna Talli [2]

[1] Center of Interdisciplinary Assessment, Counseling and Support (KE.D.A.S.Y.), 271 00 Ileia, Greece; dorabachourou@gmail.com

[2] Department of Italian Language and Literature, Aristotle University of Thessaloniki, 541 24 Thessaloniki, Greece; talli@itl.auth.gr

[3] Department of German Language and Literature, Aristotle University of Thessaloniki, 541 24 Thessaloniki, Greece; vasilikikou@del.auth.gr

* Correspondence: svoula@itl.auth.gr; Tel.: +30-2310997452

Abstract: The present study explores comparatively the effectiveness of a cognitive (verbal short-term memory (vSTM), verbal working memory (vWM)) and of a linguistic training (10-week duration each) in the diffusion of gains in cognitive abilities (vSTM and vWM) of in school-aged Greek-speaking children with developmental language disorder (DLD). To this purpose, two computerized training programs i.e., a linguistic and a cognitive one, were developed and applied to three groups (A, B, and C) of children with DLD (N = 49, in total). There were three assessments with two vSTM tasks (non-word repetition and forward digit span) and a vWM task (backward digit span): pre-therapeutically (time 1), where no significant between-group differences were found, post-therapeutically I (time 2), and post-therapeutically II (time 3) and two training phases. In phase I, group A received meta-syntactic training, whereas group B vSTM/vWM training and group C received no training. In phase II, a reversal of treatment was performed for groups A and B: group A received vSTM/vWM while group B meta-syntactic training. Again, group C received no training. Overall, the results indicated a significant performance improvement for the treatment groups and revealed beneficial far-transfer effects as language therapy can affect vSTM and vWM in addition to direct and near transfer effects. In addition, the intervention type order affected performance as follows: first, better performance on the vSTM task (non-word repetition) was shown when the linguistic treatment was delivered first; second, better performance on the vWM in Time 2 and Time 3 was shown by group B, for which the cognitive treatment was delivered first. Concluding, not only intervention type but also intervention type order can affect performance in DLD.

Keywords: developmental language disorder; linguistic training; cognitive training; intervention effects

1. Introduction

Developmental language disorder (DLD) (previously known as specific language impairment) is defined as the difficulty with the acquisition of language skills, in all or some linguistic domains, including phonology, grammar, syntax, vocabulary, and pragmatics [1]. DLD affects about 8% of children [2], a number that undoubtedly requires the efficient implementation of clinical and evidence-based studies.

Among the biological and environmental etiology factors of the disorder that are now considered, family history, socio-economic status, or peri-natal problems should also be taken into account as inclusionary criteria for DLD. In addition, the new DLD inclusionary criteria set by the CATALISE group put an end to the obligatory differentiation between verbal and non-verbal intelligence and accept the coexistence of DLD with other neurodevelopmental disorders, such as ADHD, speech sound disorder, behavioral disorders, or below average non-verbal ability [3]. Children with DLD demonstrate mostly difficulties in morphosyntax [4–6], in non-canonical word order [7], and in the comprehension or/and production of relative clauses [8–15]. Moreover, children with DLD have difficulties in cognitive domains including phonological short-term memory [16], visuospatial working memory [17], executive function [18–21], cognitive flexibility [22], and processing skills [6,23].

Intensive research in the past was performed to identify whether the core clinical marker of children DLD concerned a deficit in linguistic or cognitive abilities. Many studies revealed that the core deficit in DLD is manifested in the domain of inflectional morphology, especially past tense marking in English-speaking children [24] or in the domain of complex syntax, especially in sentences formed though the operations of A-movement or A-bar movement [8,25–27]. On the other hand, other studies suggested that deficient performance on non-word repetition, a proper measure for vSTM, constitutes a strong index for DLD [28] or provided evidence that the core deficit of the disorder is not only linguistic but also cognitive (e.g., Plym et al. [29]). More recent studies underline the coexistence of cognitive and linguistic impairment in DLD and investigate their relation. For example, Gilliam et al. [30] have suggested that syntactic comprehension in children with and without DLD is strongly linked to four cognitive processes: fluid reasoning, controlled attention, complex working memory, and language knowledge in long-term memory.

Despite controversy, the above evidence is still useful for intervention schemes. Specifically, more efficient remediation of the deficits could be planned and applied if research findings are taken into account. It should be mentioned that so far a large body of intervention studies target the language deficits in DLD and not the cognitive or processing deficits [31–33]. However, recent studies provide encouraging results for clinicians needing to target multiple domains (both language and cognitive) in their intervention methods [34–36].

The main aim of the present study is to explore the effectiveness of cognitive targeted therapy and linguistically targeted therapy on the improvement of verbal short-term memory (vSTM) and verbal working memory (vWM) of individuals with DLD. While there is much evidence indicating that cognitive intervention targeting STM and/or WM contributes to enhancing these abilities in individuals with language disorders, there are only a few studies that explore far-transfer effects, namely the impact of linguistic intervention on vSTM and vWM.

1.1. *The Effects of Cognitive Intervention on STM and WM*

Several studies with children with impaired WM skills showed improvement in WM and academic performance after WM training [37–39]. Wener and Archibald [40] in a preliminary intervention study with two therapy schemes targeting language and WM skills in nine 7- to 9-year-old children (three with DLD, two with WM impairment, four with both language and WM impairment) revealed a domain-specific effect: children with WM impairment improved their WM skills, while children with DLD improved their grammatical skills and children with both language and WM impairment showed more gains in their grammatical than WM skills. There were also cross-domain effects for some children. The authors suggest that intervention is more effective when targeting children's specific areas of weaknesses, which is why each child's underlying deficit(s) should be identified before designing the therapy program. Pauls and Archibald [41] used a single-subject design to explore near-transfer (WM) and far-transfer (language, reading, math) effects of a WM training program [42] in a group of seven children (8–11 years old) with WM impairments (three of them with additional language impairments). They found immediate and long-term (3-month follow-up) near-transfer effects in all participants but modest far-transfer effects in four of the participants. None of the children with additional language impairments showed an improvement in language skills, suggesting that WM training does not generally lead to improvement in language skills and is not effective for children with language impairments. They conclude that participant-specific characteristics, such as age, training intensity, and baseline abilities of WM, which might influence transfer response, should be taken into account [41]. Furthermore, Henry et al. [43] applied a working-memory intervention for a group of children with DLD and explored its impact on the WM and language abilities of this group compared to an active control group that did not participate in training. They found that the group (therapy group vs. active control group) was a significant predictor for gains in WM skills and language intervention, indicating that a WM intervention can impact on untrained skills.

1.2. The Effects of Linguistic Intervention on STM and WM

While there are several intervention studies targeting remediating impaired linguistic and meta-linguistic mechanisms (e.g., see reviews by Cleave et al. [44] and systematic reviews by Cirrin and Gillam [45], Law et al. [46], Rinaldi et al. [47], Storkel et al. [48], for bilingual children see Nair et al. [49], Bettelli et al. [50]), only a few studies with a linguistic intervention scheme have explored the effect of linguistic therapy on non-linguistic abilities in DLD. In particular, Pauls and Archibald [36] provided a narrative-based language intervention (three 40-min individual sessions per week for five weeks) and investigated its impact on linguistic abilities, related academic abilities (reading and math), and working memory. The results indicated heterogeneity in improvement of linguistic or working memory abilities of these individuals. Moreover, two phonological awareness language interventions found improvement in verbal STM and WM skills [51,52] (first study: four 20-min individual sessions per week for four weeks, second study: two 15-min sessions per week in small groups for two semesters). These interventions included phoneme awareness and rhyming activities. By contrast, Swanson et al. [53] reported no effects of a narrative-based language intervention on verbal STM (three 50-min individual sessions per week for six weeks). Their goal was to increase the frequency of use of complex grammatical forms (subordinating conjunctions, coordinating conjunctions, complex verbs, postmodification of nouns) through narration of novel stories. Other studies failed to show far-transfer effects of language intervention (three 60-min individual sessions per week for five weeks) to WM skills [54] or found modest far-transfer effects on WM skills, i.e., in some children only [55]. However, none of the abovementioned studies report pilot applications of their experimental materials. Despite controversial findings, it has been proposed that clinicians can improve cognitive abilities, including WM, by teaching strategies for comprehending and producing language skills [7]. The idea behind this is that therapeutic schemas incorporate and activate cognitive mechanisms, so that linguistic knowledge learning becomes meta-cognitive.

The present study builds on previous research and further investigates the linguistic and cognitive intervention effects on cognition abilities (vWM and vSTM abilities) in individuals with DLD, by providing assessment of these abilities before (at time 1: T1) and after (at times 2 and 3: T2 and T3) two therapy phases (phase I and phase II). It innovates by applying two training programs for children with DLD, one linguistic/meta-linguistic and one cognitive, being undertaken one after the other. The linguistic/meta-linguistic training program focused on syntax and on complex syntactic structures, while the cognitive training program focused on vWM and vSTM skills. In this study, computerized therapeutic materials specifically designed for Greek-speaking individuals were employed. To the best of our knowledge this is the first study in Greek employing this particular design. An experienced speech and language therapist (the first author) administered individually the two 10-week training programs, with frequency of one session per week and duration of 40 min each session (i.e., a total of 6 h and 40 min of intervention). Three groups participated, two of them were the experimental treatment groups while the third one was the control group. In particular, the experimental clinical groups received both linguistic and cognitive treatment, successively, with different order. In the first therapy phase (phase I), while the clinical group A was receiving linguistic therapy (language-first group), the clinical group B was receiving cognitive therapy (cognitive-first group). In the second therapy phase (phase II), while the clinical group A was receiving cognitive therapy, the clinical group B was receiving linguistic therapy at the same time (T2). In sum, the present study combines both linguistic and cognitive training in children with DLD and explores direct and near-transfer as well as far-transfer effects in the domain of vSTM and vWM. Furthermore, it examines the effectiveness of the order of training pattern and applies the reversal of training schemes, combining both linguistic and cognitive training.

2. Materials and Methods

This study was performed within the PhD project of the first author (T.B.) under the supervision of the second author (S.S.) in accordance with the rules and regulations of Aristotle University of Thessaloniki. Within this PhD project, the effectiveness of cognitive and linguistic therapy on the linguistic and cognitive abilities of children with DLD was investigated. The present study focuses on reporting the linguistic vs. cognitive intervention effectiveness on cognitive abilities.

2.1. Participants

These children (mean age: 9;8 years, age range 7;3–12;0) had been diagnosed by the official governmental agency in Peloponnese for diagnosing learning and speech language difficulties in Greece (Centers of Interdisciplinary Assessment, Counseling and Support, KE.D.A.S.Y.) as having a severe language impairment by multidisciplinary groups of experts, including a psychologist who performed standardized tests (WISC-III [56]), a speech–language therapist for language assessment, a teacher of special education needs for literacy assessment, and a social worker for case history. The social worker interviewed parents to collect information about the child and his/her family. More specifically, information was collected about the developmental and medical history of the child (developmental milestones, signs of communication difficulties at preschool age, hearing problems, other medical diagnoses, etc.), as well as if there were language disorders in the family (i.e., relatives with a history of speech, language, or hearing difficulties), behavioral and school history, and the family's socio-economic status. All children were monolingual, native speakers of Greek.

To assess their verbal and non-verbal IQ, the WISC-III was employed which has been standardized for Greek [56] and used in official diagnostic centers in Greece. The Greek version has been standardized for children 6 years–16 years 11 months. It consisted of two scales, non-verbal and verbal. The former includes the following subtests: picture completion, coding, picture arrangement, block design, object assembly, symbol search, and mazes; the latter includes the following subtests: vocabulary, similarities, arithmetic, information, comprehension, and digit span. The overall non-verbal IQ score for the participants with DLD was within the normal range (non-verbal IQ: >85). With respect to their verbal IQ, all participants of this study were assessed by means of digit span. They were required to repeat digits forward and backwards. The forward and backward digit repetition is considered a measure of verbal short-term memory and working memory, respectively. As described below (Section 2.2.1), these tasks were employed pre-therapeutically and post-therapeutically. All participants had a mental age at least 6 months below their chronological age in the digit span subscale of WISC-III^{GR} [56]. This performance is indicative of difficulties in the domain of short-term and working memory. All children attended mainstream classes. Furthermore, no children with pronounced hearing problems participated in the study. The children were diagnosed as having severe language problems. In addition, according to the official reports of clinicians, these individuals with DLD showed impairments in the domain of syntax as indicated by the group performances on a syntactic comprehension task (also employed in the study by Talli and Stavrakaki [15]). This test consisted of 36 sentences with relative clauses, passives sentences, and sentences with reflexive verbs. Accuracy was calculated as a percentage of correct responses to the sentences. The DLD group showed deficient performance compared to typically developing children of this age [15] as the scores obtained were below the average score of typically developing children of this age. Additional difficulties in language for these children were detected by means of the Reading Test A [57]. This is a standardized diagnostic tool that has been developed to assess reading abilities in Greek-speaking children aged 8–15 years [57]. It evaluates four subdomains: decoding, fluency, morphosyntax, and reading comprehension. In the morphosyntax subscale, expressive morphosyntactic skills are assessed as follows: (a) production of verbs (person, number, tense, and aspect marking), (b) production of morphologically complex words, (c) sentence elicitation with visual cue, and (d) sentence elicitation without visual cue. Scores are converted into percentiles, which are the normative data for this test (50th percentile is the mean score). All children of this study had a percentile of 30 or below in morphosyntax, which is 2 SD below the mean. Difficulties in the domain of morphosyntax indicate the presence of a language disorder in the participants of our study.

In addition, all participants performed non-word repetition [58,59] in which they had a percentile below 1. This task assesses verbal STM. It is a Greek adaptation from the French test battery EVALEC [60] (Greek adaptation: [58]; see [59] for normalization for Greek children and adolescents aged 7–13 years old). It includes twenty-four three- to six-syllable non-words (six non-words for each length: four with CV and two with CVC syllable structure) presented orally in increasing length order. The child had to repeat each non-word after the examiner and his/her responses were noted. Accuracy score was the total number of correctly repeated syllables, calculated as a percentage. The children had a percentile below 1 in the task, which indicates deficient performance in vSTM. This performance characterizes individuals with DLD [16].

All participants were randomly assigned to three groups: (1) group A (linguistic-first group) consisted of 16 children with DLD receiving meta-linguistic training first and then vSTM/vWM training; (2) group B (cognitive-first group) consisted of 17 children with DLD receiving vSTM/vWM training first and then meta-linguistic training; (3) group C (control group) consisted of 16 children with DLD who received only their usual training at school (support by special education teachers). Notably, all children followed their usual everyday schedule. They were receiving special education services, either at their public school (inclusion class) or at home after school with a private teacher. While the effect of these factors cannot be evaluated, it is underlined that they were in action for all the participants of this study. In Table 1, demographic and clinical characteristics of these groups are presented. The groups did not differ in chronological age and non-verbal IQ. With respect to chronological age (CA), a Kruskal–Wallis test for non-parametric data was performed, as the distribution of the cognitive first group was not normal, which showed no group effect ($x^2(2) = 2.7$, $p = 0.3$). With respect to the non-verbal IQ we performed a Kruskal–Wallis test for non-parametric data, as the distribution of the language-first group was not normal, and again the difference was not significant ($x^2(2) = 1.1$, $p = 0.6$). For syntactic comprehension, we performed a one-way ANOVA, which showed that the three groups did not differ from each other ($F(2, 46) = 0.76$, $p = 0.47$).

Table 1. Demographic characteristics of the three groups and their performance on non-verbal IQ test (WISC-III) [56] as well as in syntactic comprehension [15].

	Group A (N = 16)	**Group B (N = 17)**	**Group C (N = 16)**	**Statistics for the Comparisons among Groups**
	M (SD)	**M (SD)**	**M (SD)**	
Age (in months)	114.56 (18.10)	122.06 (12.40)	112.81 (15.80)	$x^2(2) = 2.7, p = 0.3$
Non-verbal IQ	96.19 (10.55)	100.06 (9.84)	97.88 (11.19)	$x^2(2) = 1.1, p = 0.6$
Syntactic comprehension (accuracy %)	62.50 (13.42)	57.68 (11.43)	61.81 (11.72)	$F(2, 46) = 0.76, p = 0.47$
Boys/girls	10/6	11/6	12/4	

2.2. *Assessment Material and Procedure*

2.2.1. Pre- and Post-Training Assessment Materials

There were three assessments: one pre-therapeutically and two post-therapeutically (one week after the first training and one week after the second training). This paper focuses on assessments performed by means of two verbal STM tasks and one verbal WM task. Additional detailed linguistic testing was also performed [61]. Assessments were performed by the first author (T.B.), an experienced speech and language therapist.

Verbal STM

Non-word repetition: this task assesses verbal STM. As described above (in Section 2.1), this task assesses vSTM by non-word repetition (twenty-four three- to six-syllable non-words (six non-words for each length: four with CV and two with CVC syllable structure). The task is presented orally in increasing length order. As pointed out above, the accuracy score was the total number of correctly repeated syllables (%).

Forward digit span: this task also assesses verbal STM. It is the subtest of the Greek WISC-III [56]. The child was asked to repeat a list of digits (2–9 digits) in increasing order which were presented orally by the examiner. It consists of eight lists, for each of which the child could complete two trials: every correct trial was awarded with 1 point. The interruption criterion was a score of zero in both trials of a list. Total raw scores were taken into consideration (maximum 16 points).

Verbal WM

Backward digit span: this task assesses verbal working memory and is also the subtest from the Greek WISC-III. It consists of seven lists presented in increasing length (2–8 digits) and the child had to repeat them backwards. For each list the child could complete two trials: for a correct response from the first trial s/he was awarded with 2 points and for a correct response in second trial 1 point. When the child failed in both trials of a list the administration would stop. Total raw scores were noted (maximum 14 points).

2.2.2. Intervention Material

Training Programs

There were two training phases for both groups A and B. Group C received no training. Each of the two interventions included a 10-week training program once a week (40 min each session) for each child with DLD that was delivered individually. Intervention materials were administered by means of two software applications for a tablet. Intervention was applied by the first author (TB), an experienced speech and language therapist. In phase I, group A received meta-syntactic training, whereas group B vSTM/vWM training. In phase II, a reversal of training was performed for groups A and B: group A received vSTM/vWM while group B meta-syntactic training. In Figure A1 (Appendix A), the intervention phases are schematically presented. The intervention design developed in this study, in line with other studies in the field [34,62], aimed to evaluate the direct and near-transfer as well as far-transfer effects of linguistic vs. cognitive intervention in the domain of vSTM and vWM, in addition to the effect of the order of training pattern. For each post-therapy time (T2 and T3), the short-term goals for the groups that received intervention concerned the improvement of vSTM and vWM abilities via application of different training patterns. By the reversal of the therapy scheme order, the direct/near-transfer vs. far-transfer effects can be evaluated. Assessment was performed one week after the completion of each therapy pattern (after T2 and after T3) to evaluate the short-term goals. While our study, in line with other intervention studies [34,62], was not designed to provide a follow-up to assess whether there are permanent beneficial effects on the vSTM and vWM abilities of individuals with DLD, there was a longer-term goal than the short-term one described above. This was to evaluate the effect of the overall therapy scheme by assessing whether the therapy type provided first (linguistic vs. cognitive) after

T1 affected the performances shown in T3. By so doing, our long-term aim was to assess the effect of the intervention type order on performances in T3.

Meta-Syntactic Training

Meta-syntactic training was delivered through a software application named Assisted Language Acquisition (AS.L.A. URL: https://lab7.gr/asla.html [63]) comprising both production and comprehension of relative clauses. The theoretical background for the development of the tasks was, on the one hand, tests developed in the context of complex structures in DLD [10,25] and, on the other hand, therapeutic approaches with a theoretical background, see Ebbels [33] for an overview. There was a pilot application of the software by an experienced speech and language therapist to five children with DLD (grades 2–5) for at least 4 sessions, in order to check the acceptance and applicability of the software to this population. The software that was developed, exclusively for research and academic purposes, can run both on Windows operating systems and on tablets and/or mobile devices running the Android operating system. There is also the possibility of future transfer to other platforms, such as iOS. Among the various application development solutions available, the RAD Studio XE65 application development environment from Embarcadero was chosen (URL: https://www.embarcadero.com/products/rad-studio) and, in particular, the free trial version of the environment., Notably, all children had previous experience with tablets and they were familiar with their use.

A combination of linguistic and meta-linguistic techniques was used for the training of relative clauses, both production and comprehension. For the comprehension of relative clauses, at first the child would listen to a sentence with relative clauses and then s/he had to represent the sentence visually, through a specially configured interface. Visual representation through pictures [64,65] and symbols (see [33] for an overview) helps the child to understand the knowledge of how the object relative clause is connected and how it specifies the object of the main clause. The construction of the sentence was performed by choosing the correct image among others for every term of the sentence (i.e., subject, verb, object, subordinate relative clause). Different shapes and colors were chosen (see Ebbels [33]) for every term of the sentence: a subject was represented with a blue circle, a verb was represented with a green triangle, an object with a red square, a subordinate relative clause with a red diamond. A subordinate relative clause had the same color as the object to show the link between them. Visual representation with symbols and colors of the function and the relationship of the terms of the sentence aims at the activation of the conscious representation of the syntactic relation, that is, the meta-syntactic awareness.

See (1) for an example (stimulus sentence in Greek, in IPA, glossed and with translation in English):

(1) Το αγόρι κοιτάζει τη γιαγιά που αγκαλιάζει η γυναίκα.
to aγori citazi ti jaja pu agaʎazi i jineka.
The boy$_{\text{NOM}}$ looks the grandma who$_{\text{ACC}}$ hugs the woman$_{\text{NOM}}$
'The boy looks at the grandmother, who the woman hugs'.

In case the child misrepresented the sentence, the clinician would use a recasting technique, s/he would repeat the sentence and would indicate to the child the correct representation of the sentence, again explaining to him/her how the object relative clause is connected with the main clause.

After training comprehension, the production of relative clauses was then trained through elicitation technique [25]. A script was presented to the child with two pictures who was asked to answer the clinician's question, which contained the target structure of relative clauses, with partial guided production. See (2) for an example of production of subject relative clause (the introductory sentence, the prompting questions, and the guiding utterance of the clinician are in English, the target sentence is in Greek, IPA, glossed and translated into English).

(2) **Introductory sentence:**

Here are two pictures with children. One child is cooking soup, the other child is eating ice cream.
Prompting question
Which child would you rather be?
Experimenter utterance for partial guiding:
I would rather be...
Target:
... το παιδί που τρώει παγωτό.
... to peδi pu troi paγoto
... the child$_{NOM}$ who eats ice cream.
... the child who eats ice cream.

Feedback was given by the clinician according to the correctness of the child's answers. The children received a small prize at the end of each session (stickers or snack). The child's score/performance was recorded by the clinician during the treatment and after the session from the application statistics. All sessions were recorded and could be extracted in a file for further examination.

vSTM/vWM Training

Verbal STM and WM training was delivered through an application named Boosting Memory (BOOM! [66]). Similarly to AS.L.A., BOOM! software (URL: https://lab7.gr/boom.html) was developed to run both on Windows operating systems and on tablets and/or mobile devices running the Android operating system. There is also the possibility of future transfer to other platforms, such as iOS. Training material was divided into 10 sessions of varying difficulty. Each session, delivered individually, included five different types of tasks, each focusing on an area of phonological short-term and working memory. The order that was followed in each session is as follows: a. digit span, b. backward digit span, c. word repetition, d. non-word repetition, e. sentence repetition, f. N-back task (recall of penultimate word of a sentence), g. N-back task (recall of last two words of a sentence). For the specific design, research data of clinical interventions for STM and WM training for individuals with language disorders have been taken into account (for intervention in aphasia, Salis et al. [67], Salis [68] and for intervention in children with DLD [35]). Similarly to AS.L.A., there was a pilot application of the software by an experienced speech and language therapist to five children with DLD (grades 2–5) for at least 4 sessions in order to check the acceptance and applicability of the software to this population. In the forward digit span the child must repeat aloud in the same order a series of digits increasing in length from 4 to 9, such as 4-7-2-1, 2-6-4-7-8-3-9-5-1. The level of difficulty increased from session to session. In the backward digit span the child must repeat aloud a series of digits increasing in length from 4 to 9 in the reversed order, such as 5-7-3-8, 8-3-7-5. Similarly to the forward digit span, the level of difficulty increased from session to session. In the word repetition task, the children had to repeat words increasing in length from one to five syllables, with the syllabic structure CV or CVC, e.g., mati (=eye), nero (=water), meli (=honey). In the non-word repetition task, children had to repeat non-words with increasing length from one to five syllables and syllabic structure CV or CVC, e.g., loma, taci, nati, tola, kapatoli. In the sentence repetition task, children had to repeat aloud a sentence. The length of sentences increased in number of words from session to session. See (3) and (4) for examples (sentences in Greek, in IPA, glossed and with translation in English).

(3) Ο κηπουρός κόβει ένα όμορφο τριαντάφυλλο.

O cipuros kovi ena omorfo triantafilo.
The gardener cuts a beautiful rose.
'The gardener cuts a beautiful rose'.

(4) Ο κηπουρός ποτίζει τον όμορφο μεγάλο κήπο.

O cipuros potizi ton omorfo meγalo cipo.
The gardener waters the beautiful large garden'.
'The gardener waters the beautiful large garden'.

For the N-back tasks children had to recall the penultimate and the last two words of a sentence, respectively. See (5) for an example of the task of recall of the penultimate word of a sentence (sentences in Greek, in IPA, glossed and with translation in English, the target word in bold):

(5) Η γυναίκα ζωγραφίζει τον **μεγάλο** καμβά
I jineka zoγrafizi ton **meγalo** kamva.
The woman paints the **large** canvas.
'The woman paints the large canvas'.

In each session the child follows a story script presented through graphics reminiscent of cartoons. The story goes like this: on a beach there is a treasure chest. The child listens to recorded instructions at each stage of the session, as well as sentences, words, etc., which s/he must repeat in the same order. As the examination progresses, the chest seems to open every now and then. A pirate appears at the right moment, whose purpose is to take the treasure. Depending on the child's performance and the answers s/he gives as the process progresses, the pirate will eventually open the chest and find either the treasure or a pumpkin inside. Depending on the child's answers the clinician touches the screen to the right for correct answer or to the left for wrong answer. Feedback was given by the clinician after each answer according to the correctness of the child's answers. The children received a small prize at the end of each session (stickers or snack).

3. Results

The group performances on the assessment measures in T1, T2, and T3 are presented in Table 2.

Table 2. Performance of groups A, B, and C in the NWR, Forward and Backward digit span tasks, in each of the three assessments: pre-training (T1), 1st post-training (T2), 2nd post-training (T3).

	T1 M (SD)	T2 M (SD)	T3 M (SD)
NWR			
Group A	0.45 (0.12)	0.56 (0.10)	0.74 (0.05)
Group B	0.48 (0.07)	0.66 (0.05)	0.68 (0.07)
Group C	0.53 (0.13)	0.51 (0.11)	0.52 (0.11)
Forward digit span			
Group A	7.38 (1.54)	9.69 (1.82)	10.44 (1.09)
Group B	8.00 (1.12)	9.76 (1.56)	10.12 (1.93)
Group C	8.06 (1.61)	8.44 (1.50)	8.81 (1.11)
Backward digit span			
Group A	3.00 (0.82)	4.13 (1.26)	4.88 (1.20)
Group B	2.71 (1.45)	3.76 (1.35)	4.76 (1.39)
Group C	2.69 (0.70)	3.13 (1.15)	2.88 (1.02)

3.1. Data Analysis

We performed three types of analyses. First, the data from the non-word repetition task were analyzed by means of multilevel models for mixed designs, since the design included one within-subject (testing time) and one between-subject (group) factor. For the factor group we used sliding contrasts, which means that the two treatment groups were compared to each other and the cognitive-first group was compared to the control group. For the factor testing time we set sliding contrasts, too, so that each level was compared to its adjacent level, i.e., testing time 1 was compared to testing time 2 and testing time 2 was compared to testing time 3. For these analyses we also included the following random components: an adjustment of each child's individual average (random intercept for subjects) and an adjustment for each child's testing time effect (random slope for subjects). For each analysis we followed the following procedure for choosing the best model. First, we fitted a model with only the intercept and subsequently added the main effects testing time and group and finally the interaction of the two factors. Then, we compared the models and if there was a significant difference between two models, it was considered as a significant contribution of the additional factor. Finally, we used the contrasts in order to break down the interactions.

The second type of analysis consisted of one-way repeated measures ANOVAs and concerned within-group comparisons across testing times, whereby we defined sliding contrasts for the factor testing time so that each level was compared to its adjacent level, i.e., testing time 1 was compared to testing time 2 and testing time 2 was compared to testing time 3, as in the multilevel model. Similarly, we fitted models only with the intercept and then added the main effect. A significant difference between the models was considered as a significant effect of time. These analyses were necessary because the multilevel models for mixed designs provide information about the difference between groups concerning the change from testing point to testing point and not the performance of each group across testing times. We also performed one-way ANOVAs for between-group comparisons at different testing times. We defined sliding contrasts for the factor group, so that the language-first group is compared to the cognitive-first group and the cognitive-first group is compared to the control group. The interpretation was the same as for the repeated measures ANOVA.

The third type of analysis concerned the data from the forward and backward digit span, which were analyzed by means of non-parametric tests because the distributions were not normal and no transformation could be applied to overcome this problem. In particular, we performed a non-parametric Kruskal–Wallis test for the between-group comparisons for each testing time. For the post hoc comparisons, we used the Dunn test, which is the most popular for post hoc comparisons after a significant Kruskal–Wallis, with Benjamini–Hochberg corrections to control the familywise error rate. For the pairwise comparisons we report the adjusted *p*-value. For the within-group analysis for each testing time we used the non-parametric Friedmans's chi-square test, which corresponds to the one-way ANOVA for repeated measures, and performed post hoc comparisons with the Wilcoxon test for repeated measures by adjusting the *p*-values with Bonferroni correction. All analyses were carried out with the R software [69].

3.1.1. Verbal Short-Term Memory—Non-Word Repetition Task

Figure 1 presents the percentage correct responses of the verbal short term memory task (non-word repetition) for the three groups at the different testing times; see also Table 2.

The one-way ANOVA for T1 showed no group effect in the pre-treatment time ($\beta = 3.85$, $t = 0.91$, $p = 0.37$ for the comparison between language first and control and $\beta = 4.22$, $t = 0.1$, $p = 0.32$ for the comparison of cognitive first and control). The comparison of the different models for the whole dataset revealed significant main effects of group ($x^2(6) = 10$, $p = 0.0077$) and testing time ($x^2(8) = 63$, $p < 0.0001$). Moreover, we found a significant interaction of group by testing time ($x^2(12) = 93$, $p < 0.0001$). We used the contrasts to gain insight into this interaction. Table 3 shows the β, the *t*-value, the *p*-value, and the effect size for each contrast.

The results indicate that non-word repetition improved for both treatment groups. In particular, the cognitive-first group improved by T2, i.e., after receiving the cognitive intervention, in comparison to both the language-first group (but the effect size is medium) and the control group (large effect). Concerning the change between T2 and T3, the language-first group improved significantly more than the cognitive-first group with the effect being large, whereas the cognitive-first group did not improve further after receiving the meta-linguistic intervention, i.e., it remained as stable as the control group. The one-way ANOVA for T2 showed that the cognitive-first group was significantly better than both the language-first ($\beta = 10.14$, $t = 2.99$, $p = 0.0045$) and the control group ($\beta = -15.3$, $t = -4.5$, $p < 0.001$). The post hoc *t*-test for the comparison between the control and the language-first group showed that there was no difference between them ($t(29.9) = 1.28$, $p = 0.63$) Interestingly, the one-way ANOVA for T3 showed that the language-first group was significantly better than the cognitive-first group ($\beta = -6.13$, $t = -2$, $p = 0.046$), which was better than the control ($\beta = -18$, $t = -6$, $p < 0.0001$). This finding suggests that the group that received the meta-linguistic intervention first not only improved in non-word repetition but also achieved higher performance in T3 than the treatment group that first received the meta-cognitive and afterwards the meta-linguistic intervention. The within-group comparisons showed that the language-first group improved significantly both from T1 to T2 ($\beta = 12$, $t = 5.4$, $p = 0$, $r = 0.7$, large effect) and from T2 to T3 ($\beta = 19$, $t = 8.6$, $p = 0$, $r = 0.8$, large effect). Concerning the cognitive-first group, it improved only from T1 to T2 ($\beta = 18$, $t = 10$, $p = 0$, $r = 0.87$, large effect) and not from T2 to T3 ($\beta = 3$, $t = 2$, $p = 0.1$). The control group did not improve across time, as expected.

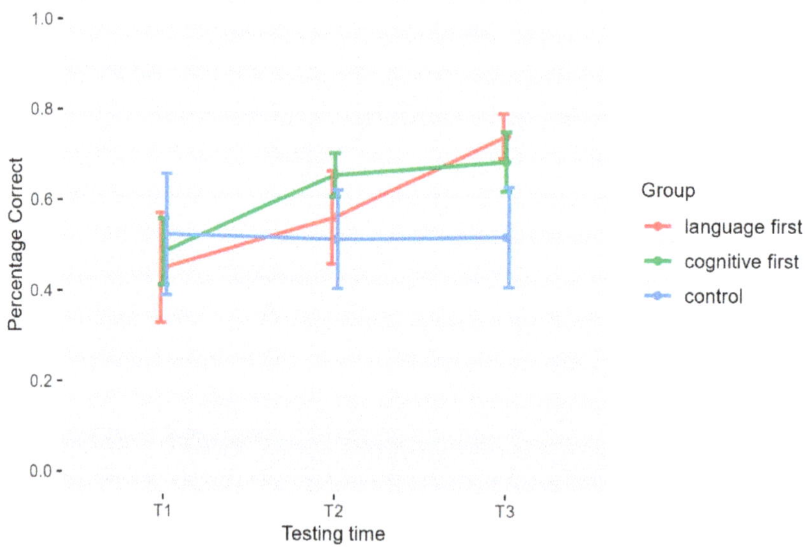

Figure 1. Percentage of the correct responses in the Verbal Short Term Memory (Non-Word Repetition Task). T1: pre-training assessment, T2: 1st post-training assessment, T3: 2nd post-training assessment, language first: group which received first meta-linguistic training and second STM/WM training, cognitive first: group which received first STM/WM training and then meta-linguistic training, control: control group which received no training.

Table 3. Contrasts, coefficients (β), *t*-values, and *p*-values as well as effect size for each interaction concerning the performance in the Non-Word Repetition task.

Contrast	β	*t*(df)	*p*-Value	Effect Size, *r*
Group 2-1: Time 2-1: Comparison between language first and cognitive first concerning the change between testing time 1 and testing time 2	6	2 (92)	0.03	0.2 small
Group 3-2: Time 2-1: Comparison between cognitive first and controls concerning the change between testing time 1 and testing time 2	−20	−7 (92)	0.0000	0.59 large
Group 2-1: Time 3-2: Comparison between language first and cognitive first concerning the change between testing time 2 and testing time 3	−16	−6 (92)	0.0000	0.53 large
Group 3-2: Time 3-2: Comparison between cognitive first and controls concerning the change between testing time 2 and testing time 3	−3	1.1 (92)	0.36	0.09 very small

3.1.2. Verbal Short-Term Memory—Forward Digit Span

Figure 2 presents the percentage of correct performances in the forward digit span, the second measure of verbal STM, for exact numbers, see Table 2. As the figure shows, the two treatment groups manifest the same pattern. This was confirmed by the statistical analysis.

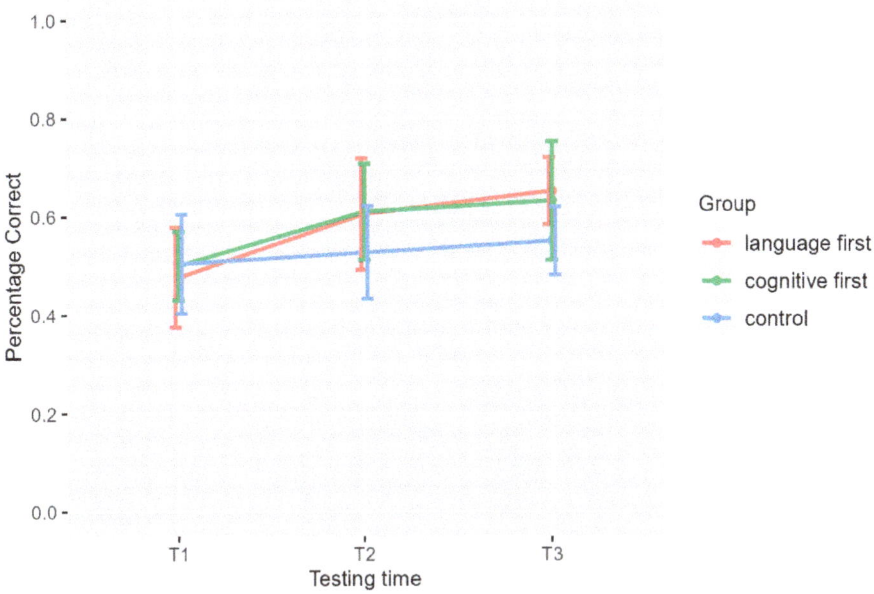

Figure 2. Percentage of correct responses in the Verbal Short-Term Memory (Forward Digit Span Task). T1: pre-training assessment, T2: 1st post-training assessment, T3: 2nd post-training assessment, language first: group which received first meta-linguistic training and second STM/WM training, cognitive first: group which received first STM/WM training and then meta-linguistic training, control: control group which received no training.

There was not a significant group effect in T1 ($x^2(2) = 0.9$, $p = 0.6$), and the effect just failed to reach significance in T2 ($x^2(2) = 6$, $p = 0.06$) but was significant in T3 ($x^2(2) = 10$, $p = 0.005$). The post hoc tests for T2 showed non-significant differences between the two treatment groups and the control group, (language first vs. control: $Z = -1.93$, adjusted $p = 0.08$ and cognitive first vs. control: $Z = 2.22$, adjusted $p = 0.08$). The difference between the two treatment groups was not significant either ($Z = 0.26$, adjusted $p = 0.8$). Post hoc comparisons for T3 showed a significant difference between language first and control ($Z = -3.14$, adjusted $p = 0.005$), with the effect being large ($r = 0.56$, large), and between cognitive first and control ($Z = 2.25$, adjusted $p = 0.037$), but the effect was moderate ($r = 0.35$). As in T2, there was no difference between the two treatment groups ($Z = -0.95$, adjusted $p = 0.34$). Summing up the between-group comparisons, we see a difference for the two treatment groups compared to the controls only in T3 after both treatment groups received both interventions. We suggest that the moderate vs. large effect is due to the large vs. small standard deviation observed in T3 for the cognitive-first and the language-first groups, respectively, although they have very similar mean performance. Within-group comparisons yielded significant effects of testing time for the two treatment groups (for the language-first group, Friedman's $x^2(2) = 21$, $p = 0$ and for the cognitive-first group, Friedman's $x^2(2) = 16$, $p = 0$). Pairwise comparisons showed a significant difference between T1 and T2 ($W = 2.5$, adjusted $p = 0.002$, $r = 0.86$, large) but not between T2 and T3 ($W = 24$, adjusted $p = 0.2$, $r = 0.46$, moderate) (Bonferroni correction) for the language-first group. A similar pattern manifested for the cognitive-first group, namely a significant difference between T1 and T2 ($W = 8$, adjusted $p = 0.006$, $r = 0.76$, large) but not between T2 and T3 ($W = 15.5$, adjusted $p = 1$, $r = 0.2$, small). Surprisingly, the Friedman's test was also significant for the control group (Friedman's $x^2(2) = 8$, $p = 0.01$). The post hoc pairwise comparisons between T1 and T2 and between T2 and T3 were not significant ($W = 15.5$, adjusted $p = 0.68$ and $W = 9$, adjusted $p = 0.66$, respectively). However, the difference between T1 and T3 was significant ($W = 0$, adjusted $p = 0.02$, $r = 0.74$, large). In sum, both treatment groups showed a large improvement in T2 compared to T1, whereas there was no substantial change between T2 and T3. Concerning the control group, there was an overall improvement from the T1 to T3.

3.1.3. Verbal Working Memory—Backward Digit Span

The percentage of correct performances in this task is shown in Figure 3 and Table 2. While the performance of both treatment groups is very low, as it remains at below 50% correct, significant improvement in the treatment group performance was found as shown by the pairwise comparisons presented below.

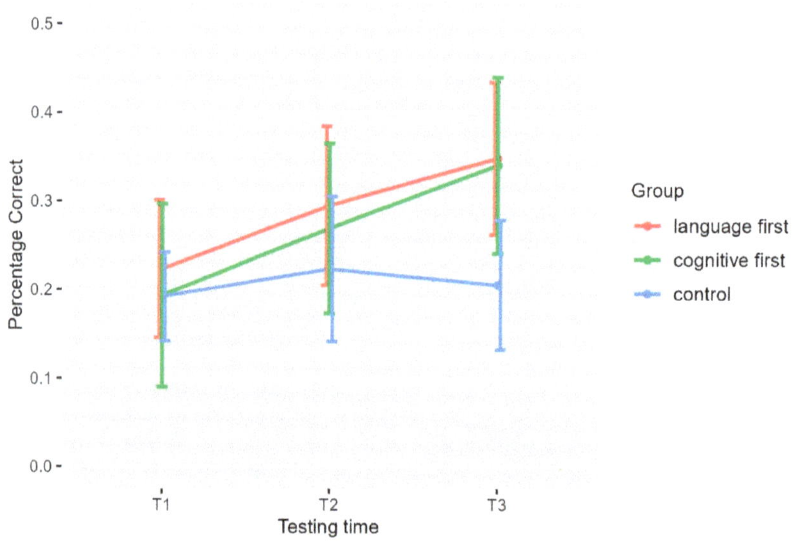

Figure 3. Percentage correct responses in the Verbal Working Memory (Backward Digit Span Task). T1: pre-training assessment, T2: 1st post-training assessment, T3: 2nd post-training assessment, language first: group which received first meta-linguistic training and second STM/WM training, cognitive first: group which received first STM/WM training and then meta-linguistic training, control: control group which received no training.

The between-group comparisons for each testing time showed no group effect for T1 ($x^2(2) = 1$, $p = 0.5$) and T2 ($x^2(2) = 5$, $p = 0.09$), whereas there was a difference in T3 ($x^2(2) = 19$, $p = 0$). The post hoc tests for T3 showed that the difference between the two treatment groups and the control group was significant and the effect sizes were large for both comparisons (language first vs. control, $Z = -3.9$, adjusted $p = 0$, effect size $r = 0.7$ and cognitive first vs. control $Z = 3.71$, adjusted $p = 0$, $r = 0.63$), whereas the difference between the two treatment groups was not significant ($Z = -0.25$, adjusted $p = 0.8$). Summarizing, neither treatment resulted in between-treatment-group differences in BWDS. In T3, when both treatment groups had received both therapies, the treatment groups were significantly better than the control and the effect was large.

Similarly to the forward digit span task, we performed the non-parametric Friedman's ANOVA for the within-group comparisons. For the language-first group there was a significant effect of testing time (Friedman's $x^2(2) = 15$, $p = 0.0007$). Pairwise comparisons showed a significant difference between T1 and T3 ($W = 10$, adjusted $p = 0.025$, $r = 0.7$, large effect), but not between T1 and T2 (W = 22, adjusted $p = 0.171$) and between T2 and T3 ($W = 9$, adjusted $p = 0.1$). This suggests a gradual improvement. There was also an effect of testing time for the cognitive-first group (Friedman's $x^2(2) = 15$, $p = 0.0006$). The pairwise comparisons showed that both the differences between T1 and T2 and between T2 and T3 were significant and the effect was large (for T1 vs. T2 $W = 14$, adjusted $p = 0.041$, $r = 0.6$ and for T2 vs. T3 $W = 7$ adjusted $p = 0.036$, $r = 0.61$). As expected, there was no effect for the control group (Friedman's $x^2(2) = 2$, $p = 0.3$). In sum, the within-group comparisons showed a gradual improvement across testing times for the language-first group. The improvement of the cognitive-first group was large for both post-treatment testing times.

4. Discussion

This study was set up to investigate the efficiency of two cross-domain training programs for children with DLD, one linguistic/meta-linguistic and one cognitive. The novelty of this study lies in the alteration of two different training programs which are evaluated in terms of (i) direct, near-, and far-transfer effects on the performance of vSTM and vWM tasks and (ii) the therapy type order, specifically, whether applying linguistic/meta-linguistic training first or applying cognitive training first is more beneficial for vSTM and vWM skills of children with DLD. The particular design of the present study was strongly motivated by the simultaneous presence of linguistic and vSTM/vWM impairments in children with DLD (see Section 1). While the bulk of research concerns the application of linguistic and cognitive methods separately in individuals with DLD, the combination of them has not received any specific attention, although the individuals with DLD might benefit from this combination as they suffer from linguistic and cognitive deficits. A further question is whether the applied therapy type could better trigger the individual response to therapy when delivered first.

With respect to vSTM, the results collectively indicated that non-word repetition improved for both treatment groups. The between-group comparisons for T2 showed that the cognitive-first group was significantly better than the language-first group. This finding can be interpreted as showing that receiving therapy targeting in the specific domain, i.e., cognitive therapy for improving vSTM, maximizes the treatment gains. However, in T3 the language-first group improved significantly compared to T2, while the performance of the cognitive-first group was not increased after receiving the meta-linguistic intervention. This finding can be interpreted as indicative of the therapy type order effect, as it shows that the group that received the meta-linguistic intervention benefited significantly. It may be the case that the meta-linguistic therapy program contributes to the enhancement of vSTM by training the ability to keep and store linguistic information as well as to perform combinational linguistic procedures on it. In our view, these results suggest that far-transfer effects can actually occur for children with DLD when cognitive training is combined with a linguistic training and, more specifically, when a meta-syntactic training is followed by a vWM/vSTM training. These findings corroborate findings from other studies that show far-transfer gains after linguistic training [70,71] and/or vWM/vSTM training [34,35,41,72–74]. What our results add to these findings is that the intervention's positive impact is affected by applying two different therapy types with the linguistic therapy to come first. By contrast, while a significant improvement was observed in the performance of the treatment groups on the forward digit span task, the overall effect of the therapy type order is hard to evaluate as no substantial change occurred between T2 and T3 for the treatment groups. In addition, the improvement of the control group from T1 to T3 indicates that the daily school routine has a positive impact on the vSTM abilities of these children.

With respect to vWM, the results collectively indicated improvement for both the treatment groups especially in T3. While the language-first group showed a gradual improvement that became significant in T3 (as shown by pairwise comparisons), the cognitive-first group showed significant improvement between T1 and T2 as well as between T2 and T3 (with large effects). These results indicate that, with respect to this particular task (BWDS), starting with cognitive therapy can effectively trigger the improvement of vWM abilities. Interestingly, the cognitive-first group starts with lower (although not significantly lower) vWM abilities in this task and shows the same performance as the language-first group in T3. Overall, these findings indicate that receiving therapy targeting a specific domain, i.e., cognitive therapy for improving vWM, positively impacts on the vWM performance as shown by the significance of the difference between T1 and T2 for the cognitive-first group.

Notably, in T3, when both treatment groups had received both therapies, the treatment groups were significantly better than the control and the effect was large. This finding indicates that applying more than one therapy type may maximize the beneficial effect. In fact, despite differences in the improvement rate (gradual improvement across testing times for the language-first group; large improvement for the cognitive-first group) at the end both the treatment groups greatly benefited.

Comparing the within-group performance over different times for the vSTM task (non-word repetition) and the vWM task, a reverse pattern is shown regarding the impact that the therapy type order has on the treatment groups' performance. Specifically, while the language-first group benefited more from receiving the language therapy first regarding vSTM, the cognitive-first group benefited more from receiving the cognitive therapy first regarding the vWM. Apparently, performance on vSTM tasks may be more closely related to linguistic demands than that on vWM. While current research indicates correlation between vSTM, vWM, and linguistic abilities [36,41,51,52], our results indicate that vSTM performance is more positively affected than vWM performance by linguistic training.

While direct comparison of the outcomes of the present mixed treatment intervention cannot be performed, as to the best of our knowledge a meta-syntactic and cognitive treatment was not simultaneously applied, it should be pointed out that beneficial effects of mixed therapy have been previously reported. In particular, Tyler et al. [74] investigated the effectiveness of a mixed therapeutic design targeting phonology and morphosyntax in children with deficient phonological and morphosyntactic abilities. They found that both treatment approaches were effective. However, they noticed that when the morphosyntactic therapy was delivered, there was a slightly better morphosyntactic improvement. While results of the present study cannot be compared directly with the results by Tyler et al., it is pointed out that the mixed therapeutic designs can reveal aspects of improvement that cannot be shown by the application of one only intervention method.

Overall, the results of this study indicate that far-transfer effects can actually occur for children with DLD (language therapy can affect vSTM and vWM) in addition to direct and near-transfer effects. Furthermore, these data show that the combination of different treatment methods and especially the treatment type order may be a significant matter in improving deficient abilities in DLD. However, this study has the following limitations. First, the combination of different methods has not been compared to the application of one only treatment method (linguistic or cognitive). Second, no assessment of the long-term therapy effects has been performed, as this study was designed to assess the short-term treatment benefits. Therefore, future studies should address the contribution of different methods in comparison to the effectiveness of one only treatment method (linguistic or cognitive) to provide more accurate assessment of the method combination outcome. In addition, the long-term therapy gains should be evaluated by providing assessment a long time after therapy completion.

5. Conclusions

Concluding, the present study provides insights into the effects of different therapy methods applied in different orders to individuals with a primary disorder in language. It evaluates their effectiveness simultaneously and assesses short-term gains. By so doing, it shows perspectives on the combination of therapy protocols and the consideration of 'what comes first' in the treatment application. There are limitations in the present study that concern the comparison of the combined methods' effects to the effects of one treatment method; in addition, no long-term assessment was provided after therapy completion. Future studies are expected to address these questions by extending the present therapy protocol to provide further comparisons of different treatment methods and assess long-term benefits, which are crucial for contributing to the well-being of individuals with language disorders.

Author Contributions: Conceptualization, S.S.; methodology, S.S.; validation, T.B. and I.T.; formal analysis, V.K.; investigation, T.B.; data curation, T.B., V.K. and I.T.; writing—original draft preparation, T.B. and I.T.; writing—review and editing, S.S., V.K. and I.T.; supervision, S.S.; project administration, S.S. and T.B. All authors have read and agreed to the published version of the manuscript.

Funding: This research received no external funding.

Institutional Review Board Statement: This study was conducted in accordance with the Declaration of Helsinki. Informed consent was attained from every parent/legal guardian of the participating children. All measures for keeping anonymity and protecting personal data were taken, while the researchers have followed the rules of Research Ethics and Deontology. Ethical review and approval were not required for the study on human participants in accordance with the local legislation and institutional requirements.

Informed Consent Statement: Written informed consent was sought and obtained from all parents of the children who participated in the study, as well as verbal assent by the children themselves.

Data Availability Statement: The data presented in this study are available upon request from the corresponding author. The data are not publicly available due to ethical reasons.

Acknowledgments: The authors would like to thank the children who participated in this therapy study and their families.

Conflicts of Interest: The authors declare no conflicts of interest.

Appendix A

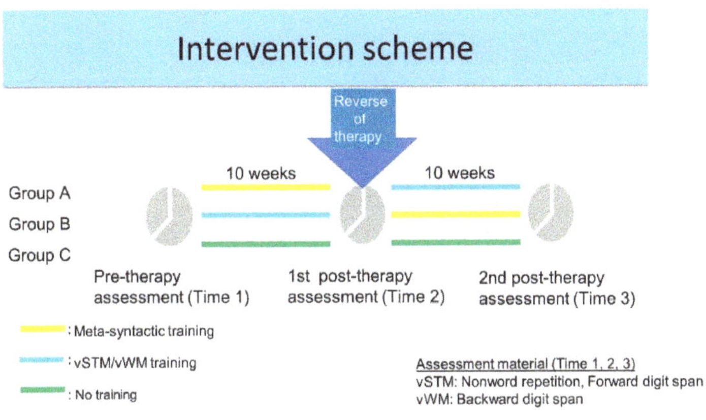

Figure A1. Overview of the intervention scheme, its duration and assessment material.

References

1. Bishop, D.V.M.; Snowling, M.J.; Thompson, P.A.; Greenhalgh, T.; CATALISE-2 Consortium. Phase 2 of CATALISE: A multinational and multidisciplinary Delphi consensus study of problems with language development: Terminology. *J. Child Psychol. Psychiatry* **2017**, *58*, 1068–1080. [CrossRef]
2. Norbury, C.F.; Gooch, D.; Wray, C.; Baird, G.; Charman, T.; Simonoff, E.; Vamvakas, G.; Pickles, A. The impact of nonverbal ability on prevalence and clinical presentation of language disorder: Evidence from a population study. *J. Child Psychol. Psychiatry* **2016**, *57*, 1247–1257. [CrossRef]

3. Bishop, D.V.M.; Snowling, M.J.; Thompson, P.A.; Greenhalgh, T.; CATALISE Consortium. CATALISE: A multinational and multidisciplinary Delphi consensus study: Identifying Language Impairments in Children. *PLoS ONE* **2016**, *11*, e0158753. [CrossRef]
4. Fey, M.E.; Catts, H.W.; Proctor-Williams, K.; Tomblin, J.B.; Zhang, X. Oral and Written Story Composition Skills of Children with Language Impairment. *J. Speech Lang. Hear. Res.* **2004**, *47*, 1301–1318. [CrossRef]
5. Georgiou, G.P.; Theodorou, E. Abilities of children with developmental language disorders in perceiving phonological, grammatical, and semantic structures. *J. Autism Dev. Disord.* **2023**, *53*, 4483–4487. [CrossRef]
6. Leonard, L.B. *Children with Specific Language Impairment*, 2nd ed.; MIT Press: Cambridge, MA, USA, 2014.
7. Montgomery, J.W.; Gillam, R.B.; Evans, J.L.; Sergeev, A.V. "Whatdunit?" Sentence Comprehension Abilities of Children with SLI: Sensitivity to Word Order in Canonical and Noncanonical Structures. *J. Speech Lang. Hear. Res.* **2017**, *60*, 2603–2618. [CrossRef]
8. Arosio, F.; Silleresi, S.; Guasti, M.T. The production of relative clauses in Italian-speaking children with DLD. *Isogloss. Open J. Roman. Linguist.* **2024**, *10*, 1–23. [CrossRef]
9. Fonteneau, E.; Van Der Lely, H.K.J. Electrical Brain Responses in Language-Impaired Children Reveal Grammar-Specific Deficits. *PLoS ONE* **2008**, *3*, e1832. [CrossRef]
10. Friedmann, N.; Novogrodsky, R. Which questions are most difficult to understand? *Lingua* **2011**, *121*, 367–382. [CrossRef]
11. Hesketh, A. The use of relative clauses by children with language impairment. *Clin. Linguist. Phon.* **2006**, *20*, 539–546. [CrossRef]
12. Hestvik, A.; Schwartz, R.G.; Tornyova, L. Relative Clause Gap-Filling in Children with Specific Language Impairment. *J. Psycholinguist. Res.* **2010**, *39*, 443–456. [CrossRef]
13. Stavrakaki, S. Comprehension of Reversible Relative Clauses in Specifically Language Impaired and Normally Developing Greek Children. *Brain Lang.* **2001**, *77*, 419–431. [CrossRef]
14. Stavrakaki, S. A-bar Movement Constructions in Greek Children with SLI. In *Clinical Linguistics: Theory and Applications in Speech Pathology and Therapy*; Fava, E., Ed.; John Benjamins: Amsterdam, The Netherlands, 2002; pp. 131–153.
15. Talli, I.; Stavrakaki, S. Short-term memory, working memory and linguistic abilities in bilingual children with Developmental Language Disorder. *First Lang.* **2020**, *40*, 437–460. [CrossRef]
16. Graf Estes, K.; Evans, J.L.; Else-Quest, N.M. Differences in the Nonword Repetition Performance of Children with and Without Specific Language Impairment: A Meta-Analysis. *J. Speech Lang. Hear. Res.* **2007**, *50*, 177–195. [CrossRef]
17. Vugs, B.; Hendriks, M.; Cuperus, J.; Knoors, H.; Verhoeven, L. Developmental Associations Between Working Memory and Language in Children with Specific Language Impairment: A Longitudinal Study. *J. Speech Lang. Hear. Res.* **2017**, *60*, 3284–3294. [CrossRef]
18. Archibald, L.M.D.; Joanisse, M.F. On the Sensitivity and Specificity of Nonword Repetition and Sentence Recall to Language and Memory Impairments in Children. *J. Speech Lang. Hear. Res.* **2009**, *52*, 899–914. [CrossRef]
19. Ebert, K.D.; Kohnert, K. Sustained Attention in Children with Primary Language Impairment: A Meta-Analysis. *J. Speech Lang. Hear. Res.* **2011**, *54*, 1372–1384. [CrossRef]
20. Im-Bolter, N.; Johnson, J.; Pascual-Leone, J. Processing Limitations in Children with Specific Language Impairment: The Role of Executive Function. *Child Dev.* **2006**, *77*, 1822–1841. [CrossRef]
21. Montgomery, J.W.; Magimairaj, B.M.; Finney, M.C. Working Memory and Specific Language Impairment: An Update on the Relation and Perspectives on Assessment and Treatment. *Am. J. Speech Lang. Pathol.* **2010**, *19*, 78–94. [CrossRef]
22. Pauls, L.J.; Archibald, L.M.D. Executive Functions in Children with Specific Language Impairment: A Meta-Analysis. *J. Speech Lang. Hear. Res.* **2016**, *59*, 1074–1086. [CrossRef]
23. Ullman, M.T.; Pierpont, E.I. Specific Language Impairment is not Specific to Language: The Procedural Deficit Hypothesis. *Cortex* **2005**, *41*, 399–433. [CrossRef]
24. Rice, M.L.; Wexler, K.; Marquis, J.; Hershberger, S. Acquisition of Irregular Past Tense by Children with Specific Language Impairment. *J. Speech Lang. Hear. Res.* **2000**, *43*, 1126–1144. [CrossRef]
25. Novogrodsky, R.; Friedmann, N. The production of relative clauses in syntactic SLI: A window to the nature of the impairment. *Adv. Speech Lang. Pathol.* **2006**, *8*, 364–375. [CrossRef]
26. Stavrakaki, S. Developmental perspectives on Specific Language Impairment: Evidence from the production of wh-questions by Greek SLI children over time. *Adv. Speech Lang. Pathol.* **2006**, *8*, 384–396. [CrossRef]
27. Van Der Lely, H.K.J.; Jones, M.; Marshall, C.R. Who did Buzz see someone? Grammaticality judgement of wh-questions in typically developing children and children with Grammatical-SLI. *Lingua* **2011**, *121*, 408–422. [CrossRef]
28. Archibald, L.M.D.; Gathercole, S.E. Short-term and working memory in specific language impairment. *Int. J. Lang. Commun. Disord.* **2006**, *41*, 675–693. [CrossRef]
29. Plym, J.; Lahti-Nuuttila, P.; Smolander, S.; Arkkila, E.; Laasonen, M. Structure of Cognitive Functions in Monolingual Preschool Children with Typical Development and Children with Developmental Language Disorder. *J. Speech Lang. Hear. Res.* **2021**, *64*, 3140–3158. [CrossRef]
30. Gillam, R.B.; Montgomery, J.W.; Evans, J.L.; Gillam, S.L. Cognitive predictors of sentence comprehension in children with and without developmental language disorder: Implications for assessment and treatment. *Int. J. Speech Lang. Pathol.* **2019**, *21*, 240–251. [CrossRef]
31. Balthazar, C.H.; Ebbels, S.; Zwitserlood, R. Explicit Grammatical Intervention for Developmental Language Disorder: Three Approaches. *Lang. Speech Hear. Serv. Sch.* **2020**, *51*, 226–246. [CrossRef]

32. Ebbels, S. Teaching grammar to school-aged children with specific language impairment using Shape Coding. *Child Lang. Teach. Ther.* **2007**, *23*, 67–93. [CrossRef]
33. Ebbels, S. Improving grammatical skill in children with specific language impairment. In *Understanding Developmental Language Disorders: From Theory to Practice*; Norbury, C.F., Tomblin, J.B., Bishop, D.V.M., Eds.; Psychology Press: London, UK, 2008; pp. 149–174.
34. Stanford, E.; Durrleman, S.; Delage, H. The Effect of Working Memory Training on a Clinical Marker of French-Speaking Children with Developmental Language Disorder. *Am. J. Speech Lang. Pathol.* **2019**, *28*, 1388–1410. [CrossRef]
35. Delage, H.; Stanford, E.; Durrleman, S. Working memory training enhances complex syntax in children with Developmental Language Disorder. *Appl. Psycholinguist.* **2021**, *42*, 1341–1375. [CrossRef]
36. Pauls, L.J.; Archibald, L.M. Cognitive and linguistic effects of narrative-based language intervention in children with Developmental Language Disorder. *Autism Dev. Lang. Impair.* **2021**, *6*, 23969415211015867. [CrossRef]
37. Alloway, T.P.; Bibile, V.; Lau, G. Computerized working memory training: Can it lead to gains in cognitive skills in students? *Comput. Hum. Behav.* **2013**, *29*, 632–638. [CrossRef]
38. Dunning, D.L.; Holmes, J.; Gathercole, S.E. Does working memory training lead to generalized improvements in children with low working memory? A randomized controlled trial. *Dev. Sci.* **2013**, *16*, 915–925. [CrossRef]
39. Holmes, J.; Gathercole, S.E. Taking working memory training from the laboratory into schools. *Educ. Psychol.* **2014**, *34*, 440–450. [CrossRef]
40. Wener, S.E.; Archibald, L.M. Domain-specific treatment effects in children with language and/or working memory impairments: A pilot study. *Child Lang. Teach. Ther.* **2011**, *27*, 313–330. [CrossRef]
41. Pauls, L.J.; Archibald, L.M.D. Cognitive and Linguistic Effects of Working Memory Training in Children with Corresponding Deficits. *Front. Educ.* **2022**, *6*, 812760. [CrossRef]
42. Klingberg, T.; Fernell, E.; Olesen, P.J.; Johnson, M.; Gustafsson, P.; Dahlström, K.; Gillberg, C.G.; Forssberg, H.; Westerberg, H. Computerized Training of Working Memory in Children with ADHD—A Randomized, Controlled Trial. *J. Am. Acad. Child Adolesc. Psychiatry* **2005**, *44*, 177–186. [CrossRef]
43. Henry, L.A.; Christopher, E.; Chiat, S.; Messer, D.J. A Short and Engaging Adaptive Working-Memory Intervention for Children with Developmental Language Disorder: Effects on Language and Working Memory. *Brain Sci.* **2022**, *12*, 642. [CrossRef]
44. Cleave, P.L.; Becker, S.D.; Curran, M.K.; Van Horne, A.J.O.; Fey, M.E. The Efficacy of Recasts in Language Intervention: A Systematic Review and Meta-Analysis. *Am. J. Speech Lang. Pathol.* **2015**, *24*, 237–255. [CrossRef]
45. Cirrin, F.M.; Gillam, R.B. Language Intervention Practices for School-Age Children with Spoken Language Disorders: A Systematic Review. *Lang. Speech Hear. Serv. Sch.* **2008**, *39*, S110–S137. [CrossRef]
46. Law, J.; Garrett, Z.; Nye, C. The Efficacy of Treatment for Children with Developmental Speech and Language Delay/Disorder: A Meta-Analysis. *J. Speech Lang. Hear. Res.* **2004**, *47*, 924–943. [CrossRef]
47. Rinaldi, S.; Caselli, M.C.; Cofelice, V.; D'amico, S.; De Cagno, A.G.; Della Corte, G.; Di Martino, M.V.; Di Costanzo, B.; Levorato, M.C.; Penge, R.; et al. Efficacy of the Treatment of Developmental Language Disorder: A Systematic Review. *Brain Sci.* **2021**, *11*, 407. [CrossRef]
48. Storkel, H.L.; Komesidou, R.; Fleming, K.K.; Romine, R.S. Interactive Book Reading to Accelerate Word Learning by Kindergarten Children with Specific Language Impairment: Identifying Adequate Progress and Successful Learning Patterns. *Lang. Speech Hear. Serv. Sch.* **2017**, *48*, 108–124. [CrossRef]
49. Nair, V.K.; Clark, G.T.; Siyambalapitiya, S.; Reuterskiöld, C. Language intervention in bilingual children with developmental language disorder: A systematic review. *Int. J. Lang. Commun. Disord.* **2023**, *58*, 576–600. [CrossRef]
50. Bettelli, G.; Guasti, M.T.; Ajmone, P.F.; Tenca, E.; Arosio, F. The effect of a priming-based training on the production of object clitic pronouns in Italian speaking children with DLD. *Clin. Linguist. Phon.* **2023**, 105404. [CrossRef]
51. Park, J.; Ritter, M.; Lombardino, L.J.; Wiseheart, R.; Sherman, S. Phonological awareness intervention for verbal working memory skills in school-age children with specific language impairment and concomitant word reading difficulties. *Int. J. Res. Stud. Lang. Learn.* **2013**, *3*, 3–22. [CrossRef]
52. Van Kleeck, A.; Gillam, R.B.; Hoffman, L.M. Training in phonological awareness generalizes to phonological working memory: A preliminary investigation. *J. Speech Lang. Pathol. Appl. Behav. Anal.* **2006**, *1*, 228–243. [CrossRef]
53. Swanson, L.A.; Fey, M.E.; Mills, C.E.; Hood, L.S. Use of Narrative-Based Language Intervention with Children Who Have Specific Language Impairment. *Am. J. Speech Lang. Pathol.* **2005**, *14*, 131–141. [CrossRef]
54. Shahmahmood Toktam, M.; Zahra, S.; AliPasha, M.; Ali, M.; Shahin, N. Cognitive and language intervention in primary language impairment: Studying the effectiveness of working memory training and direct language intervention on expansion of grammar and working memory capacities. *Child Lang. Teach. Ther.* **2018**, *34*, 235–268. [CrossRef]
55. Pauls, L. Exploring Associations between Language and Working Memory Abilities in Children with Specific or Combined Impairments in Language and Working Memory. Ph.D. Thesis, The University of Western Ontario, London, ON, Canada, 2018.
56. Georgas, J.; Paraskevopoulos, I.; Bezevegis, I.; Giannitsas, N. *WISC-III GR: Wechsler Intelligence Scale for Children*; Ellinika Grammata: Athens, Greece, 1997.
57. Padeliadu, S.; Antoniou, F. *Reading Test (Test-A)*; Program EPEAEK—Greek Ministry of Education: Athens, Greece, 2007.
58. Talli, I. Linguistic Abilities in Developmental Dyslexia and Specific Language Impairment: A Comparative and Cross-Linguistic Approach. Ph.D. Thesis, Paris Descartes University Aristotle University of Thessaloniki, Thessaloniki, Greece, 2010.

59. Talli, I.; Kotsoni, P.; Stavrakaki, S.; Sprenger-Charolles, L. Assessing phonological short-term memory in Greek: Reliability and validity of a non-word repetition test. *Front. Psychol.* **2023**, *13*, 904268. [CrossRef]
60. Sprenger-Charolles, L.; Colé, P.; Béchennec, D.; Kipffer-Piquard, A. French normative data on reading and related skills from EVALEC, a new computerized battery of tests (end Grade 1, Grade 2, Grade 3, and Grade 4). *Eur. Rev. Appl. Psychol.* **2005**, *55*, 157–186. [CrossRef]
61. Bachourou, T. Γνωστικές Και Γλωσσικές Μέθοδοι Κλινικής Παρέμβασης Στην Ειδική Γλωσσική Διαταραχή. Μια Συγκριτική Μελέτη Της Αποτελεσματικότητάς Τους. (Cognitive and Language Methods of Clinical Intervention in Special Language Impairment: A Comparative Study of Their Efficiency). Ph.D. Thesis, Aristotle University of Thessaloniki, Thessaloniki, Greece, 2021.
62. Delage, H.; Stanford, E.; Baratti, C.; Durrleman, S. Working memory training in children with developmental language disorder: Effects on complex syntax in narratives. *Front. Rehabil. Sci.* **2023**, *3*, 1068959. [CrossRef]
63. Bachourou, T.; Bachourou, A. Assisted Language Acquisition (AS.L.A.). Software. 2015. Available online: https://lab7.gr/asla.html (accessed on 1 June 2024).
64. Bryan, A. Colourful semantics: Thematic role therapy. In *Language Disorders in Children and Adults*; Chiat, S., Marshall, J., Eds.; Whurr: London, UK, 1997; pp. 143–161.
65. Spooner, L. Addressing expressive language disorder in children who also have severe receptive language disorder: A psycholinguistic approach. *Child Lang. Teach. Ther.* **2002**, *18*, 289–313. [CrossRef]
66. Bachourou, T.; Bachourou, A. BOOM! (Boosting Memory). Software. 2015. Available online: https://lab7.gr/boom.html (accessed on 1 June 2024).
67. Salis, C.; Kelly, H.; Code, C. Assessment and treatment of short-term and working memory impairments in stroke aphasia: A practical tutorial: Memory and aphasia. *Int. J. Lang. Commun. Disord.* **2015**, *50*, 721–736. [CrossRef]
68. Salis, C. Short-term memory treatment: Patterns of learning and generalisation to sentence comprehension in a person with aphasia. *Neuropsychol. Rehabil.* **2012**, *22*, 428–448. [CrossRef]
69. R Core Team. R: A Language and Environment for Statistical Computing. 2023. Available online: https://www.R-project.org/ (accessed on 1 June 2024).
70. Gillam, R.B.; Loeb, D.F.; Hoffman, L.M.; Bohman, T.; Champlin, C.A.; Thibodeau, L.; Widen, J.; Brandel, J.; Friel-Patti, S. The Efficacy of Fast ForWord Language Intervention in School-Age Children with Language Impairment: A Randomized Controlled Trial. *J. Speech Lang. Hear. Res.* **2008**, *51*, 97–119. [CrossRef]
71. Stevens, C.; Fanning, J.; Coch, D.; Sanders, L.; Neville, H. Neural mechanisms of selective auditory attention are enhanced by computerized training: Electrophysiological evidence from language-impaired and typically developing children. *Brain Res.* **2008**, *1205*, 55–69. [CrossRef]
72. Ebert, K.D.; Kohnert, K.; Pham, G.; Rentmeester Disher, J.; Payesteh, B. Three Treatments for Bilingual Children with Primary Language Impairment: Examining Cross-Linguistic and Cross-Domain Effects. *J. Speech Lang. Hear. Res.* **2014**, *57*, 172–186. [CrossRef]
73. Ebert, K.D.; Rentmeester-Disher, J.; Kohnert, K. Nonlinguistic cognitive treatment for bilingual children with primary language impairment. *Clin. Linguist. Phon.* **2012**, *26*, 485–501. [CrossRef]
74. Tyler, A.A.; Lewis, K.E.; Haskill, A.; Tolbert, L.C. Efficacy and cross-domain effects of a morphosyntax and a phonology intervention. *Lang. Speech Hear. Ser. Sch.* **2002**, *33*, 52–66. [CrossRef]

brain sciences

MDPI

Systematic Review

Efficacy of rTMS Combined with Cognitive and Language Training in People Living with Alzheimer's Disease: A Systematic Review

Eleni-Nefeli Georgopoulou [1], Anastasia Nousia [2], Maria Martzoukou [3], Nefeli K. Dimitriou [1], Ioannis Liampas [4], Lambros Messinis [3] and Grigorios Nasios [1,*]

1 Department of Speech and Language Therapy, School of Health Sciences, University of Ioannina, 45332 Ioannina, Greece; e.georgopoulou@uoi.gr (E.-N.G.); nefelikdimitriou@gmail.com (N.K.D.)
2 Department of Speech and Language Therapy, University of Peloponnese, 24100 Kalamata, Greece; sialogo@gmail.com
3 Laboratory of Neuropsychology and Behavioral Neuroscience, School of Psychology, Aristotle University of Thessaloniki, 54124 Thessaloniki, Greece; mmartzo@lit.auth.gr (M.M.); lmessinis@psy.auth.gr (L.M.)
4 Department of Neurology, Faculty of Medicine, School of Health Sciences, University of Thessaly, 41500 Larissa, Greece; liampasioannes@gmail.com
* Correspondence: grigoriosnasios@gmail.com

Abstract: Repetitive transcranial magnetic stimulation (rTMS) is a non-invasive brain stimulation method that has been suggested as a possible treatment method for cognitive impairment in patients with Alzheimer's Disease (pwAD), similar to multidomain cognitive training (CT). The effectiveness, however, of combining these techniques for pwAD remains controversial due to the variability in rTMS parameters, differences in CT protocol designs—many of which neglect the language domain—and the inclusion of patients at various stages of Alzheimer's Disease (AD) and with different types of dementia. The current review aims to evaluate the cognitive benefits of combining rTMS with CT, including language training, for individuals with mild to moderate AD. An extensive literature search was conducted in PubMed, Google Scholar, and the Cochrane Library with relevant terms, resulting in nine studies with a total of 290 participants [190 in the Active Group (AG) and 100 in the Control Group (CG)]. The comprehensive review of the articles revealed that the combined treatment improved global cognitive function, as well as neurocognitive, neuropsychiatric, and quality of life in the AG. Nevertheless, these results should be interpreted cautiously, given the relatively small number of existing studies on this specific combination.

Keywords: rTMS; cognitive and language training Alzheimer; Alzheimer; language rehabilitation; cognitive training and rTMS

1. Introduction

Alzheimer's disease (AD) is a leading neurodegenerative condition. It is linked to the buildup of abnormal proteins in the brain, including beta-amyloid plaques and tau tangles, which result in the loss of brain cells. AD is a form of dementia leading to cognitive deficits, personality and behavioral changes, and motor impairments [1]. AD is associated with a gradual decline in memory, cognitive abilities, attention, and executive functions, as well as deterioration in thinking and behavioral skills [2,3]. It also affects language, reasoning, social conduct, verbal and auditory naming, and the capacity to perform basic tasks, all of which are significantly impaired by the underlying neurodegenerative mechanisms [4]. Available medications, such as cholinesterase inhibitors (donepezil, rivastigmine, and galantamine) [5] and the glutamate receptor antagonist memantine [6], are used for symptomatic treatment. These treatments, however, have limited efficacy, do not prevent disease progression, and are not well-tolerated by all patients [7–9].

Therefore, it is critically important to expand the range of therapeutic options available for treating AD. One such option is cognitive training (CT) programs, which involve structured tasks/exercises presented either via computer or by paper and pencil. These programs range in difficulty level and target specific cognitive domains. They have shown promising results in AD, with positive effects on global and individual-domain cognitive outcomes [10]. Such interventions seem to be much more effective in a multidomain rather than a single-domain design base [11]. The results are still suboptimal, but they have a limited long-term beneficial impact. It is worth noting that CT programs in AD are often designed primarily to enhance memory, while language and other executive functions are often neglected, sometimes not trained at all, or not properly assessed [12]. Furthermore, despite numerous studies asserting the inclusion of language training in their CT programs, none of them offered detailed information regarding the specific language domains targeted or the content of the language tasks employed [10].

Among the non-pharmacological treatment modalities that have emerged in recent years is repetitive transcranial magnetic stimulation (rTMS). This treatment uses rhythmic electromagnetic induction (magnetic pulses) to non-invasively stimulate the brain, modulating cortical excitability and neural activity [13,14]. These pulses are generated by a coil placed on the scalp and can modulate neural activity in targeted brain regions. rTMS is considered a safe, low-cost, and effective treatment for patients with AD. Additionally, it is a promising alternative for the treatment of certain prevalent non-cognitive symptoms among patients with AD, including depression [15], as well as for the improvement of cognitive and motor functions in otherwise healthy older adults [16]. In addition, the combination of rTMS with neurophysiological techniques such as motor-evoked potential (MEP) allows for the reliable and convenient measurement of cortical plasticity [17]. Recent studies have reported that rTMS may specifically improve global cognitive function, language performance, and memory in patients with AD (pwAD), especially at the mild or early stage of the disease [18].

To date, there are four reviews and meta-analyses exploring the beneficial effects of rTMS on cognitive function in AD [19–23]. These articles, however, vary in terms of rTMS stimulation parameters (i.e., single site, multisite, etc.) and include both studies utilizing rTMS alone and studies combining rTMS with CT. They lack analytical information about the CT training programs (specific tasks, domains, etc.), and it appears that the CT programs used did not always include language in the training sessions. Furthermore, these reviews and analyses did not focus on specific clinical groups, as they analyzed studies examining patients with Mild Cognitive Impairment (MCI) as well as those with different stages of AD and other types of dementia.

Considering the well-established effects of rTMS and CT (with or without language included) on AD [24,25], we speculated that the combination of rTMS and CT but with language included might produce additional beneficial effects. Therefore, and due to the aforementioned literature gap, we conducted a systematic review carefully designed to summarize evidence from studies that implemented CT but specifically included language in the training, along with multisite rTMS for pwAD exclusively. Furthermore, analytical information regarding their CT programs is provided. In particular, the objectives of the present systematic review are as follows: the present systematic review aimed to (a) explore the effectiveness of combining high-frequency rTMS with CT, including language for patients with mild to moderate AD, and (b) determine which domains benefit most from the intervention.

2. Method

2.1. Search Strategy

A comprehensive literature search was performed using PubMed, Google Scholar, and the Cochrane Library. The following search terms were employed (entered as free text): (a) "rTMS cognitive and language training" and "Alzheimer"; (b) "Alzheimer" and "language rehabilitation" and "cognitive training" and "rTMS". The final literature search was performed on 30th January 2024. The initial search yielded a total of 1538 studies published up to January 2024. Specifically, 546 studies were retrieved from PubMed, 848 studies from Google Scholar, and 144 studies from Cochrane.

2.2. Eligibility Criteria

The following eligibility criteria were considered: (1) use of rTMS combined with CT including language, (2) inclusion of participants with AD, (3) availability of pre- and post-intervention cognitive data, and (4) inclusion of at least 8 participants in the rTMS combined with CT (including language) arm. Studies were excluded if they (1) lacked pre- or post-intervention cognitive data, (2) involved different kinds of interventions (e.g., physical training, CT not including language, rTMS without CT including language), (3) included participants with other neurological conditions (e.g., stroke, Parkinson's disease), or (4) were systematic review articles and meta-analyses. Conclusions were based on randomized controlled trials (RCTs). Further information regarding the procedure of the selection for eligible studies is presented in Figure 1 [26].

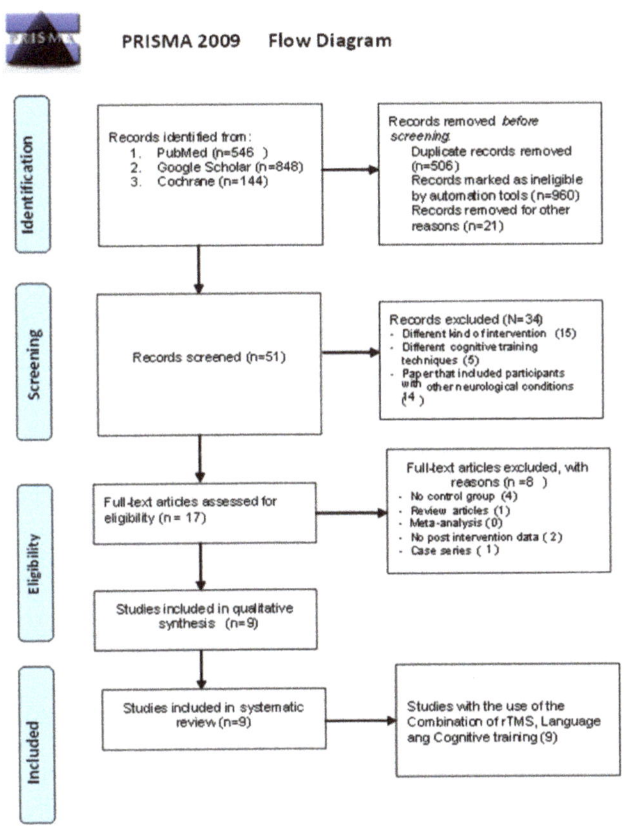

Figure 1. Caption [27–35].

2.3. Data Extraction

The data listed below were exported from the retrieved studies: author, year of publication, number of participants, AD stage, targeted domains for intervention, study design, outcome measures, intervention duration and frequency, and outcomes. The PRISMA guidelines for reporting systematic reviews were followed (Appendix A) [26]. The flow diagram illustrates the progression of information through the different phases of the systematic review, mapping the number of records identified, included, and excluded, as well as the reasons for exclusions [36]. Two independent reviewers (A.N. and N.G.) conducted the search of the literature, performed the quality evaluation, and independently extracted the data. Any potential discrepancies were resolved by a third author (I.L.).

2.4. Risk of Bias Assessment Tool

Risk of bias was evaluated using the risk of bias (RoB) tool from Cochrane for systematic reviews of interventions (Figure 2). Five methodological domains were assessed: (1) randomization process (including sequence generation, allocation concealment, and baseline differences between groups), (2) deviations from intended interventions (such as blinding of participants and personnel, and appropriateness of analysis), (3) missing outcome data (data availability and reasons for missing data), (4) measurement of outcomes (methods of measurement for both groups), and (5) selective reporting (prespecified protocols and multiple analyses). Each domain was categorized as "low risk of bias", "high risk of bias", or "some concerns" based on the methodological features and reporting of the reviewed studies.

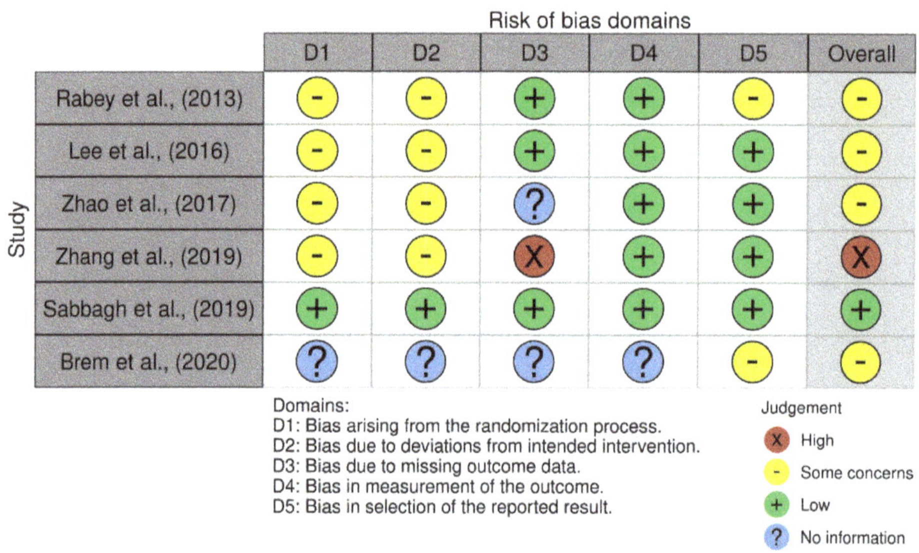

Figure 2. Caption [28,30–34].

3. Results

3.1. Study Characteristics

Study characteristics are included in Table 1. Regarding the study settings, three studies were conducted in Israel [27–29], one study in both Israel and the USA [30], one study in Italy and the USA [31], one study in North Korea [32], two studies in China [33,34], and one study in France [35]. Two of them were pilot studies [27,33], and one was a clinical trial [29]. Three studies were randomized, double-blinded, and placebo-controlled [30–33]. One was randomized and double-blinded without placebo control [28]; one was randomized, single-blind, and placebo-controlled [31]; and, finally, one was a non-randomized open-label study [34].

Table 1. General information of the retrieved studies.

Study	Country	NIBS	Study Design	Parallel Arm	Groups	Participants per Group
Bentwich et al. (2011) [27]	Israel	rTMS/Computer CT, including language	Pilot	-	AC	8
Rabey et al. (2013) [28]	Israel	rTMS/Computer CT, including language	Randomized, double-blind	YES	AG	7
					SG	8
Lee et al. (2016) [32]	North Korea	rTMS/Computer CT, including language	Randomized, double-blind, placebo-controlled	YES	AG	18
					SG	8
Rabey and Dobronevsky (2016) [29]	Israel	rTMS/ Computer CT, including language	Clinical trial	YES	AG	25
					AG2	5
Zhao et al. (2017) [34]	China	rTMS/ CT, including language	Prospective randomized, double-blind, placebo-controlled trial	YES	AG	17
					SG	13
Nguyen et al. (2017) [35]	France	rTMS/Computer CT, including language	Prospective open-label	-	AG	10
Sabbagh et al. (2019) [30]	USA + Israel	rTMS/Computer CT, including language	Prospective randomized, double-blind, sham-controlled	YES	AG	59
					SG	50
Zhang et al. (2019) [33]	China	rTms/ Computer CT, including language	Pilot	YES	AG	15
					SG	13
Brem et al. (2020) [31]	USA + Italy	rTMS/Computer CT, including language	Randomized, sham controlled	YES	AG1	16
					AG2	10
					SG (Sc + ST)	8

CT: Cognitive training; rTMS: repetitive transcranial magnetic stimulation; AG: active group; SG: sham group.

Six of the studies [28,30–34] included both an active group (AG) and a sham group (SG), whereas all participants in the other three studies [27,29,35] received active stimulation. In total, 290 participants were included, with 190 in the AG and 100 in the SG. The treatment group received rTMS sessions combined with computer-based CT, including language in all studies except for one [34], which did not specify whether the training was computer-based or paper-pencil.

3.2. Participants' Characteristics

Patients' characteristics are presented in Table 2. In total, 290 patients were included. More specifically, the AG consisted of 190 individuals, 100 males and 90 females, while the SG included 100 participants, 52 males and 48 females [27–35]. The mean age of the participants ranged from 67.5 to 76.9 years old; Rabey and Dobronevsky [29] did not provide information about the age of the participants. There were no significant differences in age and gender between the patient groups in the studies of Brem et al. [31], Lee et al. [32], Rabey et al. [28], Sabbagh et al. [30], Zhang et al. [33] and Zhao et al. [34], whereas Rabey and Dobronevsky [29] did not give specific information regarding the comparison in age and gender between groups. The participants in the other two studies [27,35] were not separated into groups.

Table 2. Demographics of study participants of the retrieved studies.

Study	N	AD Stage	AG1(r/r)		AG2(r/s)		SG(s/s)	
			Gender (Males/Total)	Mean Age in Years (Sd)/Range in y	Gender (Males/Total)	Mean Age in y (Sd)/Range in Years	Gender (Males/Total)	Mean Age in y (Sd)/Range in Years
Bentwich et al. (2011) [27]	8	Mild to moderate	7/8	75.4 (4.4)	--	--	--	--
Rabey et al. (2013) [28]	15	Mild to moderate	5/7	72.6 (8.9)	--	--	5/8	75.4 (9.07)
Lee et al. (2016) [32]	26	Mild	8 /18	72.1 (7.6)	--	--	3/ 8	70.3 (4.8)
Rabey and Dobronevsky (2016) [29]	30	Mild to moderate	17/30	N/A	N/A/5	N/A	--	--
Zhao et al. (2017) [34]	30	Mild to moderate	7/17	69.3 (5.8)	--	--	7/13	71.4 (5.2)
Nguyen et al. (2017) [35]	10	Mild	5/10	73 (7.2)	--	--	--	--
Sabbagh et al. (2019) [30]	109	Mild to moderate	41/59	76.9 (NA)	--	--	29/50	76.7 (NA)
Zhang et al. (2019) [33]	28	Mild	3/15	69 (8.19)	--	--	3/13	68.54 (7.93)
Brem et al. (2020) [31]	34	Mild to moderate	4/16	69.25 (6.80)	5/10	69.10 (5.24)	5/ 8	67.50 (10.27)

N: number of participants; AD: Alzheimer's Disease; AG: active group; SG: sham group; AG2: 2nd active group; Y: years; Sd: standard deviation; N/A: not available.

In terms of disease severity, all nine studies included patients who were classified as having mild to moderate disease [27–35]. Five studies used the Diagnostic and Statistical Manual of Mental Disorders, 4th Edition (DSM-IV) criteria [27,28,30,32,34] to evaluate patients' disease stage, and one study used the 5th edition's criteria (DSM-5) combined with the National Institute on Aging and Alzheimer's Association (NIA-AA) criteria [31].

One study fulfilled the National Institute of Neurological and Communicative Disorders and Stroke and the Alzheimer's Disease and Related Disorders Association (NINCDS-ADRDA) criteria [33], and two studies provided no further information about the diagnostic process [29,35]. Moreover, five of the studies [27–29,32,34] supported the diagnosis with magnetic resonance imaging (MRI), and one study [32] used MRI to exclude cerebrovascular disease.

As a measure of baseline cognitive function, seven of the studies presented participants' Mini Mental State Examination (MMSE) scores. Specifically, in five studies [27,28,30–32], the MMSE score ranged from 18 to 24; in one study [34], the score ranged from 18 to 26, and one study [29] did not provide the exact score. In addition, five studies used the Clinical Dementia Rate (CDR) scale; in two of these studies [27,28], the CDR score was 1, and in the other three studies [30,32,34], the CDR score was either 1 or 2. Moreover, Rabey and Dobronevsky [29] and Sabbagh et al. [30] used the Alzheimer Disease Assessment Scale—Cognitive (ADAS-Cog). Finally, Zhang et al. [33] conducted a baseline neuropsychological evaluation using CDR, MMSE, activities of daily living (ADL), and the Neuropsychiatric Inventory (NPI).

Three of the studies did not provide details about the concomitant pharmacological treatment administered to patients [27,29,35]. While Rabey et al. [28] noted that patients did not have serious metabolic or cardiac diseases, Zhao et al. [34], both with Lee et al. [32], reported no drug changes two months before and throughout the treatment duration, and Sabbagh et al. [30] and Zhang et al. [33] mentioned that if some patients were medicated for AD, they should have been taking the medication for at least three months at a stable dosage.

3.3. Intervention Characteristics

In all studies, participants underwent treatment programs with repetitive transcranial magnetic stimulation (rTMS) and CT, including language, which lasted for a period ranging between four weeks and three months, with session frequency varying from once per day to twice per week. Each session lasted from 30 to 60 min. The Neuro AD Neuronix protocol was applied in all studies except for two [33,34].

The majority of the studies stimulated six brain areas. Specifically, in most of the studies [27–32,35], the coil was placed on the Broca area (left frontal part of the temporal lobe), the Wernicke area (left posterior part of the temporal lobe), the right and left dorsolateral prefrontal cortex (R-dlPFC and L-dlPFC), and the right and left parietal somatosensory association cortex (R-pSAC + L-pSAC). In one study [34], the coil was placed on the parietal P3/P4 and the posterior temporal T5/T6 areas, and in another study [33], it was placed on the DLPFC and the lateral temporal lobe (LTL). Tasks were designed to train syntax, grammar, lexical meaning, categorization, action naming, object naming, spatial memory, and spatial attention. All tasks were associated with the stimulation of the corresponding brain areas. Analytical information is shown in Table 3.

Table 3. Summary of the procedural characteristics and main findings of the retrieved studies.

Study	N	Coil Placement	Diagnostic Criteria/Inclusion Criteria	Cognitive Domains Targeted	Stimulation Parameters	Number of Sessions	Outcome Measures	Results
Bentwich et al. (2011) [27]	8	Neuro AD Neuronix Broca (left frontal part of the temporal lobe) Wernicke (left posterior part of the temporal lobe) R-dlPFC + L-dlPFC (right and left prefrontal cortex) R-pSAC + L-pSAC (right and left parietal somatosensory association cortex)	DSM-IV criteria MMSE score18–24 CDR (1) Hebrew language MRI supporting probable AD Having a caregiver	Language functions Judgment Executive function Long-term memory Spatial and topographical orientation and praxis	45 min/session 3 brain areas stimulated/session 20 trains of rTMS (2 s of 10 Hz/train, 20 pulses/train.) Followed by 1–4 cog. tasks during a period of 20–40 s 20 repetitions = Each area stimulated with 400 pulses during a 7–15 min.	Int-rTMS-COG 1/day for 6 weeks = 30 sessions and maint-rTMS-COG 2/week for 3 months = 24 sessions 54 total	ADAS-cog CGIC MMSE ADAS-ADL HAMIL-TON NPI	ADAS-Cog (average) improved by approximately 4 points after both 6 weeks and 4.5 months of treatment and CGIC by 1.0 and 1.6 points, respectively. MMSE, ADAS-ADL, and HAMILTON improved, but without statistical significance. NPI did not change.
Rabey et al. (2013) [28]	15	Neuro AD Neuronix Broca (left frontal part of the temporal lobe) Wernicke (left posterior part of the temporal lobe) R-dlPFC + L-dlPFC (right and left prefrontal cortex) R-pSAC + L-pSAC	DSM-IV criteria MMSE score18–24 CDR (1) MRI indicating probable AD No serious metabolic or cardiac diseases Having a caregiver at least for 10 h/week Hebrew or Russian language	Language functions Judgment Executive function Long-term memory Spatial and topographical orientation and praxis	45–60 min/session 3 brain areas stimulated/session 20–25 trains of rTMS (2 s of 10 Hz/region, 20 pulses/train.) Followed by 1–4 cog. tasks during a period of 20–40 s 20 repetitions = Each area stimulated with 400–500 pulses during a 7–15 min.	Int-rTMS-COG 1/day for 6 weeks = 30 sessions and maint-rTMS-COG 2/week for 3 months = 24 sessions 54 total	ADAS-cog CGIC NPI	There was an improvement in the average ADAS-cog score of 3.76 points after 6 weeks in the treatment group compared to 0.47 in the placebo group and 3.52 points after 4.5 months of treatment, compared to a worsening of 0.38 in the placebo. There was also an improvement in the average CGIC score of 3.57 (after 6 weeks) and 3.67 points (after 4.5 months), compared to 4.25 and 4.29 in the placebo group (mild worsening). NPI improved non-significantly.

Table 3. *Cont.*

Study	N	Coil Placement	Diagnostic Criteria/Inclusion Criteria	Cognitive Domains Targeted	Stimulation Parameters	Number of Sessions	Outcome Measures	Results
Lee et al. (2016) [32]	26	Neuro AD protocol Neuronix Broca (left frontal part of the temporal lobe) Wernicke (left posterior part of the temporal lobe) R-dlPFC + L-dlPFC (right and left prefrontal cortex) R-pSAC + L-pSAC (right and left parietal somatosensory association cortex	DSM-IV criteria MMSE score18–24 CDR(1 or 2) Having a caregiver Read and write Korean MRI	Language functions Judgment Executive function Long-term memory Spatial and topographical orientation and praxis	60 min/session 3 brain areas stimulated/session 20 trains of rTMS (2 s of 10 Hz/train, 20 pulses/train.) Followed by 1–4 cog. tasks during a period of 20–40 s 20 repetitions = Each area stimulated with 400 pulses during a 7–15 min.	1/day for 6 weeks = 30 sessions	ADAS-cog CGIC MMSE GDS	The ADAS-cog score in the treatment group was significantly improved compared to the sham group (4.28 and 5.39 in the treatment group vs. 1.75 and 2.88 in the sham group immediately and 6 weeks after treatment, respectively). The MMSE and CGIC scores were also improved in the treatment group. Based on subgroup analysis, the effect of rTMS-COG was superior for the mild group compared to the total patients, especially in the domains of memory and language.
Rabey and Do-bronevsky, (2016) [29]	30	Neuro AD Neuronix Broca (left frontal part of the temporal lobe) Wernicke (left posterior part of the temporal lobe) R-dlPFC + L-dlPFC (right and left prefrontal cortex) R-pSAC + L-pSAC (right and left parietal somatosensory association cortex	Mild to moderate AD diagnosis ADAS-Cog MMSE MRI	Language functions Judgment Executive function Long-term memory Spatial and topographical orientation and praxis	60 min/session 3 brain areas stimulated/session 20 trains of rTMS (2 s of 10 Hz/train, 400 pulses/train) for 3/4 paradigms and 5 trains (2 s of 10 Hz/train, 100 pulses/train) for the 4th paradigm	1/day for 6 weeks = 30 sessions Second treatment course for 5 participants after 10.2 months	ADAS-Cog MMSE	The effect of rTMS-COG treatment was statistically significant regarding both ADAS-Cog (−2.4 point improvement) and MMSE (1.7 points improvement) scores.

Table 3. *Cont.*

Study	N	Coil Placement	Diagnostic Criteria/Inclusion Criteria	Cognitive Domains Targeted	Stimulation Parameters	Number of Sessions	Outcome Measures	Results
Zhao et al. (2017) [34]	30	Parietal P3/P4 Posterior temporal T5/T6	DSM-IV criteria MMSE 18–26 CDR 1 or 2 Caregiver for >72 h/week Capable of Reading and writing Chinese language MRI No drug change 2 months before and throughout the duration	Memory Language Executive function	1 h/session 3 brain areas each session 10 min of rTMS (10 s of 20 Hz/train, 20 s intermediate/train) followed by 2–4 cognitive tasks over the course of 20–40 s/brain area	1/day, 5days/week for 6 weeks = 30 sessions No maintenance intervention	ADAS-Cog MMSE MoCA WHO-UCLAAVLT	The ADAS-cog, MMSE, and WHO-UCLA AVLT score in the rTMS group was significantly improved compared with baselines at 6 weeks after treatment (all $p < 0.05$). Meanwhile, MoCA scores were also obviously ameliorated in the mild AD patients with rTMS. Additionally, subgroup analysis showed that the effect of rTMS on the memory and language of mild AD patients were superior to those of moderate AD patients.
Nguyen et al. (2017) [35]	10	Neuro AD Neuronix Broca (left frontal part of the temporal lobe) Wernicke (left posterior part of the temporal lobe) R-dlPFC + L-dlPFC (right and left prefrontal cortex) R-pSAC + L-pSAC (right and left parietal somatosensory association cortex	Probable AD diagnosis	Language functions Judgment Executive function Long-term memory Spatial and topographical orientation and praxis	30–45 min/session 3 brain areas stimulated/session 20 trains of rTMS (2 s of 10 Hz/ train, 20 pulses/train.) 20 repetitions = Each area stimulated with 400 pulses	1/day for 5 weeks= 25 sessions	ADAS-Cog MMSE Dubois FAB Stroop Locomotor Apathy score Zarit score Dependence score	The primary endpoint was the improvement of the ADAS-Cog score. Six months after the end of the treatment, the ADAS-Cog score returned to baseline value, except for the best responders, who remained significantly improved. The other main result was the improvement of apathy and dependence scores for the entire series of patients.

Table 3. *Cont.*

Study	N	Coil Placement	Diagnostic Criteria/Inclusion Criteria	Cognitive Domains Targeted	Stimulation Parameters	Number of Sessions	Outcome Measures	Results
Sabbagh et al. (2019) [30]	109	Neuro AD Neuronix Broca (left frontal part of the temporal lobe) Wernicke (left posterior part of the temporal lobe) R-dlPFC + L-dlPFC (right and left prefrontal cortex) R-pSAC + L-pSAC (right and left parietal somatosensory association cortex	DSM-IV criteria MMSE score18–24 CDR(1 or 2) Age 60–90 ADAS-Cog >17 Reliable caregiver No deficits in hearing or vision Good English or Hebrew >8 th grade education Stable dose >90 days (if medicated for AD)	Language functions Judgment Executive function Long-term memory Spatial and topographical orientation and praxis	60 min/session 3 brain areas stimulated/session trains of rTMS (2 s of 10 Hz/train, 20 pulses/train.) 20 repetitions = Each area stimulated with 400 pulses	1/day, 5 days/week for 6 weeks = 30 sessions	ADAS-Cog CGIC MRI	Subjects with baseline ADAS-Cog showed a statistically significant benefit favoring active over sham. Responder analysis showed 31.7% participants in the active group with ≤ -4 point improvement on ADAS-Cog versus 15.4% in the sham group.

Table 3. *Cont.*

Study	N	Coil Placement	Diagnostic Criteria/Inclusion Criteria	Cognitive Domains Targeted	Stimulation Parameters	Number of Sessions	Outcome Measures	Results
Zhang et al. (2019) [33]	28	DLPFC LTL	NINCDS-ADRDA Tests for: vitamin B12, folate, and thyroid hormone levels; quantitative hepatitis C antibody, quantitative human immunodeficiency virus antibody, and syphilis serology E epsilon 4 (APOEε4) genotype MRI to exclude cerebrovascular disease Before therapy, patients had taken the medicine for at least 3 months at a stable dosage. neuropsychological evaluation.	Memory Mathematical calculations Language Logic thinking	60 min/session 2 brain areas Repetition of 10 Hz for 5 s and intermittent for 25 s 20 trains = 1000 pulses Approximately 10 min/encephalic region	1/day, 5 days/week for 4 weeks 20 total No maintenance sessions	ADAS-Cog CDR MMSE ADL NPI ACE-III	After 4 weeks of rTMS-CT treatment, the patients' scores improved in the ADAS-cog, MMSE, ACE-III, in the subscales of word recall memory in the ADAS-cog, and in the domains of attention, memory, visual–spatial function in the ACE-III. The effects lasted for at least 4 weeks

Table 3. *Cont.*

Study	N	Coil Placement	Diagnostic Criteria/Inclusion Criteria	Cognitive Domains Targeted	Stimulation Parameters	Number of Sessions	Outcome Measures	Results
Brem et al. (2020) [31]	34	Neuro AD Neuronix Broca (left frontal part of the temporal lobe) Wernicke (left posterior part of the temporal lobe) R-dlPFC + L-dlPFC (right and left prefrontal cortex) R-pSAC + L-pSAC (right and left parietal somatosensory association cortex	DSM-V criteria NIA-AA Criteria MMSE score 18–24 Age 55–90 Normal/corrected ability to see and hear English or Italian as primary language	Language functions Judgment Executive function Long-term memory Spatial and topographical orientation and praxis	60 min/session 3 brain areas stimulated/session 20 trains of rTMS (2 s of 10 Hz/ region, 20 pulses/train.) Followed by 1–4 cog. tasks during a period of 20–40 s 15 repetitions/region	1/day for 6 weeks = 30 sessions	ADAS-cog GDS CDR ADAS-ADL MMSE CGIC	The real/real group showed significant cognitive improvement compared to the sham/sham, but not the real/sham group.

N: number of participants; MMSE: Mini Mental Stage Test; DSM-IV: Diagnostic and Statistical Manual of Mental Disorders, 4th Edition; DSM-V: Diagnostic and Statistical Manual of Mental Disorders, 5th Edition; rTMS: repetitive transcranial magnetic stimulation; CDR: Clinical Dementia Rating; ADAS-cog: Alzheimer Disease Assessment Scale—Cognitive; CGIC: Clinical Global Impression of Change scale; ADAS-ADL: Alzheimer Disease Assessment Scale—Activities of Daily Living; WHO-UCLAAVLT: World Health Organization university of California—Los Angeles, Auditory Verbal Learning Test; NIA- AA:National Institute on Aging—Alzheimer's Association criteria; COG: Cognitive Training; Dubois: the 5 word test; Hamilton Rating Scale for Depression; MoCA: Montreal Cognitive Assessment NPI: Neuropsychiatric Inventory test; ICF, intracortical facilitation; LICI, long-interval intracortical inhibition; rMT/aMT, resting/active motor threshold; SICI, short-interval intracortical inhibition; GDS: Geriatric Depression Scale; FAB: Frontal Assessment Battery; Dubois score; Stroop color test; R-pSAC: parietal somatosensory association cortex; L-pSAC: left parietal somatosensory association cortex; R-dlPFC: right prefrontal cortex; L-dlPFC: left prefrontal cortex); ACE-III: Addenbrooke's Cognitive Examination III; NINCDS-ADRDA: National Institute of Neurological and Communicative Disorders and Stroke—the Alzheimer's Disease and Related Disorders Association criteria.

3.4. Cognitive Including Language Outcomes

The targeted domains were language functions, judgment, executive function, long-term memory, spatial and topographical orientation, and praxis in almost all studies [27–32,35], except for the studies of Zhao et al. [34], who targeted memory, language and executive function, and Zhang et al. [33], who aimed at the domains of memory, mathematical calculations, language, and logical thinking.

The comprehensive review of the articles revealed the benefits of rTMS and CT, including language treatment, which are summarized in Table 4. Two studies [29,30] used only the Alzheimer Disease Assessment Scale—Cognitive (ADAS-cog) to evaluate the outcomes of the therapy sessions; both showed a statistically significant benefit favoring the active group over the sham group. Four studies [27,28,31,32] evaluated the language and cognitive profile before and after treatment with the ADAS-cog and the Mini Mental State Examination (MMSE). Specifically, Rabey and Dobronevsky [29] found statistically significant improvement in both tests after the combination of rTMS CT, including language treatment, whereas patients in the studies of Bentwich et al. [27] and Lee et al. [32] showed significant improvement only in ADAS-cog. In the latter research, however, the MMSE score improved significantly between baseline and six weeks after treatment in the mild AD group. Brem et al. [31] noticed that the group with active stimulation and real CT, including language treatment, showed significant improvement compared to the sham stimulation and sham CT/language group but not compared to the real CT/language and sham stimulation group.

The other three studies [33–35] assessed participants' outcomes with additional tests. In particular, Zhao et al. [34] found significantly improved performance in the ADAS-cog; the MMSE; and the World Health Organization University of California—Los Angeles, Auditory Verbal Learning Test (WHO-UCLA AVLT) score for the AG, with notably improved Montreal Cognitive Assessment (MoCA) scores in mild AD patients of the AG. Nguyen et al. [35] used the ADAS-cog, the MMSE, the Stroop color test, and Dubois' five-word testing (5WT), of which only the ADAS-cog indicated a significantly improved score. Finally, Zhang et al. [33] reported improved scores in the ADAS-cog (specifically in the word recall memory subscale), the MMSE, and Addenbrooke's Cognitive Examination III (ACE-III, in the domains of attention, memory, visual–spatial function), with these effects lasting for at least four weeks.

Table 4. Detailed results (neuropsychological measures, global cognitive function, and mood) from the studies.

Study	Outcome Measures	AG-PRE	AG-POST 6 Weeks			AG-POST 6 Weeks				
		Mean (Sd)	**Mean (Sd)**		***p*-value**	**Mean (Sd)**		***p*-value**		
Bentwich et al. (2011) [27]	**ADAS-Cog**	22.5 (5.6)	18.3 (6.8)		--	18.5 (3.8)		--		
	CGIC	3.0 (1.0)	2.4 (0.7)		--	N/A		--		
	MMSE	22.9 (1.7)	24.1 (2.2)		$p = 0.049$	22.1 (2.1)		--		
	ADAS-ADL	61.4 (14.6)	66.3 (15.1)		--	63.0 (17.8)		--		
	HAMILTON	9.7 (4.2)	5.6 (1.8)		$p = 0.053$	8.3 (2.7)		--		
	NPI	5.2 (7.8)	3.8 (8.9)		--	7.1 (10.0)		--		
		AG-PRE	**AG-POST 6 weeks**	**AG-POST 4.5 months**		**SG-PRE**	**SG-POST 6 weeks**	**SG-POST 4.5 months**		
		Mean (Sd)	**Mean (Sd)**	**Mean (Sd)**	***p*-value**	**Mean (Sd)**	**Mean (Sd)**	**Mean (Sd)**		***p*-value**
Rabey et al. (2013) [28]	**MMSE**	22(8.9)	--	--		22 (1.41)				
	ADAS-Cog	24.09	20.33	20.57	$p = 0.04$	24.09	23.19	24.47		$p = 0.05$
	GCIC		3.57	3.67	$p = 0.05$		4.25	4.29		$p = 0.05$
	NPI		+3.43				−1.38			
		Group (n)	**Pre (Mean, Sd)**		**Post (Mean, Sd)**	**6weeks after (Mean, Sd)**		***p*-value (Pre) vs. post**	***p*-value (Pre) vs. 6 week after**	**Time group**
Lee et al. (2016) [32]	**ADAS-Cog**	**Treatment**	23.61 (6.40)		19.33 (8.30)	18.22 (8.85)		0.0018	0.002	0.208
		Sham	22.88 (6.20)		21.12 (7.66)	20.00 (9.49)		0.238	0.46	
		Mild treatment	21.77 (5.09)		16.31 (6.40)	14.92 (7.43)		0.035	0.005	0.111
		Mild sham	20.83 (3.76)		18.17 (4.54)	17.33 (4.93)		0.396	0.056	
		Moderate treatment	28.4 (7.54)		27.2 (7.95)	26.8 (6.30)		1	0.264	0.966
		Moderate sham	29.0 (9.90)		30.0 (9.90)	28.0 (18.39)		NA	NA	
	MMSE	**Treatment**	22.39 (2.87)		23.89 (4.44)	24.39 (4.57)		0.139	0.058	0.729
		Sham	22.75 (2.49)		24.50 (4.90)	25.75 (4.56)		0.769	0.213	
		Mild treatment	23.77 (2.01)		25.62 (3.33)	26.46 (2.93)		0.058	0.015	0.785
		Mild sham	23.83 (1.71)		26.67 (2.16)	27.5 (2.51)		0.461	0.204	
		Moderate treatment	8.8 (0.84)		19.4 (3.98)	19.0 (3.54)		1	1	0.784
		Moderate sham	19.5 (0.71)		18.0 (5.66)	20.5 (6.36)		NA	NA	
		Treatment	10.50 (6.41)		7.89 (5.71)	7.50 (6.44)		0.635	0.452	0.77

Table 4. *Cont.*

Study	Outcome Measures	AG-PRE	AG-POST 6 Weeks		AG-POST 6 Weeks			
	GDS	Sham	11.63 (8.93)	9.38 (6.99)	8.00 (6.97)	0.656	0.020	
		Mild treatment	11.62 (4.82)	9.92 (5.25)	10.00 (5.85)	1	1	0.754
		Mild sham	11.83 (8.45)	9.17 (6.97)	7.50 (6.16)	1	0.148	
		Moderate treatment	7.60 (8.71)	2.60 (2.70)	1.00 (0.71)	1	1	0.484
		Moderate sham	11.00 (14.14)	10.00 (9.90)	9.50 (12.02)	NA	NA	
			Changes in scores mean (SD)		*p*-value			
	Group	Cognitive domains	Δ immediate after treatment	Δ 6 weeks after	Baseline vs. immediate after	Baseline vs. 6 weeks after	time	
Lee et al. (2016) [32]	Treatment	Memory	2.00 (2.81)	2.56 (3.05)	0.040	0.006	0.054	
		Language	1.28 (1.53)	1.44 (1.92)	0.004	0.003	0.009	
		Executive function	0.67 (1.24)	0.56 (1.50)	0.117	0.413	0.995	
	Mild treatment	Memory	1.77 (3.09)	2.69 (3.71)	0.275	0.031	0.178	
		Language	1.23 (1.79)	1.23 (1.96)	0.04	0.019	0.002	
		Executive function	0.92 (1.12)	0.92 (1.50)	0.031	0.093	0.2	
	Moderate Treatment	Memory	2.60 (2.07)	2.20 (3.03)	0.627	0.688	0.397	
		Language	1.40 (0.55)	2.00 (1.87)	0.243	1.000	0.947	
		Executive function	0.00 (1.41)	0.40 (1.14)	1.000	0.969	0.084	

Study	Parameter/measurement		Mean	SE	Min	Median	Max	N	*p*-value
Rabey and Dobronevsky, (2016) [29]	ADAS-Cog	Pre-treatment	20.5	1.3	10.6	18.8	37.3	30	
		Post-treatment	18.1	1.4	8.3	16.2	33.6	30	
		Difference	-2.4	0.6	--	--	--	30	<0.001
	MMSE-Cog	Pre-treatment	22.2	0.5	16.8	2.9	27.0	30	
		Post-treatment	23.9	0.5	18.0	24.5	28.0	30	
		Difference	1.7	0.3	--	--	--	30	<0.001

Table 4. *Cont.*

Study	Outcome Measures	AG-PRE	AG-POST 6 Weeks			AG-POST 6 Weeks		
	Neuropsychological assessments	Group (N)	Mean scores (SD)			*p*-value		
			Baseline [B]	Immediate after treatment [6]	6 weeks after treatment [12]	[B] vs. [6]	[B] vs. [12]	Time * group
Zhao et al. (2017) [34]	ADAS-cog	Treatment (17)	22.6 (5.9)	18.5 (5.4)	16.8 (6.9)	0.042 *	0.013 *	0.332
		Sham (13)	24.2 (6.1)	22.9 (8.9)	21.2 (8.6)	0.668	0.315	
		Mild treat. (12)	20.6 (5.2)	16.4 (4.4)	14.2 (6.8)	0.044 *	0.017 *	0.263
		Mild sham (8)	21.7 (4.6)	20.3 (5.6)	19.4 (6.8)	0.593	0.441	
		Moderate treat. (5)	23.8 (6.6)	20.3 (6.5)	18.9 (6.7)	0.423	0.278	0.878
		Moder sham (5)	27.5 (5.9)	24.2 (8.6)	23.5 (8.5)	0.499	0.413	
	MMSE	Treatment (17)	22.2 (2.8)	23.9 (2.5)	25.5 (4.6)	0.071	0.017 *	0.557
		Sham (13)	22.8 (2.3)	23.1 (3.3)	24.2 (4.1)	0.790	0.294	
		Mild treat. (12)	25.6 (2.1)	27.1 (4.1)	29.7 (4.5)	0.147	0.042	0.639
		Mild sham (8)	25.8 (2.3)	26.2 (3.5)	28.1 (3.5)	0.791	0.395	
		Moderate treat. (5)	19.2 (2.5)	20.4 (3.3)	21.7 (4.3)	0.535	0.294	0.812
		Moderate sham.(5)	19.5 (1.9)	20.4 (2.5)	21.5 (2.1)	0.539	0.153	
	MoCA	Treatment (17)	17.5 (6.2)	19.8 (6.5)	21.5 (5.9)	0.299	0.063	0.552
		Sham (13)	18.1 (7.3)	19.3 (6.7)	20.1 (6.6)	0.666	0.471	
		Mild treat. (12)	18.6 (5.1)	21.1 (4.3)	23.1 (5.3)	0.208	0.046 *	0.799
		Mild sham (8)	19.7 (7.5)	20.9 (7.1)	21.4 (7.8)	0.747	0.664	
		Moderate treat. (5)	16.6 (6.2)	17.9 (5.8)	19.2 (5.5)	0.741	0.503	0.517
		Moder sham (5)	17.5 (6.8)	18.4 (5.9)	19.5 (6.6)	0.829	0.650	
	WHO-UCLA AVLT	Treatment (17)	32.5 (7.9)	35.8 (7.8)	38.7 (8.9)	0.229	0.039 *	0.667
		Sham (13)	34.1 (8.1)	35.8 (7.4)	37.8 (8.7)	0.582	0.273	
		Mild treat.(12)	35.6 (5.6)	37.9 (6.5)	41.8 (6.6)	0.363	0.021 *	0.524
		Mild sham (8)	35.8 (6.7)	36.6 (6.7)	38.7 (4.5)	0.815	0.327	
		Moderate treat.(5)	30.5 (7.6)	33.5 (2.3)	35.6 (6.3)	0.423	0.281	0.550
		Moder sham (5)	30.6 (6.7)	33.9 (5.4))	36.5 (4.9)	0.416	0.151	
	Group	Cognitive domains	Changes in score, mean (SD)			*p*-value		
			Δ immediate after treatment		Δ 6 weeks after treatment	[B] vs. [6]	[B] vs. [12]	Time * group
	Treatment	Memory	2.04 (2.25)		2.62 (3.66)	0.002 *	0.009 *	0.575
		Language	1.17 (1.36)		1.39 (1.67)	0.003 *	0.003 *	0.676

Table 4. *Cont.*

Study	Outcome Measures	AG-PRE	AG-POST 6 Weeks		AG-POST 6 Weeks		
Zhao et al. (2017) [34]	Mild treatment	Executive function	0.86 (1.89)	0.49 (1.67)	0.079	0.244	0.549
		Memory	2.56 (2.33)	2.88 (2.87)	0.003 *	0.005 *	0.767
		Language	1.36 (1.56)	1.54 (1.67)	0.012 *	0.009 *	0.788
	Moderate treatment	Executive function	0.92 (1.12)	0.92 (1.50)	0.016 *	0.058	1.000
		Memory	1.56 (2.32)	2.41 (3.12)	0.207	0.159	0.638
		Language	0.98 (0.89)	1.15 (1.56)	0.070	0.175	0.838
		Executive function	0.74 (0.99)	0.00 (1.11)	0.170	1.000	0.298
Nguyen et al. (2017) [35]			Baseline	Post	Follow-up 6 months		*p*
		ADAS-Cog	20.1 (3.1)	17.2 (2.5)	20.1 (3.4)		0.0165
		ADAS-Cog word recognition	17.8 (1.4)	18.5 (1.7)	19.0 (1.3)		0.1738
		ADAS-Cog word recall	2.7 (0.5)	2.8 (0.5)	2.7 (0.5)		0.9592
		MMSE	18.8 (1.9)	19.7 (1.4)	17.8 (1.5)		0.1168
		MMSE language	6.4 (0.4)	6.6 (0.3)	6.6 (0.4)		0.5945
		Dubois	5.0 (0.8)	4.3 (0.85)	4.5 (1.1)		0.2466
		FAB	11.5 (1.3)	11.7 (1.4)	11.4 (1.5)		0.9155
		Stroop color test	39.3 (7.3)	36.3 (6.1)	40.6 (8.2)		0.8357
		Locomotor	26.2 (0.6)	27.0 (0.4)	27.0 (0.5)		0.0478
		Apathy	17.4 (2.7)	10.8 (2)	9.4 (1.8)		0.0125
		Caregiver (Zarit)	4.1 (0.4)	3.6 (0.5)	3.7 (0.4)		0.2385
		Dependence	48.4 (5.5)	36.8 (5)	34.7 (4.4)		0.0085
Sabbagh et al. (2019) [30]		Baseline		Post (week 7)		Post (week 12)	
		AG	SG	AG	SG	AG	SG
	ADAS	23.6	24.4	−0.7	−0.7	−2.11	24.4
	MMSE	21.7	21.3	N/A	N/A	N/A	N/A
	Neuropsychological assessment		Sham rTMS-CT (Mean ± SE)	Real rTMS-CT (Mean ± SE)	F value		*p*-value

Table 4. *Cont.*

Study	Outcome Measures	AG-PRE	AG-POST 6 Weeks	AG-POST 6 Weeks	
			MMSE		
	[4]-[B]	0.75 ± 0.49	3.27 ± 0.56	17.411	<0.01 *
	[8]-[B]	0.80 ± 0.81	2.571 ± 0.45	5.894	0.024 *
			ACE-III		
	[4]-[B]	2.18 ± 1.43	11.77 ± 1.32	25.817	<0.01 *
	[8]-[B]	0.69 ± 1.63	7.70 ± 1.51	10.692	<0.01 *
			Attention		
	[4]-[B]	0.07 ± 0.49	2.19 ± 0.44	19.308	<0.01 *
	[8]-[B]	−0.14 ± 0.53	1.88 ± 0.40	22.590	<0.01 *
			Memory		
Zhang et al. (2019) [33]	[4]-[B]	0.29 ± 1.07	3.87 ± 0.82	7.798	0.009 *
	[8]-[B]	−0.36 ± 1.22	1.87 ± 0.63	4.021	0.055
			Verbal fluency		
	[4]-[B]	0.21 ± 0.42	0.94 ± 0.42	2.951	0.097
	[8]-[B]	0.14 ± 0.43	0.69 ± 0.36	3.208	0.084
			Language		
	[4]-[B]	1.00 ± 0.92	2.31 ± 0.89	3.310	0.080
	[8]-[B]	1.71 ± 0.79	2.25 ± 0.44	2.697	0.113
			Visuospatial function		
	[4]-[B]	0.79 ± 0.65	1.56 ± 0.45	5.472	0.027 *
	[8]-[B]	0.57 ± 0.61	1.56 ± 0.43	7.416	0.011 *
		PRE	23.00 (2.48)	25.10 (3.06)	23.61 (3.99)
		POST	20.79 (2.81)	24.7 (3.23)	24.11 (3.31)

AG: active group; SG: sham group; Sd: standard deviation; N/A: not available; N: number of participants; MMSE: Mini Mental Stage Test; ADAS-cog: Alzheimer Disease Assessment Scale—Cognitive; CGIC: Clinical Global Impression of Change scale; ADAS-ADL: Alzheimer Disease Assessment Scale—Activities of Daily Living; Hamilton Rating Scale for Depression; NPI: Neuropsychiatric Inventory test; ICF, intracortical facilitation; FAB: Frontal Assessment Battery; Dubois score; WHO-UCLAAVLT: World Health Organization University of California—Los Angeles, Auditory Verbal Learning Test; ACE-III: Addenbrooke's Cognitive Examination III.

3.5. Neurocognitive, Neuropsychiatric, and Quality of Life Outcomes

Apart from language and cognitive domains, seven studies [27,28,30–33,35] also evaluated non-cognitive factors. Three of these studies [27,31,33] assessed the impact of the treatment on patients' quality of life. For this purpose, Bentwich et al. [27] and Brem et al. [31] used the Alzheimer Disease Assessment Scale—Activities of Daily Living (ADAS-ADL) with positive outcomes, and Zhang et al. [33] demonstrated a tendency toward improvement in the Activities of Daily Living (ADL). Depression was measured either with the Hamilton Rating Scale for Depression [27] or with the Geriatric Depression Scale (GDS) [31,32]. The study by Bentwich et al. [27] showed an improved score but without statistical significance, whereas the study by Lee et al. [32] found that the GDS score did not improve significantly in the treatment group. Brem et al. [31] evaluated depression only at baseline.

Five of the studies [27,28,30–32] used the Clinical Global Impression of Change scale (CGIC), all of which indicated better results. In addition, three studies assessed the presence and severity of neuropsychiatric symptoms using the Neuropsychiatric Inventory test (NPI); two of them [28,33] found an improvement after treatment, while the other one [27] noted that the NPI scores did not change after treatment. Moreover, two studies [31,33] assessed patients using the Clinical Dementia Rating (CDR) at the baseline phase. Lastly, Nguyen et al. [35] utilized the Frontal Assessment Battery (FAB), locomotor, apathy score, Zarit score, and dependence score, showing an improvement in apathy and dependence scores. His section may be divided by subheadings. It should provide a concise and precise description of the experimental results, their interpretation, and the experimental conclusions that can be drawn.

4. Discussion

To the best of our knowledge, the present review is the first to focus on the impact of combining rTMS with CT, including language, for pwAD, on their overall language and cognitive profile. The critical examination of the retrieved studies revealed that this combinatory treatment led to language and cognitive improvement as assessed with the ADAS-Cog, WHO-UCLAAVLT, MMSE, MoCA, and ACE-III. In contrast, using rTMS or CT, including language alone, did not result in notable improvements.

Our findings are consistent with those of Cheng et al. [37], who concluded in their meta-analysis that combining rTMS with CT enhances cognitive abilities in Alzheimer's disease more effectively than rTMS alone. It should be noted, however, that their analysis included studies where CT did not necessarily train language along with other cognitive domains. Furthermore, Lin et al. [19], in their review, did not reveal significantly improved results when combining rTMS with CT but observed that the complementary use of CT yielded greater overall enhancement than either treatment alone. Similar findings were observed in the reviews and meta-analyses by Xie et al. [20], Wei et al. [21], and Liu [23]. In contrast to our findings, Yan et al. [38] found in their systematic review and meta-analysis that the additional application of CT did not show significant improvement in the cognitive status of the participants. Their sample, however, included both Mild Cognitive Impairment (MCI) participants and individuals with different stages of pwAD.

Specifically, beneficial outcomes were observed for language and overall cognitive profile [28,31–35], as well as for memory and overall cognitive profile [27,32–34], and for quality of life [27,31,33]. Furthermore, outcomes regarding neuropsychiatric symptoms and quality of life were conflicting and lacked statistical significance. Specifically, while quality of life outcomes showed a tendency towards improvement [27,31,34], improvements in neuropsychiatric symptoms did not reach significance [27,28,35]. It is worth noting that all studies reported beneficial effects for their active groups on overall cognitive profile [27–35].

The studies that provided stronger evidence were primarily those offering outcome data for both language and executive functions [28,31–35]. These results, however, should be interpreted cautiously due to several factors. Firstly, participant nationality varied across the studies, leading to potential differences in language tasks. Additionally, sample sizes varied, further influencing the robustness of the findings. Moreover, there remains an insufficient number of studies focusing on pwAD and the impact of combining rTMS with CT, including language intervention, to draw definite conclusions. Another significant limitation is the lack of comprehensive analytical evaluation beyond the language sections of assessment tools such as ADAS-Cog and ACE-III, both before and after interventions, which hampers the ability to assess the specific impact on the language domain distinct from other cognitive and executive functions.

It has been suggested that combining rTMS with CT merges "exogenous" and "endogenous" stimulation, thus enhancing neuroplasticity. rTMS may pre-activate the neural system, while subsequent CT interacts with the ongoing brain activation to amplify or extend the neural effects [39–41]. This synergistic interaction may allow CT to modulate the impact of rTMS, potentially explaining why studies involving both rTMS and CT predominantly yield more consistent results than those without cognitive training. Multidomain or holistic CT, including language, would be the optimal approach. This review and the data derived from the aggregation of the results of the included studies demonstrate originality, as it is the first time the effectiveness of rTMS combined with CT and language training has been investigated. Additionally, it contributes to the current literature and holds significant clinical relevance, as the role of language (and its training) in AD seems to be critically important and should always be taken into consideration.

Furthermore, our review fully supports the relevance of non-pharmacological multidomain interventions for pwAD and provides the groundwork for future studies to build upon. When it comes to the effectiveness of these interventions, there is growing evidence (supported even by the present review) that a multidomain design, which targets a variety of domains and brain regions, seems to be more efficient and effective [11,42–45]. Many questions, however, remain to be answered, including how language and other cognitive improvements would translate into improved patients' quality of life and whether neurophysiological measures could act as potential biomarkers for designing interventions in AD. Future research should assess in greater detail the potential that language rehabilitation might play a more important role in the improvement of the overall everyday living profile of patients and their caregivers/families.

Author Contributions: Conceptualization, E.-N.G. and A.N.; methodology, E.-N.G., A.N. and N.K.D.; data analysis I.L. and M.M.; writing—original draft preparation, E.-N.G. and A.N.; writing—review and editing, M.M. and L.M.; supervision, G.N. All authors have read and agreed to the published version of the manuscript.

Funding: This research received no external funding.

Institutional Review Board Statement: Not applicable.

Informed Consent Statement: Not applicable.

Data Availability Statement: No new data were created.

Conflicts of Interest: The authors declare no conflicts of interest.

Appendix A

Section and Topic	Item #	Checklist Item	Location Where Item Is Reported
TITLE			
Title	1	Identify the report as a systematic review.	1
ABSTRACT			
Abstract	2	See the PRISMA 2020 for Abstracts checklist.	1
INTRODUCTION			
Rationale	3	Describe the rationale for the review in the context of existing knowledge.	1, 2
Objectives	4	Provide an explicit statement of the objective(s) or question(s) the review addresses.	1, 2
METHODS			
Eligibility criteria	5	Specify the inclusion and exclusion criteria for the review and how studies were grouped for the syntheses.	2
Information sources	6	Specify all databases, registers, websites, organizations, reference lists, and other sources searched or consulted to identify studies. Specify the date when each source was last searched or consulted.	3
Search strategy	7	Present the full search strategies for all databases, registers, and websites, including any filters and limits used.	3
Selection process	8	Specify the methods used to decide whether a study met the inclusion criteria of the review, including how many reviewers screened each record and each report retrieved, whether they worked independently, and, if applicable, details of automation tools used in the process.	4
Data collection process	9	Specify the methods used to collect data from reports, including how many reviewers collected data from each report, whether they worked independently, any processes for obtaining or confirming data from study investigators, and, if applicable, details of automation tools used in the process.	4
Data items	10a	List and define all outcomes for which data were sought. Specify whether all results that were compatible with each outcome domain in each study were sought (e.g., for all measures, time points, analyses), and if not, the methods used to decide which results to collect.	NA
	10b	List and define all other variables for which data were sought (e.g., participant and intervention characteristics, funding sources). Describe any assumptions made about any missing or unclear information.	NA
Study risk of bias assessment	11	Specify the methods used to assess the risk of bias in the included studies, including details of the tool(s) used, how many reviewers assessed each study and whether they worked independently, and if applicable, details of automation tools used in the process.	4, 5
Effect measures	12	Specify for each outcome the effect measure(s) (e.g., risk ratio, mean difference) used in the synthesis or presentation of results.	NA

Section and Topic	Item #	Checklist Item	Location Where Item Is Reported
Synthesis methods	13a	Describe the processes used to decide which studies were eligible for each synthesis (e.g., tabulating the study intervention characteristics and comparing against the planned groups for each synthesis (item #5)).	2
	13b	Describe any methods required to prepare the data for presentation or synthesis, such as handling of missing summary statistics or data conversions.	NA
	13c	Describe any methods used to tabulate or visually display the results of individual studies and syntheses.	
	13d	Describe any methods used to synthesize results and provide a rationale for the choice(s). If meta-analysis was performed, describe the model(s), method(s) to identify the presence and extent of statistical heterogeneity, and software package(s) used.	NA
	13e	Describe any methods used to explore possible causes of heterogeneity among study results (e.g., subgroup analysis, meta-regression).	NA
	13f	Describe any sensitivity analyses conducted to assess the robustness of the synthesized results.	NA
Reporting bias assessment	14	Describe any methods used to assess the risk of bias due to missing results in a synthesis (arising from reporting biases).	NA
Certainty assessment	15	Describe any methods used to assess certainty (or confidence) in the body of evidence for an outcome.	NA
RESULTS			
Study selection	16a	Describe the results of the search and selection process, from the number of records identified in the search to the number of studies included in the review, ideally using a flow diagram.	NA
	16b	Cite studies that might appear to meet the inclusion criteria but which were excluded, and explain why they were excluded.	NA
Study characteristics	17	Cite each included study and present its characteristics.	9
Risk of bias in studies	18	Present assessments of risk of bias for each included study.	NA
Results of individual studies	19	For all outcomes, present, for each study, (a) summary statistics for each group (where appropriate) and (b) an effect estimate and its precision (e.g., confidence/credible interval), ideally using structured tables or plots.	17–25
Results of syntheses	20a	For each synthesis, briefly summarise the characteristics and risk of bias among contributing studies.	NA
	20b	Present results of all statistical syntheses conducted. If meta-analysis was performed, present for each the summary estimate and its precision (e.g., confidence/credible interval) and measures of statistical heterogeneity. If comparing groups, describe the direction of the effect.	NA
	20c	Present results of all investigations of possible causes of heterogeneity among study results.	NA
	20d	Present results of all sensitivity analyses conducted to assess the robustness of the synthesized results.	NA
Reporting biases	21	Present assessments of risk of bias due to missing results (arising from reporting biases) for each synthesis assessed.	NA
Certainty of evidence	22	Present assessments of certainty (or confidence) in the body of evidence for each outcome assessed.	NA

Section and Topic	Item #	Checklist Item	Location Where Item Is Reported
DISCUSSION			
Discussion	23a	Provide a general interpretation of the results in the context of other evidence.	26, 27
	23b	Discuss any limitations of the evidence included in the review.	26, 27
	23c	Discuss any limitations of the review processes used.	27
	23d	Discuss the implications of the results for practice, policy, and future research.	27
OTHER INFORMATION			
Registration and protocol	24a	Provide registration information for the review, including the register name and registration number, or state that the review was not registered.	NA
	24b	Indicate where the review protocol can be accessed or state that a protocol was not prepared.	NA
	24c	Describe and explain any amendments to information provided at registration or in the protocol.	NA
Support	25	Describe sources of financial or non-financial support for the review and the role of the funders or sponsors in the review.	27
Competing interests	26	Declare any competing interests of review authors.	27
Availability of data, code, and other materials	27	Report which of the following are publicly available and where they can be found: template data collection forms; data extracted from included studies; data used for all analyses; analytic code; any other materials used in the review.	27

References

1. Andrade-Guerrero, J.; Santiago-Balmaseda, A.; Jeronimo-Aguilar, P.; Vargas-Rodríguez, I.; Cadena-Suárez, A.R.; Sánchez-Garibay, C.; Soto-Rojas, L.O. Alzheimer's Disease: An Updated Overview of Its Genetics. *Int. J. Mol. Sci.* **2023**, *24*, 3754. [CrossRef] [PubMed]
2. Cappa, S.F.; Binetti, G.; Pezzini, A.; Padovani, A.; Rozzini, L.; Trabucchi, M. Object and Action Naming in Alzheimer's Disease and Frontotemporal Dementia. *Neurology* **1998**, *50*, 351–355. [CrossRef]
3. Almor, A.; Aronoff, J.M.; MacDonald, M.C.; Gonnerman, L.M.; Kempler, D.; Hintiryan, H.; Hayes, U.L.; Arunachalam, S.; Andersen, E.S. A Common Mechanism in Verb and Noun Naming Deficits in Alzheimer's Patients. *Brain Lang.* **2009**, *111*, 8–19. [CrossRef] [PubMed]
4. De Almeida, R.G.; Mobayyen, F.; Antal, C.; Kehayia, E.; Nair, V.P.; Schwartz, G. Category-Specific Verb-Semantic Deficits in Alzheimer's Disease: Evidence from Static and Dynamic Action Naming. *Cogn. Neuropsychol.* **2021**, *38*, 1–26. [CrossRef] [PubMed]
5. Birks, J.; Craig, D. Galantamine for Vascular Cognitive Impairment. *Cochrane Database Syst. Rev.* **2006**, *2006*, CD004746. [CrossRef] [PubMed]
6. McShane, R.; Westby, M.J.; Roberts, E.; Minakaran, N.; Schneider, L.; Farrimond, L.E.; Maayan, N.; Ware, J.; Debarros, J. Memantine for Dementia. *Cochrane Database Syst. Rev.* **2019**, *3*, CD003154. [CrossRef] [PubMed]
7. Alzheimer's Association National Plan Milestone Workgroup; Fargo, K.N.; Aisen, P.; Albert, M.; Au, R.; Corrada, M.M.; DeKosky, S.; Drachman, D.; Fillit, H.; Gitlin, L.; et al. 2014 Report on the Milestones for the US National Plan to Address Alzheimer's Disease. *Alzheimers Dement.* **2014**, *10*, S430–S452. [CrossRef] [PubMed]
8. Lane, C.A.; Hardy, J.; Schott, J.M. Alzheimer's Disease. *Eur. J. Neurol.* **2018**, *25*, 59–70. [CrossRef]
9. Anand, A.; Patience, A.A.; Sharma, N.; Khurana, N. The Present and Future of Pharmacotherapy of Alzheimer's Disease: A Comprehensive Review. *Eur. J. Pharmacol.* **2017**, *815*, 364–375. [CrossRef]
10. Bahar-Fuchs, A.; Martyr, A.; Goh, A.M.; Sabates, J.; Clare, L. Cognitive Training for People with Mild to Moderate Dementia. *Cochrane Database Syst. Rev.* **2019**, *3*, CD013069. [CrossRef]
11. Nousia, A.; Siokas, V.; Aretouli, E.; Messinis, L.; Aloizou, A.-M.; Martzoukou, M.; Karala, M.; Koumpoulis, C.; Nasios, G.; Dardiotis, E. Beneficial Effect of Multidomain Cognitive Training on the Neuropsychological Performance of Patients with Early-Stage Alzheimer's Disease. *Neural Plast.* **2018**, *2018*, 2845176. [CrossRef] [PubMed]
12. Kallio, E.-L.; Öhman, H.; Kautiainen, H.; Hietanen, M.; Pitkälä, K. Cognitive Training Interventions for Patients with Alzheimer's Disease: A Systematic Review. *J. Alzheimers Dis.* **2017**, *56*, 1349–1372. [CrossRef]
13. Rossi, S.; Hallett, M.; Rossini, P.M.; Pascual-Leone, A. Safety, Ethical Considerations, and Application Guidelines for the Use of Transcranial Magnetic Stimulation in Clinical Practice and Research. *Clin. Neurophysiol.* **2009**, *120*, 2008–2039. [CrossRef] [PubMed]
14. Valero-Cabré, A.; Amengual, J.L.; Stengel, C.; Pascual-Leone, A.; Coubard, O.A. Transcranial Magnetic Stimulation in Basic and Clinical Neuroscience: A Comprehensive Review of Fundamental Principles and Novel Insights. *Neurosci. Biobehav. Rev.* **2017**, *83*, 381–404. [CrossRef] [PubMed]
15. Health Quality Ontario. Repetitive Transcranial Magnetic Stimulation for Treatment-Resistant Depression: A Systematic Review and Meta-Analysis of Randomized Controlled Trials. *Ont. Health Technol. Assess. Ser.* **2016**, *16*, 1–66.

16. Zimerman, M.; Hummel, F.C. Non-Invasive Brain Stimulation: Enhancing Motor and Cognitive Functions In Healthy Old Subjects. *Front. Aging Neurosci.* **2010**, *2*, 149. [CrossRef]
17. Lefaucheur, J.-P. Transcranial Magnetic Stimulation. In *Handbook of Clinical Neurology*; Elsevier: Amsterdam, The Netherlands, 2019; Volume 160, pp. 559–580. [CrossRef]
18. Lefaucheur, J.-P.; Aleman, A.; Baeken, C.; Benninger, D.H.; Brunelin, J.; Di Lazzaro, V.; Filipović, S.R.; Grefkes, C.; Hasan, A.; Hummel, F.C.; et al. Evidence-Based Guidelines on the Therapeutic Use of Repetitive Transcranial Magnetic Stimulation (rTMS): An Update (2014–2018). *Clin. Neurophysiol.* **2020**, *131*, 474–528. [CrossRef]
19. Lin, Y.; Jiang, W.-J.; Shan, P.-Y.; Lu, M.; Wang, T.; Li, R.-H.; Zhang, N.; Ma, L. The Role of Repetitive Transcranial Magnetic Stimulation (rTMS) in the Treatment of Cognitive Impairment in Patients with Alzheimer's Disease: A Systematic Review and Meta-Analysis. *J. Neurol. Sci.* **2019**, *398*, 184–191. [CrossRef]
20. Xie, Y.; Li, Y.; Nie, L.; Zhang, W.; Ke, Z.; Ku, Y. Cognitive Enhancement of Repetitive Transcranial Magnetic Stimulation in Patients with Mild Cognitive Impairment and Early Alzheimer's Disease: A Systematic Review and Meta-Analysis. *Front. Cell Dev. Biol.* **2021**, *9*, 734046. [CrossRef]
21. Wei, Z.; Fu, J.; Liang, H.; Liu, M.; Ye, X.; Zhong, P. The Therapeutic Efficacy of Transcranial Magnetic Stimulation in Managing Alzheimer's Disease: A Systemic Review and Meta-Analysis. *Front. Aging Neurosci.* **2022**, *14*, 980998. [CrossRef]
22. Zhang, T.; Sui, Y.; Lu, Q.; Xu, X.; Zhu, Y.; Dai, W.; Shen, Y.; Wang, T. Effects of rTMS Treatment on Global Cognitive Function in Alzheimer's Disease: A Systematic Review and Meta-Analysis. *Front. Aging Neurosci.* **2022**, *14*, 984708. [CrossRef]
23. Liu, G.; Xue, B.; Guan, Y.; Luo, X. Effects of Repetitive Transcranial Magnetic Stimulation Combined with Cognitive Training on Cognitive Function in Patients with Alzheimer's Disease: A Systematic Review and Meta-Analysis. *Front. Aging Neurosci.* **2024**, *15*, 1254523. [CrossRef]
24. Sitzer, D.I.; Twamley, E.W.; Jeste, D.V. Cognitive Training in Alzheimer's Disease: A Meta-analysis of the Literature. *Acta Psychiatr. Scand.* **2006**, *114*, 75–90. [CrossRef]
25. Li, X.; Qi, G.; Yu, C.; Lian, G.; Zheng, H.; Wu, S.; Yuan, T.-F.; Zhou, D. Cortical Plasticity Is Correlated with Cognitive Improvement in Alzheimer's Disease Patients after rTMS Treatment. *Brain Stimulat.* **2021**, *14*, 503–510. [CrossRef]
26. Moher, D.; Liberati, A.; Tetzlaff, J.; Altman, D.G.; The PRISMA Group. Preferred Reporting Items for Systematic Reviews and Meta-Analyses: The PRISMA Statement. *PLoS Med.* **2009**, *6*, e1000097. [CrossRef] [PubMed]
27. Bentwich, J.; Dobronevsky, E.; Aichenbaum, S.; Shorer, R.; Peretz, R.; Khaigrekht, M.; Marton, R.G.; Rabey, J.M. Beneficial Effect of Repetitive Transcranial Magnetic Stimulation Combined with Cognitive Training for the Treatment of Alzheimer's Disease: A Proof of Concept Study. *J. Neural Transm.* **2011**, *118*, 463–471. [CrossRef]
28. Rabey, J.M.; Dobronevsky, E.; Aichenbaum, S.; Gonen, O.; Marton, R.G.; Khaigrekht, M. Repetitive Transcranial Magnetic Stimulation Combined with Cognitive Training Is a Safe and Effective Modality for the Treatment of Alzheimer's Disease: A Randomized, Double-Blind Study. *J. Neural Transm.* **2013**, *120*, 813–819. [CrossRef] [PubMed]
29. Rabey, J.M.; Dobronevsky, E. Repetitive Transcranial Magnetic Stimulation (rTMS) Combined with Cognitive Training Is a Safe and Effective Modality for the Treatment of Alzheimer's Disease: Clinical Experience. *J. Neural Transm.* **2016**, *123*, 1449–1455. [CrossRef] [PubMed]
30. Sabbagh, M.; Sadowsky, C.; Tousi, B.; Agronin, M.E.; Alva, G.; Armon, C.; Bernick, C.; Keegan, A.P.; Karantzoulis, S.; Baror, E.; et al. Effects of a Combined Transcranial Magnetic Stimulation (TMS) and Cognitive Training Intervention in Patients with Alzheimer's Disease. *Alzheimers Dement.* **2019**, *16*, 641–650. [CrossRef] [PubMed]
31. Brem, A.-K.; Di Iorio, R.; Fried, P.J.; Oliveira-Maia, A.J.; Marra, C.; Profice, P.; Quaranta, D.; Schilberg, L.; Atkinson, N.J.; Seligson, E.E.; et al. Corticomotor Plasticity Predicts Clinical Efficacy of Combined Neuromodulation and Cognitive Training in Alzheimer's Disease. *Front. Aging Neurosci.* **2020**, *12*, 200. [CrossRef] [PubMed]
32. Lee, J.; Choi, B.H.; Oh, E.; Sohn, E.H.; Lee, A.Y. Treatment of Alzheimer's Disease with Repetitive Transcranial Magnetic Stimulation Combined with Cognitive Training: A Prospective, Randomized, Double-Blind, Placebo-Controlled Study. *J. Clin. Neurol.* **2016**, *12*, 57. [CrossRef]
33. Zhang, F.; Qin, Y.; Xie, L.; Zheng, C.; Huang, X.; Zhang, M. High-Frequency Repetitive Transcranial Magnetic Stimulation Combined with Cognitive Training Improves Cognitive Function and Cortical Metabolic Ratios in Alzheimer's Disease. *J. Neural Transm.* **2019**, *126*, 1081–1094. [CrossRef]
34. Zhao, J.; Li, Z.; Cong, Y.; Zhang, J.; Tan, M.; Zhang, H.; Geng, N.; Li, M.; Yu, W.; Shan, P. Repetitive Transcranial Magnetic Stimulation Improves Cognitive Function of Alzheimer's Disease Patients. *Oncotarget* **2017**, *8*, 33864–33871. [CrossRef] [PubMed]
35. Nguyen, J.-P.; Suarez, A.; Kemoun, G.; Meignier, M.; Le Saout, E.; Damier, P.; Nizard, J.; Lefaucheur, J.-P. Repetitive Transcranial Magnetic Stimulation Combined with Cognitive Training for the Treatment of Alzheimer's Disease. *Clin. Neurophysiol.* **2017**, *47*, 47–53. [CrossRef]
36. Distiller, S.R. What Is a PRISMA Flow Diagram. *DistillerSR*. Available online: https://www.distillersr.com/resources/systematic-literature-reviews/what-is-a-prisma-flow-diagram (accessed on 23 August 2024).
37. Cheng, C.P.W.; Wong, C.S.M.; Lee, K.K.; Chan, A.P.K.; Yeung, J.W.F.; Chan, W.C. Effects of Repetitive Transcranial Magnetic Stimulation on Improvement of Cognition in Elderly Patients with Cognitive Impairment: A Systematic Review and Meta-analysis. *Int. J. Geriatr. Psychiatry* **2018**, *33*, e1–e13. [CrossRef] [PubMed]

38. Yan, Y.; Tian, M.; Wang, T.; Wang, X.; Wang, Y.; Shi, J. Transcranial Magnetic Stimulation Effects on Cognitive Enhancement in Mild Cognitive Impairment and Alzheimer's Disease: A Systematic Review and Meta-Analysis. *Front. Neurol.* **2023**, *14*, 1209205. [CrossRef] [PubMed]
39. Miniussi, C.; Rossini, P.M. Transcranial Magnetic Stimulation in Cognitive Rehabilitation. *Neuropsychol. Rehabil.* **2011**, *21*, 579–601. [CrossRef]
40. Miniussi, C.; Vallar, G. Brain Stimulation and Behavioral Cognitive Rehabilitation: A New Tool for Neurorehabilitation? Neuropsychological. *Rehabilitation* **2011**, *21*, 553–559. [CrossRef]
41. Bagattini, C.; Zanni, M.; Barocco, F.; Caffarra, P.; Brignani, D.; Miniussi, C.; Defanti, C.A. Enhancing Cognitive Training Effects in Alzheimer's Disease: rTMS as an Add-on Treatment. *Brain Stimul.* **2020**, *13*, 1655–1664. [CrossRef]
42. Beishon, L.; Intharakham, K.; Swienton, D.; Panerai, R.B.; Robinson, T.G.; Haunton, V.J. Neuroimaging Outcomes in Studies of Cognitive Training in Mild Cognitive Impairment and Early Alzheimer's Disease: A Systematic Review. *Curr. Alzheimer Res.* **2020**, *17*, 472–486. [CrossRef]
43. Castro, C.B.; Costa, L.M.; Dias, C.B.; Chen, J.; Hillebrandt, H.; Gardener, S.L.; Sohrabi, H.R. Multi-Domain Interventions for Dementia Prevention–A Systematic Review. *J. Nutr. Health Aging* **2023**, *27*, 1271–1280. [CrossRef] [PubMed]
44. Noach, S.; Witteman, B.; Boss, H.M.; Janse, A. Effects of Multidomain Lifestyle Interventions on Cognitive Decline and Alzheimer's Disease Prevention: A Literature Review and Future Recommendations. *Cereb. Circ. Cogn. Behav.* **2023**, *4*, 100166. [CrossRef] [PubMed]
45. Heath, A.; Taylor, J.L.; McNerney, M.W. rTMS for the Treatment of Alzheimer's Disease: Where Should We Be Stimulating? *Expert Rev. Neurother.* **2018**, *18*, 903–905. [CrossRef] [PubMed]

brain sciences

MDPI

Review

The Effectiveness of Lee Silverman Voice Treatment (LSVT LOUD) on Children's Speech and Voice: A Scoping Review

Angelos Papadopoulos [1,2], Louiza Voniati [3], Nafsika Ziavra [4] and Dionysios Tafiadis [4,*]

1 School of Health Rehabilitation Sciences, University of Patras, 26500 Patras, Greece; angelospapadopoulos@gmail.com
2 General Children's Hospital of Patras "Karamandaneio", 26331 Patras, Greece
3 Department of Health Sciences, Speech and Language Therapy, European University, Nicosia 22006, Cyprus; l.voniati@euc.ac.cy
4 Department of Speech and Language Therapy, School of Health Sciences, University of Ioannina, GR45500 Ioannina, Greece; nziavra@uoi.gr
* Correspondence: tafiadis@uoi.gr

Abstract: Background: This scoping review had as a primary goal a review of the literature and the an analysis of the possible effectiveness of the LSVT LOUD approach in children with voice and speech deficits. Methods: A search was conducted in the Scopus and PubMed databases in May of 2024. Eleven articles were obtained from the search. The standards of PRISMA recommendations were used for scoping reviews and the PCC framework was used for the eligibility criteria. Furthermore, the study used the instructions in the Cochrane Handbook for a quality assessment. The Mendeley Reference Manager software collected the studies and removed duplicates. Results: The reviewed studies employed formal and informal measures to assess voice and speech abilities in the children. Regarding the sample's characteristics, the studies mostly included children with Cerebral Palsy (CP) and also those with Down Syndrome (DS). All the studies reported that children with CP and DS undertook a total dose of the LSVT LOUD treatment. Significant post-treatment findings indicated increased speech function and sound pressure level, regarding the auditory–perceptual ratings of voice and speech improvement. In many studies, parents' and expert listeners' ratings of voice, perception of vocal loudness, speech, and communication indicated improvement. Conclusions: The majority of the included studies provide positive evidence for the LSVT as an approach. However, the small sample size that featured in the studies, as well as their limitations, made these conclusions uncertain. Moreover, the study's findings provided recommendations that speech language therapists and other clinicians need to follow when setting a treatment plan with children with CP and DS.

Keywords: Lee Silverman Voice Treatment; LSVT LOUD; children; speech; voice; scoping review

1. Introduction

Numerous studies have documented the frequency of voice disorder in preschool- [1,2] and school-aged children [3–7]. Additionally, a variety of studies report prevalence rates in terms of different countries [3,5,6] and cultural backgrounds [1]. In general, the prevalence of children's Voice Disorders (chVD) ranges from 1.4% to 6.7% [1,7]. Furthermore, many studies have documented the prevalence of chVD and other speech and language disorders [5,6] and in relation to specific medical conditions such as cerebral palsy (CP) and down syndrome (DS). Approximately 40% to 90% of children diagnosed with CP experience motor speech impairments, which have a detrimental impact on their ability to communicate clearly [8,9]. In addition, the motor speech characteristics of children with DS have traditionally been classified as childhood dysarthria [10]. Specifically, 37.8% had childhood dysarthria, and the dysarthria subtype that was most commonly observed was ataxia. As indicated above, dysarthria is common in children with CP and DS diagnosis. Accordingly, the necessity of screening and assessment procedures [4,11], as well-documented therapy for chVD, is of high priority for speech-language pathologists [12,13].

Regarding children with CP, the literature [14] has reported that their vocal quality is altered. These changes refer to a reduced ability to control loudness and an altered voice quality [14–16], and, according to Nordberg et al., those who listen to the voices of children with CP rated them as having a strained–strangled, harsh, wet hoarseness [15]. Something that was noticed by a study [17] is that the voice quality of these children differs, and many times changes are present during a part of a sentence rather than there being consistency throughout the sentence. Furthermore, the voice quality differences result from motor control deficits of the laryngeal subsystem and/or coordination of the laryngeal and respiratory subsystems [14]. Specifically, a study concluded that children have presented vocal quality issues, and their vocal quality is characterized as constricted and creaky as a result of lower spectral tilt and greater noise. Worse vocal fold articulation was also demonstrated when producing single vowels, and low vowels produced less constriction compared to high vowels.

The children with DS also exhibited vocal quality issues, such as harsh, husky, monotonous, low-pitched, and ventricular voices [18,19]. These are a result of the squeezing of the larynx during phonation [18]. Moreover, children with DS have increased rates of laryngeal issues when compared to the population of children without DS [20]. The literature [20,21] has reported that these children are at a higher risk of subglottic stenosis, vocal fold paralysis, and laryngomalacia, and that the most common etiology of dysphonia in children with DS is vocal fold immobility and hypomobility.

In recent research, there has been a growing interest in treating pediatric voice disorders [22]. Voice therapists and researchers are aware that many of the well-established adult voice treatment techniques can be adjusted to accommodate children's developmental stages, but there is still much to learn about the essential elements and techniques needed for positive results [22,23], as well as several factors (e.g., parental perceptions of chVD, patient demographic data, etc.), such as the duration of therapy [12,24], that have been set as barriers to a successful outcome of voice therapy (VT) [25]. In their study, Braden and Thibeault (2020) concluded that even though voice therapy is a gold standard and established as the intervention for harmless tumors in adults, it has become more widely recommended for children as well. They have also underlined the fact that there is still debate, as in other studies [12,25,26], regarding the effectiveness of therapy for children with harmless tumors. There is conflict about what therapy should consist: only of vocal hygiene instruction [27], or whether it should also incorporate a direct intervention [23]. As an outcome of their study, they stated that children with benign vocal fold lesions and dysphonia following voice therapy had statistically significant improvements in phonation cut-off pressure and the perceptual quality of the voice. They also suggested that voice therapy is beneficial for enhancing children's voice quality [23], which was in agreement with other studies as well [12].

VT is divided into direct and indirect interventions and can be delivered via extrinsic and/or intrinsic approaches [28–31]. These interventions are organized in detailed programs/protocols with specific therapy tasks or components [28]. The seven well-known VT protocols are as follows: (i) Manual Circumlaryngeal Therapy (MCT) [32,33], (ii) Laryngeal Manual Therapy (LMT) [34,35], (iii) Vocal Function Exercises (VFE) [36,37], (iv) Resonant Voice Therapy (RVT) [38,39], (v) Accent Method [40–42], (vi) Confidential Voice Therapy (CVT) [43], and (vii) the Lee Silverman Voice Treatment (LSVT) [44,45].

Dr. Lorraine Ramig developed the LSVT method which has been a research focus for over 25 years [44,45]. LSVT has the scope to increase vocal fold adduction and loudness. This could be achieved by high-effort, intensive stimulation [46]. The literature features several randomized control trials [45,47–49] which have proved that LSVT can improve vocal intelligibility, speech issues, and quality of life [50]. Currently, LSVT has standardized research methods, and the reliability of outcome indicators has improved [46,47,51,52].

LSVT uses a simple but intensive approach while promoting a healthy increase in vocal loudness. The aforementioned indicated improvements are not only in voice production but also in speech articulation and intelligibility [50]. Studies have reported that LSVT LOUD positively affects swallowing, facial expression, and neural function, as revealed by PET [50,53,54] in people with PD. It has been implemented in individuals with Parkinson's, multiple sclerosis, ataxia, traumatic brain injury, DS, and CP. The implementation was made without modifying the primary treatment protocol [50]. LSVT trains Parkinson's disease (PD) patients at moderate stages of the disease and other atypical Parkinsonisms to employ their vocal cords at a more normal loudness in different settings. The treatment is delivered via several sessions under the guidance of a certified SLP four times a week for four weeks [55].

The LSVT method was also applied in children with voice and speech disorders [56–59]. According to Fox and Boliek's (2012) study, which examined the impact of LSVT on children with dysarthria and spastic cerebral palsy (CP), their preliminary data showed that children with spastic CP not only were able to deal with accurate voice treatment but also demonstrated improvement in a few specific areas of vocal functioning. Another study by Langlois et al. (2020) evaluated the treatment effects of LSVT on (1) vowel duration, (2) acoustic vowel space, and (3) the ratio of F2/i/ to F2/u/, which were analyzed using acoustic data from children with secondary motor speech disorders. The authors concluded that it may foster articulation improvements for children with CP changes, indicated in the duration of vowels and acoustic vowel space, and for children diagnosed with DS, observed in alterations in acoustic vowel space. Furthermore, Reed et al. (2017), in their research on eight children with motor speech disorders and CP, compared to an age- and sex-matched group of peers whose speech has developed without motor speech disorders and CP, administered LSVT in a 12-week program to examine any changes in white matter integrity. The researchers implied that the long-term practice of skills during therapy improved the integrity of the white matter neural networks of speech production.

Finally, Boliek et al. (2022), in their phase I outcomes for LSVT for children with Down syndrome, suggested that phase II treatment trials employing LSVT with a bigger population of children with DS will set the foundation for the method of obtaining treatment evidence data. They have also reported that their early findings demonstrated some significant therapeutic improvements in children with DS as well those not experiencing any high levels of intolerance.

To our knowledge, this could cover a research gap in the literature regarding the lack of such a review in pediatric samples. LSVT has been implemented and standardized in several randomized clinical trial studies in adult samples over the years [46]. Until now, there have only been systematic reviews of adults that systematically compared the effectiveness of LSVT with speech therapy or with no intervention for dysarthria in the adult population [46,60]. Given what has just been mentioned, the current scoping review aims to explore and analyze the literature on the LSVT LOUD approach's effect on diagnosis for children with CP and DS with voice and speech problems arising from dysarthria secondary to CP and DS.

The Aim of the Study

This scoping review was conducted to enhance the current body of literature by addressing the following research questions: is the LSVT LOUD approach effective and practical for children regarding their voice and speech issues that arise from various disorders?

Moreover, the primary objective of this scoping review is to supply researchers with an up-to-date and complete information resource on the subject, promoting its use in clinical practice. Moreover, it would be considered a valuable manuscript since it consisted of useful pieces of the literature. In addition, this scoping review provides a comprehensive view of the impact of the LSVT LOUD approach in speech and voice treatment.

2. Materials and Methods

2.1. Databases and Search

The authors conducted the research in the Scopus and PubMed databases in May 2024. The articles were selected from relevant journals. The following keywords were used in both databases: "LSVT", "LOUD", and "CHILDREN". Following the search, the researchers read the abstracts to verify that the studies included the LSVT LOUD implementation in children. Specifically, after we completed searches of the literature, and identified studies, we conducted screening of the titles and abstracts of each article to determine if they met the criteria for inclusion. These criteria for titles and abstracts were pre-defined and were as follows:

- Date range of publication: the dataset for this study included all available literature without any time restrictions.
- Study design type: only research articles and no review studies.
- Study focuses on a specific disease, condition, or patient population: only children.
- Study focuses mainly on a specific intervention: LSVT LOUD
- Study took place in a certain country/hospital or other context: no restriction of the context.
- Study written in a specific language: we only included studies written in English. However, all the available studies were written in English.

Furthermore, Figure 1 shows the process flow diagram for this scoping review, which is based on the standards of PRISMA guidelines for scoping reviews (PRISMA-ScR) [61]. Moreover, the researchers used the PCC framework as guidance to construct the criteria for this scoping review (Table 1) [62].

The electronic research strategy for Scopus was as follows: (TITLE-ABS-KEY (LSVT LOUD) AND TITLE-ABS-KEY (CHILDREN)), and for PubMed was as follows: ((LSVT) AND (LOUD)) AND (CHILDREN).

Table 1. The PCC framework.

PCC Element	
Population	Children
Concept	The use of the LSVT LOUD approach
Context	Every disorder that affects children

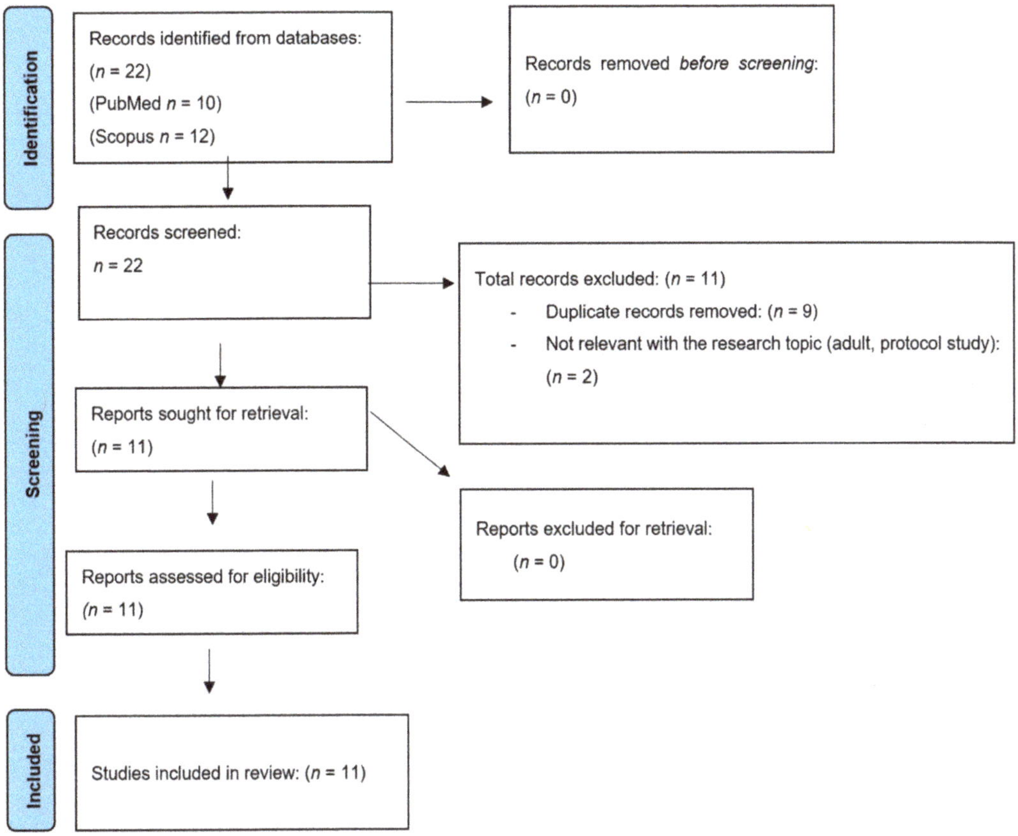

Figure 1. PRISMA ScR flow diagram.

2.2. Eligibility Criteria

The current review included only articles that met specific qualifying criteria. The criteria were as follows: the article (1) investigates children treated with the LSVT LOUD approach, (2) focuses on addressing deficiencies in speech and voice issues, (3) is written in the English language, (4) utilizes either standardized or non-standardized measures for the speech and voice assessment process, (5) is only a research article, with no review studies allowed, and that (6) every county/hospital or other place would be included with no restriction of the context.

On the other hand, the criteria for exclusion were as follows: (1) data such as letters to the editor or other review studies were excluded, and (2) studies that were part of a thesis or dissertation (grey literature) were excluded.

The database screening resulted from 12 Scopus and 10 PubMed abstracts that were reviewed to confirm if they met the inclusion criteria for the scoping review study. At this point, it needs to be mentioned that two (2) independent researchers (A.P. and N.Z.) evaluated the eligibility of the abstracts, and two reasons arose for exclusion: (a) nine (9) duplicates were removed, and (b) one (1) study was not related to the research topic as it featured an adult sample [63].

The next step was for the researchers to retrieve the complete texts. The two independent researchers aimed to ensure that the studies met the eligibility criteria for the scoping review. There were no conflicts between the authors. As the next step, all the studies' characteristics were gathered and presented in Table 2. The researchers used a review matrix in excel to record the extracted data and the Mendeley reference manager software, Version 2.116.1, to collect the studies and remove duplicates. After 11 articles were excluded for the abovementioned reasons, the authors finally gathered 11 studies from the search. In the current scoping review, a thematic analysis was used as we focused on identifying and interpreting patterns within data.

Table 2. Data extraction and synthesis (sample characteristics and essential findings of the articles included in the review).

Study	Sample Gender (M/F)	Age [Range] (Years)	Disorder of the Child	Assessment	Duration	Interventions	Findings
Levy et al., 2013 [64]	2 [LSVT] 1 [Control] 3 (F)	7.03 [3.3–9.7]	CP	-Test of Auditory Comprehension of Language-3 (TACL-3). -Kaufman Brief Intelligence Test-2 (KBIT-2). -Informal assessment.	Four days per week (total 4 weeks) where the LSVT LOUD intervention was administered for 50 to 60 min plus 10 min of homework.	1. LSVT LOUD. 2. Speech intervention with sound theoretical motivations.	• Speech function and sound pressure level increased in children with dysarthria, although success may vary across linguistic levels and children.
Watts 2013 [65]	5 [LSVT] 5 [Control]	[5–7]	CP	-Informal assessment.	Sixteen treatment sessions across 4 consecutive weeks, with four treatment sessions per week.	LSVT LOUD blinded (participants were randomly assigned).	• Positive results for voice and speech production in children with CP. • Auditory–perceptual ratings of voice and speech revealed improvement in CP. • Parent's ratings of voice, speech, and communication indicated improvement.
Fortin et al., 2023 [66]	1 (F)	5 years old	CP	-Functional communication measures of the American Speech–Language–Hearing Association (1997). -Clinical evaluation. -Assessment of vocal medical status to verify potential vocal fold pathology through otorhinolaryngologic examination (video-laryngostroboscopy). -Acoustic measures.	As specified in the LSVT LOUD protocol, the girl received 16 individual 1 h long therapy sessions four times per week over a period of 4 weeks.	Standard LSVT LOUD acoustic measures that included average vocal intensity during sustained vowel phonations and sentence repetitions and the maximum duration of sustained vowel phonations were collected pre- and post-treatment.	• Significant post-treatment increases in average vocal intensity during sustained vowel phonations and sentence repetitions for the preschooler. • Significant increases in the maximum duration of sustained vowel phonation pre- to post-treatment for the preschooler. • Perceptual ratings revealed improvements in communicative effectiveness, participation, and speech.
Reed et al. [59]	8 [LSVT] 3 (F) 5 (M) 8 [Control]	11.6 [7–16]	CP	A pediatric neurologist made the diagnosis; the GMFCS expanded and a revised grade was determined by a physical therapist, and a licensed speech-language pathologist determined speech/voice status. -Test of Children's Speech (TOCS+). -Loudness (dB SPL) and duration (s) were derived from the maximum duration phonation task. -Loudness, speaking rate (syllables per s), and pitch variability (F 0 in Hz) were derived from the vowel segments of the phrase repetition task. -The DDK sequential motion rate task measured the MMRtri repetition rate (syllables/s).	Each participant with CP received a total dose of LSVT LOUD provided by a certified speech-language pathologist.	-A certified speech-language pathologist provides LSVT LOUD for children with CP. -Control children received no additional speech or language activities outside their typical daily home and school routines.	Immediately following treatment and after the 12-week maintenance program the following were observed: • Increased fractional anisotropy in several motor and association tracts. • Acoustic data on untrained tasks correlated with FA changes.

Table 2. *Cont.*

Study	Sample Gender (M/F)	Age [Range] (Years)	Disorder of the Child	Assessment	Duration	Interventions	Findings
Bakhtiari et al., 2017 [67]	8 [LSVT] 3 (F) 5 (M) 8 [Control]	11.6 [7–16]	CP with dysarthria diagnosis (with motor speech disorders)	-Test of Children's Speech Plus (TOCS+). -The DDK task. -Pre-treatment: Children were asked to overtly produce phonation at conversational loudness, cued-phonation at perceived twice-conversational loudness, a series of single words, and a prosodic imitation task while being scanned using fMRI.	Four weeks of LSVT LOUD, followed by a 12-week maintenance program.	-Children with CP: intensive neuroplasticity-principled voice treatment protocol; LSVT LOUD. -Control children did not receive treatment or any similar training throughout the course of the study.	• Post-treatment changes in connectivity between the left supramarginal gyrus, the right supramarginal gyrus, and the left precentral gyrus for children with cerebral palsy. • LSVT LOUD enhanced the contributions of the feedback system in the speech production network instead of having a high reliance on the feedforward control system and the somatosensory target map for regulating vocal effort.
Levy 2014 [68]	1 (M) [LSVT] 1 (F) [SSIT]	7 13	CP with dysarthria diagnosis	Test for Auditory Comprehension of Language—3rd Edition. -Kaufman Brief Intelligence Test—2nd Edition. -Audiological screening.	A total of 1 h for 4 days per week for 4 weeks (16 day sessions for each intervention).	-LSVT. -Speech Systems Intelligibility Treatment (SSIT).	• Better acoustic vowel space following LSVT LOUD and SSIT. • Increased articulatory accuracy and post-treatment stimuli were preferred, and they were judged more intelligible than pre-treatment stimuli.
Boliek and Fox 2014 [69]	1 (M) 1 (F)	10:9 for both children	CP with dysarthria diagnosis	-Acoustic and perceptual data were collected according to standardized protocols. -Measurements of untrained and trained tasks. -The standardized TOCS +.	Four individual 1 h sessions, 4 days per week for 4 weeks, delivered by certified LSVT clinicians.	LSVT	• Individual and environmental features affected fast-phase and long-phase responses to LSVT LOUD. • Intensity (dose) is necessary but not sufficient for change. • Weak responders may require a more extended treatment phase, better timing (e.g., developmentally and socially), and a more prominent desire to communicate successfully during daily activities. • Strong responders appear to benefit from treatment intensity and saliency and intrinsic and extrinsic rewards for using trained skills for everyday communication.
Boliek and Fox 2017 [70]	7 (5 F, 2 M)	[6–10]	Spastic CP with dysarthria diagnosis	-Auditory–perceptual measures (listener task). -Acoustic measures. -Parent ratings. -Visual analog scales, perceptual ratings of single-word intelligibility, and parent interviews.	Four individual 1 h sessions, 4 days per week for 4 weeks.	LSVT	• Expert listeners preferred voice quality and the articulatory precision of children with CP at follow-up compared to pre-treatment. • Acoustic data indicated improvements in select measures of vocal functioning immediately post-treatment, with some maintenance at follow-up. • Single-word intelligibility improved immediately post-treatment. • Parents rated positive changes in voice and speech characteristics and qualitative changes in communication.

Table 2. *Cont.*

Study	Sample Gender (M/F)	Age [Range] (Years)	Disorder of the Child	Assessment	Duration	Interventions	Findings
Langlois et al., 2020 [57]	17 (CP) 8 (M) 9 (F) 9 (DS) 1 (M) 8 (F)	10.6 (CP) [6–16] 6.8 (DS) [4–8]	CP and DS with dysarthria diagnosis	-Recordings. -Test of Children's Speech Plus (TOCS+) for the CP group. -The Goldman–Fristoe Test of Articulation 2 (GFTA) for the DS group and the sentences.	CP and DS groups: 16 one-hour sessions delivered over four weeks (four days per week) and daily homework assignments (one per day on treatment days and two per day on non-treatment days).	LSVT	• Significant changes in vowel duration and acoustic vowel space occurred from pre-treatment to 12 weeks post-treatment in those with CP. • Increased acoustic vowel space was observed in five participants with Down syndrome. • Findings provide preliminary evidence of intensive voice treatment spreading effects on the articulatory system in some children with CP and DS.
Fox and Bolek 2012 [56]	5 3 (M) 2 (F)	6.5 [5–7]	CP with dysarthria diagnosis	-Brief voice and speech screening. -An assessment of abilities to follow directions related to the study tasks. -A hearing screening (500 Hz, 1000 Hz, 2000 Hz, and 4000 Hz at 25 dB HL). -Parent rating forms. Visual analog scale.	Sixteen treatment sessions (four sessions a week for 4 consecutive weeks), two recording sessions 1 week immediately following treatment, and two recording sessions 6 weeks after the conclusion of treatment.	LSVT	• Findings support that the children with CP tolerated intensive voice treatment and showed improvement in select aspects of vocal functioning. • Parents reported an improved perception of vocal loudness immediately following treatment. • Listeners preferred the speech samples taken immediately post-treatment over those taken during the baseline phase.
Bolek et al. 2022, [58]	9 8 (F) 1 (M)	[4;6 and 8;10]	DS with motor speech disorders (dysarthria diagnosis).	-PPVT-R. -Expressive One-Word Picture Vocabulary Test-R. -Expressive language without modeling. Ratings of single-word intelligibility.	All participants completed the total dose of LSVT LOUD: four individual 1 h sessions, 4 days per week for 4 weeks.	LSVT	• Children with DS tolerated intensive voice treatment without adverse effects and made select meaningful therapeutic gains. • Patterns of post-treatment improvements were not consistent across participants but were more frequently observed on trained maximum performance tasks compared to tasks reflecting the general treatment skillset. • Some participants exhibited a more robust response to treatment, whereas others showed a mixed or weaker response. • Parents perceived benefits from intensive intervention and indicated they would strongly recommend LSVT LOUD to other parents who have children with DS and motor speech disorders.

LSVT LOUD: Lee Silverman Voice Treatment; SPL: sound pressure level; CP: cerebral palsy; DS: Down syndrome; FA: fractional anisotropy; TOCS+: Test of Children's Speech Plus; GFTA: Goldman–Fristoe Test of Articulation 2; PPVT-R: Peabody Picture Vocabulary Test-Revised.

2.3. *Quality Assessment for Scoping Review Study*

The literature on quality assessment in scoping review studies supports the idea that it is not mandatory [71–75]. However, this study used the instructions in the Cochrane Handbook and included a quality assessment identifying the possibility of bias in each study [74–76]. Table 2 presents factors such as limitations, funding sources, and declarations of interest.

3. Results

3.1. Quality Assessment

The small sample size was a frequent limitation, as revealed in this scoping review (Table 2). Furthermore, four studies did not report relevant information for funding and support. Three of the included studies [57,59,67] were supported by the Cerebral Palsy International Research Foundation, four studies [57,58,69,70] from the Stollery Children's Hospital Foundation, and three studies [58,69,70] by the Edmonton Oilers Community Foundation. One study [56] was supported, in part, by the National Multipurpose Research and Training Center of the National Institute of Deafness and Other Communication Disorders and by the Final Project Fund from the Graduate College at the University of Arizona. Moreover, this study is based on a doctoral dissertation at the University of Arizona. In addition, the Coleman Institute for Cognitive Disabilities supported two studies [57,58].

Regarding the declaration of interest/disclosure, no relevant information was found in two studies. Moreover, in two studies [65,68], the author reported no conflict of interest; the remaining seven declare that at least one author has a conflict of interest. All the studies were published in journals relevant to the field of the studies. Regarding the journals, three studies [56–58] were published in the International Journal of Speech-Language Pathology.

3.2. Presentation of Studies

As shown in Table 2, of the studies published from 2012 onwards, six were published in the last 6 years. Regarding the sample's characteristics, nine studies included only children with CP, one study included only children with Down syndrome, and one study included children with both a CP and DS diagnosis. All the studies included small sample sizes. Specifically, a higher sample number was featured in two (2) studies that recruited eight (8) children for implementation of the LSVT, and one (1) study featured seven children. In addition, 54 children with CP were involved in LSVT LOUD, and 18 with DS were, while 42 children were females, and 27 were males (one study that used LSVT LOUD as an intervention [65] did not give information about the sample's gender). The ages of the children ranged from 3.3 to 16 years old.

Direct and indirect measures were employed to evaluate the children's abilities. Furthermore, four studies used the Test of Children's Speech (TOCS+) to measure how well children can make their spoken messages understood by listeners. Two studies used the Kaufman Brief Intelligence Test-2 (KBIT-2) to briefly measure verbal and nonverbal intelligence, while two used the DDK task to measure maximum speech rate (syllables per second). One study used fMRI to evaluate the children's abilities.

The aspects and skills assessed included functional communication, acoustic measures (auditory screening), language (expression and comprehension), speech and articulation, and voice; vocal medical status was used to verify potential vocal fold pathology through otorhinolaryngology examination (video-laryngostroboscopy); and loudness (dB SPL), duration (s), and speaking rate were derived from the maximum duration phonation task, along with diadochokinetic skills (DDK). Regarding the duration of the intervention, all the studies reported that children with CP and DS received a total dose of the LSVT LOUD. As specified in the LSVT LOUD protocol, 16 individual 60-minute therapy sessions, at a frequency of four times per week over 4 weeks, were conducted. Additionally, the data show that LSVT LOUD positively affected children with CP and DS with a diagnosis of dysarthria.

Regarding the methodology, two studies used another approach to compare the effectiveness of LSVT LOUD. Specifically, one study [64] used traditionally used speech interventions with sound theoretical motivations and another study [68] used the speech systems intelligibility treatment (SSIT). Two studies consisted of control samples of children [64,68]. One study compared children with CP and DS [57]. Most of the studies, in terms of the methodological aspects, were conducted using a consistent methodological intervention approach. All of the studies satisfied the ethical requirements as they received approval from the Research Ethics Committee.

4. Discussion

This scoping review had as its primary goal an overview of the literature via a search of highly regarded databases (Scopus and PubMed) and analyzed the possible effectiveness of the LSVT LOUD approach in children with voice and speech issues arising from CP and DS. Of the 11 studies, 9 only included children with CP, 1 included children with DS, and 1 more included children with both CP and DS diagnoses.

The analysis of the studies in Table 2 indicated positive and encouraging evidence about the implementation of LSVT LOUD in the pediatric population with CP and DS diagnoses. It is a very important tool for the clinicians and all those professionals who are involved in the therapy process with children with CP and DS. Approximately 40% to 90% of children diagnosed with CP experience motor speech impairments, which have a detrimental impact on their ability to communicate clearly [8,9]. The prominent speech characteristics of dysarthria in children with CP exhibit significant variation among individuals. Nevertheless, it is frequently noticed that there are decreases in the area of vowel space, speech rate, and vocal intensity [8,9]. In addition, the motor speech characteristics of children with DS have traditionally been classified as childhood dysarthria [10]. In a study [77], out of all the participants, 97.8% fulfilled the criteria for speech disorders and the same percentage met the criteria for motor speech disorders and 37.8% had childhood dysarthria. The dysarthria subtype that was most commonly observed was ataxia. As indicated above, dysarthria is common in children with CP and DS diagnoses.

All the studies that were examined used both formal and informal methods to evaluate the voice and speech abilities of the children. A very common measure was the parent and expert listeners' baseline and post-treatment ratings of LSVT. More specifically, parents and expert listeners stated that there was an improvement from baseline to post-treatment for LSVT LOUD in voice, speech, and communication [55,65,66,68,70]. As it is well-understood in clinical contexts, direct assessment only occurs once [75,78]. In contrast, assessment tools parents typically fill out offer a longitudinal reflection in real-life situations, as mentioned in other studies [75,78].

Regarding the findings, the outcomes of two studies in children with CP diagnosis revealed that LSVT LOUD increased speech function and vocal loudness [sound pressure level] [64,67] and this is consistent with other studies conducted on individuals with Parkinson's disease [44,79]. The acoustic data about voice and speech production and intelligibility in children with spastic CP showed positive results in several studies [64–66,68,70] and suggested that LSVT should be a promising intervention approach.

The very recent study of Langlois et al. [57] that included CP and DS samples recorded notable statistical changes in vowel duration and acoustic vowel space in the CP group, from before therapy to 12 weeks after treatment. Additionally, an increase in acoustic vowel space was seen in five participants with DS, and this is supported by the literature. [80–82]. Moreover, the above results provided initial evidence of the wide-ranging impact of intensive voice treatments on the articulatory system of children with CP and DS.

There were two studies [59,67] that an offered indication of activity-dependent neuroplasticity following LSVT LOUD in children with a CP diagnosis. Similar findings were found in the study of Narayana et al. [83], in individuals with Parkinson's disease with increased activation in the right dorsolateral prefrontal cortex. Furthermore, in the studies of Reed et al. and Bakhtiari et al. [59,67], there was evidence of the positive performance of the brain regarding white matter and brain activity after the intensive implementation of LSVT in children with cerebral palsy [59,67]. Another study, conducted by Bakhtiari et al., examined the effects of LSVT LOUD on children with CP. The above study used graphical models and analyzed changes in connectivity between specific brain regions, namely the left supramarginal gyrus, right supramarginal gyrus, and left precentral gyrus, and the results indicated that LSVT LOUD improved the functioning of the feedback system in the speech production network [67]. Specifically, the results of the study [59] showed increased integrity in association areas following LSVT and slow-phase changes were observed in the left and right cingulum and the posterior corpus callosum in children with CP. As mentioned in the study [59], the left and right corpus callosum comprise fibers projecting from the cingulate gyrus to the medial temporal and parietal lobes. The cingulum has been linked to monitoring, encoding, correcting cognitive and motor processes, and behavioral memory. Moreover, fibers in the posterior aspects of the corpus callosum which project to the occipital and temporal lobes are integral to communication between the right and left hemispheres. The aforementioned connections improve opportunities for auditory and somatosensory feedback commands in speech production. Furthermore, as also referred in the study [59], the slow-phase changes observed in the white matter integrity of the cingulate gyrus and potentially the posterior corpus callosum in children with CP are encouraging and in line with establishing an internal calibration of appropriate vocal effort for speaking. According to the literature [84,85], LSVT LOUD follows the concept of activity-dependent neuroplasticity (e.g., sensory feedback, repetitive practice). In addition, a study [86] showed that learning to move expertly changes how the brain is wired to handle complex movements, including speech and voice.

The study by Boliek and Fox [69] was the only study that investigated the individual and environmental aspects. The results indicated that both individual and environmental factors influenced the fast-phase and long-phase responses to LSVT LOUD. More precisely, those who respond weakly to treatment may need a longer duration of treatment, more appropriate timing in terms of their developmental and social needs, and a greater readiness to effectively communicate in their everyday activities [69]. Furthermore, those who respond well to treatment seem to gain advantages from the high level of treatment intensity and relevance, as well as from both internal and external incentives for utilizing the acquired abilities in their daily interactions [69].

Regarding the age of the children that were involved in the LSVT approach, half of studies featured children aged below 8 years old. The above is quite important because according to a study's outcomes the period between 5 and 7 years of age is crucial for the development of connected speech. During this time, nearly all children experience notable improvements in speech intelligibility and the number of intelligible words spoken per minute [87]. Implementing approaches such as LSVT for children at an earlier stage yields improved results.

At this point, it should be mentioned that three studies [56,58,64] reported that children with CP and DS diagnoses were able to participate in intense voice treatment without experiencing any negative effects, and this can be compared to findings of the study of Mahler and Jones [88] in two adults with dysarthria secondary to DS. Regarding the long-term sustainability and maintenance over time of the treatment effects and the improvements in the voice and speech of the children, the data that were extracted from the studies referred to the immediate period after treatment so it is not possible to present any conclusions.

4.1. Key Findings

- Speech function and sound pressure level increased.
- Auditory–perceptual ratings of voice and speech revealed improvement.
- Parents' and expert listeners' ratings of voice, perception of vocal loudness, speech, and communication indicated improvement.
- Significant post-treatment increases in average vocal intensity during sustained vowel phonations and sentence repetitions were observed for the preschoolers.
- Increased fractional anisotropy in several motor and association tracts was observed.
- Acoustic data on untrained tasks correlated with FA changes detected.
- Post-treatment changes in connectivity between the left supramarginal gyrus, the right supramarginal gyrus, and the left precentral gyrus for children with cerebral palsy were observed.

- Treatment enhanced the contributions of the feedback system in the speech production network.
- Increased acoustic vowel space was found.
- Articulatory accuracy and single-word intelligibility improved immediately post-treatment.
- Individual and environmental features affected fast-phase and long-phase responses to LSVT LOUD.
- Intensity (dose) is necessary but not sufficient for change.
- Weak responders may require a more extended treatment phase, better timing, and a more prominent desire to communicate successfully. In contrast, strong responders benefit from treatment intensity and saliency and intrinsic and extrinsic rewards for using the trained skills for everyday communication.
- Children with DS tolerated intensive voice treatment without adverse effects and made select meaningful therapeutic gains.

4.2. Strengths and Limitations

In our study, we acknowledged that the first strong point of the studies was the clear methodology, and the second was that the participants were appropriately diagnosed with CP and DS. However, the study has limitations, and it is crucial to evaluate whether the intervention outcomes observed in these studies represent enough of those observed in the larger population of children with CP or DS. Furthermore, this study could not assess the possible effects of gender or other socio-demographic factors in response to the intervention because the studies did not include relative data.

Finally, in our review study, all included studies show evidence that LSVT LOUD improves children's skills, as indicated by their findings and conclusions. However, a study design should focus on a comparison with other therapies with a larger sample size or aim to research the long-term effects of the LSVT LOUD approach. Nevertheless, findings like these may prompt more researchers to design studies with a larger sample size of children with CP or DS.

5. Conclusions

Moreover, findings revealed that the children with CP and DS showed positive reactions when involved in the full dose of LSVT, and this was clear when comparing the baseline assessment for the post-treatment evaluation. The included studies confirmed the effectiveness of LSVT for dysarthria in children with DS and CP, compared with either other speech interventions or no intervention. Parents of these children perceived benefits from intensive intervention and indicated they would recommend LSVT LOUD to other parents. However, the evidence so far cannot be generalizable due to the limited sample size. Furthermore, it would be worthwhile for the researchers and clinicians (e.g., speech and language therapists) to undertake future randomized controlled trials to implement LVST within a larger sample size of children with CP and DS, as it is essential to evaluate further the effectiveness of LSVT for dysarthria in children with CP and DS.

In conclusion, the evidence is sufficient and conclusive based on our comprehensive analysis. Our review highlights the key finding that all included studies prove that LSVT LOUD improves children's voice and speech skills (vowel duration, acoustic vowel space, and increased articulatory accuracy) and LSVT could be suggested as an effective intervention approach for children with CP and DS diagnosis.

Author Contributions: Conceptualization, A.P., L.V., N.Z. and D.T.; methodology, A.P., L.V., N.Z. and D.T.; writing—original draft preparation, A.P., L.V., N.Z. and D.T.; writing—review and editing, A.P., L.V., N.Z. and D.T.; supervision, N.Z. and D.T. All authors have read and agreed to the published version of the manuscript.

Funding: This research received no external funding.

Institutional Review Board Statement: Not applicable.

Informed Consent Statement: Not applicable.

Data Availability Statement: Not applicable.

Conflicts of Interest: The authors declare no conflicts of interest.

References

1. Duff, M.C.; Proctor, A.; Yairi, E. Prevalence of voice disorders in African American and European American preschoolers. *J. Voice* **2004**, *18*, 348–353. [CrossRef] [PubMed]
2. Possamai, V.; Hartley, B. Voice disorders in children. *Pediatr. Clin. N. Am.* **2013**, *60*, 879–892. [CrossRef] [PubMed]
3. Mohammadzadeh, A.; Sandoughdar, N. Prevalence of voice disorders in iranian primary school students. *J. Voice* **2017**, *31*, 263.e13–263.e18. [CrossRef]
4. Campano, M.; Cox, S.R.; Caniano, L.; Koenig, L.L. A Review of voice disorders in school-aged children. *J. Voice* **2023**, *37*, 301.e1–301.e7. [CrossRef] [PubMed]
5. McKinnon, D.H.; McLeod, S.; Reilly, S. The prevalence of stuttering, voice, and speech-sound disorders in primary school students in Australia. *Lang. Speech Hear. Serv. Sch.* **2007**, *38*, 5–15. [CrossRef] [PubMed]
6. Mohamadi, O.; Rahimi-Madiseh, M.; Sedehi, M. The prevalence of stuttering, voice disorder, and speech sound disorders in preschoolers in Shahrekord, Iran. *Int. J. Child Youth Fam. Stud.* **2016**, *7*, 456. [CrossRef]
7. Bhattacharyya, N. The prevalence of pediatric voice and swallowing problems in the U nited S tates. *Laryngoscope* **2015**, *125*, 746–750. [CrossRef]
8. Levy, E.S.; Chang, Y.M.; Ancelle, J.A.; McAuliffe, M.J. Acoustic and perceptual consequences of speech cues for children with dysarthria. *J. Speech Lang. Hear. Res.* **2017**, *60*, 1766–1779. [CrossRef]
9. Mei, C.; Reilly, S.; Reddihough, D.; Mensah, F.; Morgan, A. Motor speech impairment, activity, and participation in children with cerebral palsy. *Int. J. Speech Lang. Pathol.* **2014**, *16*, 427–435. [CrossRef]
10. Rupela, V.; Velleman, S.L.; Andrianopoulos, M.V. Motor speech skills in children with Down syndrome: A descriptive study. *Int. J. Speech Lang. Pathol.* **2016**, *18*, 483–492. [CrossRef]
11. Simões-Zenari, M.; Nemr, K.; Behlau, M. Voice disorders in children and its relationship with auditory, acoustic and vocal behavior parameters. *Int. J. Pediatr. Otorhinolaryngol.* **2012**, *76*, 896–900. [CrossRef] [PubMed]
12. Adriaansen, A.; Meerschman, I.; Van Lierde, K.; D'Haeseleer, E. Effects of voice therapy in children with vocal fold nodules: A systematic review. *Int. J. Lang. Commun. Disord.* **2022**, *57*, 1160–1193. [CrossRef] [PubMed]
13. McAllister, A.; Sjölander, P. Children's voice and voice disorders. In *Seminars in Speech and Language*; Thieme Medical Publishers: New York, NY, USA, 2013; pp. 71–79.
14. Nip, I.S.B.; Garellek, M. Voice quality of children with cerebral palsy. *J. Speech Lang. Hear. Res.* **2021**, *64*, 3051–3059. [CrossRef] [PubMed]
15. Nordberg, A.; Miniscalco, C.; Lohmander, A. Consonant production and overall speech characteristics in school-aged children with cerebral palsy and speech impairment. *Int. J. Speech Lang. Pathol.* **2014**, *16*, 386–395. [CrossRef] [PubMed]
16. Nip, I.S.B.; Arias, C.R.; Morita, K.; Richardson, H. Initial observations of lingual movement characteristics of children with cerebral palsy. *J. Speech Lang. Hear. Res.* **2017**, *60*, 1780–1790. [CrossRef]
17. Allison, K.M.; Hustad, K.C. Impact of sentence length and phonetic complexity on intelligibility of 5-year-old children with cerebral palsy. *Int. J. Speech Lang Pathol.* **2014**, *16*, 396–407. [CrossRef]
18. Novak, A. The voice of children with Down's syndrome. *Folia Phoniatr. Logop.* **1972**, *24*, 182–194. [CrossRef]
19. Moura, C.P.; Cunha, L.M.; Vilarinho, H.; Cunha, M.J.; Freitas, D.; Palha, M.; Pueschel, S.M.; Pais-Clemente, M. Voice parameters in children with down syndrome. *J. Voice* **2008**, *22*, 34–42. [CrossRef]
20. Hseu, A.F.; Spencer, G.P.; Jo, S.; Clark, R.; Nuss, R.C. Laryngeal pathologies in dysphonic children with Down Syndrome. *Int. J. Pediatr. Otorhinolaryngol.* **2022**, *157*, 111118. [CrossRef]
21. Chin, C.J.; Khami, M.M.; Husein, M. A general review of the otolaryngologic manifestations of Down Syndrome. *Int. J. Pediatr. Otorhinolaryngol.* **2014**, *78*, 899–904. [CrossRef]
22. Braden, M.; Abbott, K.V. Advances in pediatric voice therapy. *Perspect. ASHA Spéc. Interest Groups* **2018**, *3*, 68–76. [CrossRef]
23. Braden, M.; Thibeault, S.L. Outcomes of voice therapy in children with benign vocal fold lesions. *Int. J. Pediatr. Otorhinolaryngol.* **2020**, *136*, 110121. [CrossRef] [PubMed]
24. Fujiki, R.B.; Thibeault, S.L. Pediatric voice therapy: How many sessions to discharge? *Am. J. Speech Lang. Pathol.* **2022**, *31*, 2663–2674. [CrossRef] [PubMed]
25. Hseu, A.F.; Spencer, G.; Woodnorth, G.; Kagan, S.; Kawai, K.; Nuss, R.C. Barriers to voice therapy in dysphonic children. *J. Voice* **2023**, *37*, 410–414. [CrossRef] [PubMed]
26. Trani, M.; Ghidini, A.; Bergamini, G.; Presutti, L. Voice therapy in pediatric functional dysphonia: A prospective study. *Int. J. Pediatr. Otorhinolaryngol.* **2007**, *71*, 379–384. [CrossRef] [PubMed]
27. Salderay, Z.E.; Yılmaz, M.; Altınyay, Ş.; Gölaç, H.; Gökdoğan, Ç. The effect of an indirect voice therapy approach on the voice of children with vocal fold nodules: A prospective cohort study. *J. Voice* **2022**, *38*, 858–863. [CrossRef]
28. Van Stan, J.H.; Roy, N.; Awan, S.; Stemple, J.; Hillman, R.E. A taxonomy of voice therapy. *Am. J. Speech Lang. Pathol.* **2015**, *24*, 101–125. [CrossRef]

29. Behrman, A.; Rutledge, J.; Hembree, A.; Sheridan, S. Vocal hygiene education, voice production therapy, and the role of patient adherence: A treatment effectiveness study in women with phonotrauma. *J. Speech Lang. Hear. Res.* **2008**, *51*, 350–366. [CrossRef]
30. Carding, P.N.; Horsley, I.A.; Docherty, G.J. A study of the effectiveness of voice therapy in the treatment of 45 patients with nonorganic dysphonia. *J. Voice* **1999**, *13*, 72–104. [CrossRef]
31. Holmberg, E.B.; Hillman, R.E.; Hammarberg, B.; Södersten, M.; Doyle, P. Efficacy of a behaviorally based voice therapy protocol for vocal nodules. *J. Voice* **2001**, *15*, 395–412. [CrossRef]
32. Dromey, C.; Nissen, S.L.; Roy, N.; Merrill, R.M. Articulatory changes following treatment of muscle tension dysphonia: Preliminary acoustic evidence. *J. Speech Lang. Hear. Res.* **2008**, *51*, 196–208. [CrossRef] [PubMed]
33. Roy, N.; Bless, D.M.; Heisey, D.; Ford, C.N. Manual circumlaryngeal therapy for functionaldysphonia: An evaluation of short- and long-term treatment outcomes. *J. Voice* **1997**, *11*, 321–331. [CrossRef] [PubMed]
34. Mathieson, L.; Hirani, S.; Epstein, R.; Baken, R.; Wood, G.; Rubin, J. Laryngeal Manual Therapy: A Preliminary Study to Examine its Treatment Effects in the Management of Muscle Tension Dysphonia. *J. Voice* **2009**, *23*, 353–366. [CrossRef] [PubMed]
35. Mathieson, L. The evidence for laryngeal manual therapies in the treatment of muscle tension dysphonia. *Curr. Opin. Otolaryngol. Head Neck Surg.* **2011**, *19*, 171–176. Available online: https://journals.lww.com/co-otolaryngology/fulltext/2011/06000/the_evidence_for_laryngeal_manual_therapies_in_the.8.aspx (accessed on 22 August 2024). [CrossRef] [PubMed]
36. Sabol, J.W.; Lee, L.; Stemple, J.C. The value of vocal function exercises in the practice regimen of singers. *J. Voice* **1995**, *9*, 27–36. [CrossRef]
37. Stemple, J.C.; Lee, L.; D'Amico, B.; Pickup, B. Efficacy of vocal function exercises as a method of improving voice production. *J. Voice* **1994**, *8*, 271–278. [CrossRef]
38. Verdolini, K.; Druker, D.G.; Palmer, P.M.; Samawi, H. Laryngeal adduction in resonant voice. *J. Voice* **1998**, *12*, 315–327. [CrossRef]
39. Yiu, E.M.-L.; Lo, M.C.; Barrett, E.A. A systematic review of resonant voice therapy. *Int. J. Speech Lang. Pathol.* **2017**, *19*, 17–29. [CrossRef]
40. Kotby, M.N.; El-Sady, S.R.; Basiouny, S.E.; Abou-Rass, Y.A.; Hegazi, M.A. Efficacy of the accent method of voice therapy. *J. Voice* **1991**, *5*, 316–320. [CrossRef]
41. Bassiouny, S. Efficacy of the Accent Method of Voice Therapy. *Folia Phoniatr. Logop.* **1998**, *50*, 146–164. [CrossRef]
42. Kotby, M.N.; Fex, B. The accent method: Behavior readjustment voice therapy. *Logoped. Phoniatr. Vocol.* **1998**, *23*, 39–43.
43. Verdolini-Marston, K.; Burke, M.K.; Lessac, A.; Glaze, L.; Caldwell, E. Preliminary study of two methods of treatment for laryngeal nodules. *J. Voice* **1995**, *9*, 74–85. [CrossRef] [PubMed]
44. Ramig, L.O.; Countryman, S.; Thompson, L.L.; Horii, Y. Comparison of two forms of intensive speech treatment for parkinson disease. *J. Speech Lang. Hear. Res.* **1995**, *38*, 1232–1251. [CrossRef]
45. Ramig, L.O.; Sapir, S.; Fox, C.; Countryman, S. Changes in vocal loudness following intensive voice treatment (LSVT) in individuals with Parkinson's disease: A comparison with untreated patients and normal age-matched controls. *Mov. Disord.* **2001**, *16*, 79–83. [CrossRef] [PubMed]
46. Yuan, F.; Guo, X.; Wei, X.; Xie, F.; Zheng, J.; Huang, Y.; Huang, Z.; Chang, Z.; Li, H.; Guo, Y.; et al. Lee Silverman Voice Treatment for dysarthria in patients with Parkinson's disease: A systematic review and meta-analysis. *Eur. J. Neurol.* **2020**, *27*, 1957–1970. [CrossRef] [PubMed]
47. Sapir, S.; Ramig, L.O.; Spielman, J.L.; Fox, C. Formant Centralization ratio: A proposal for a new acoustic measure of dysarthric speech. *J. Speech Lang. Hear. Res.* **2010**, *53*, 114–125. [CrossRef] [PubMed]
48. Saffarian, A.; Shavaki, Y.A.; Shahidi, G.A.; Hadavi, S.; Jafari, Z. Lee Silverman voice treatment (LSVT) mitigates voice difficulties in mild Parkinson's disease. *Med. J. Islam. Repub. Iran.* **2019**, *33*, 5. [CrossRef]
49. Constantinescu, G.; Theodoros, D.; Russell, T.; Ward, E.; Wilson, S.; Wootton, R. Treating disordered speech and voice in Parkinson's disease online: A randomized controlled non-inferiority trial. *Int. J. Lang. Commun. Disord.* **2011**, *46*, 1–16. [CrossRef]
50. Spielman, J.; Mahler, L.; Halpern, A.; Gilley, P.; Klepitskaya, O.; Ramig, L. Intensive voice treatment (LSVT®LOUD) for Parkinson's disease following deep brain stimulation of the subthalamic nucleus. *J. Commun. Disord.* **2011**, *44*, 688–700. [CrossRef]
51. Dumer, A.I.; Oster, H.; McCabe, D.; Rabin, L.A.; Spielman, J.L.; Ramig, L.O.; Borod, J.C. Effects of the lee silverman voice treatment (LSVT®LOUD) on hypomimia in Parkinson's disease. *J. Int. Neuropsychol. Soc.* **2014**, *20*, 302–312. [CrossRef]
52. Theodoros, D.G.; Hill, A.J.; Russell, T.G. Clinical and quality of life outcomes of speech treatment for Parkinson's disease delivered to the home via telerehabilitation: A noninferiority randomized controlled trial. *Am. J. Speech-Lang. Pathol.* **2016**, *25*, 214–232. [CrossRef] [PubMed]
53. El Sharkawi, A.; Ramig, L.; Logemann, J.A.; Pauloski, B.R.; Rademaker, A.W.; Smith, C.H.; Pawlas, A.; Baum, S.; Werner, C. Swallowing and voice effects of Lee Silverman Voice Treatment (LSVT®): A pilot study. *J. Neurol. Neurosurg. Psychiatry* **2002**, *72*, 31–36. [CrossRef] [PubMed]
54. Spielman, J.L.; Borod, J.C.; Ramig, L.O. The effects of intensive voice treatment on facial expressiveness in Parkinson disease: Preliminary data. *Cogn. Behav. Neurol.* **2003**, *16*, 177–188. [CrossRef] [PubMed]
55. Fox, C.M.; Ramig, L.O.; Ciucci, M.R.; Sapir, S.; McFarland, D.H.; Farley, B.G. The science and practice of LSVT/LOUD: Neural plasticity-principled approach to treating individuals with Parkinson disease and other neurological disorders. *Semin. Speech Lang.* **2006**, *27*, 283–299. [CrossRef] [PubMed]
56. Fox, C.M.; Boliek, C.A. Intensive voice treatment (LSVT LOUD) for children with spastic cerebral palsy and dysarthria. *J. Speech Lang. Hear. Res.* **2012**, *55*, 930–945. [CrossRef] [PubMed]

57. Langlois, C.; Tucker, B.V.; Sawatzky, A.N.; Reed, A.; Boliek, C.A. Effects of an intensive voice treatment on articulatory function and speech intelligibility in children with motor speech disorders: A phase one study. *J. Commun. Disord.* **2020**, *86*, 106003. [CrossRef]
58. Boliek, C.A.; Halpern, A.; Hernandez, K.; Fox, C.M.; Ramig, L. Intensive Voice Treatment (Lee Silverman Voice Treatment [LSVT LOUD]) for Children With Down Syndrome: Phase I Outcomes. *J. Speech Lang. Hear. Res.* **2022**, *65*, 1228–1262.
59. Reed, A.; Cummine, J.; Bakhtiari, R.; Fox, C.M.; Boliek, C.A. Changes in White Matter Integrity following Intensive Voice Treatment (LSVT LOUD®) in Children with Cerebral Palsy and Motor Speech Disorders. *Dev. Neurosci.* **2017**, *39*, 460–471. [CrossRef]
60. McDonnell, M.N.; Rischbieth, B.; Schammer, T.T.; Seaforth, C.; Shaw, A.J.; Phillips, A.C. Lee Silverman Voice Treatment (LSVT)-BIG to improve motor function in people with Parkinson's disease: A systematic review and meta-analysis. *Clin. Rehabil.* **2018**, *32*, 607–618. [CrossRef]
61. Tricco, A.C.; Lillie, E.; Zarin, W.; O'Brien, K.K.; Colquhoun, H.; Levac, D.; Moher, D.; Peters, M.D.; Horsley, T.; Weeks, L.; et al. PRISMA extension for scoping reviews (PRISMA-ScR): Checklist and explanation. *Ann. Intern. Med.* **2018**, *169*, 467–473. [CrossRef]
62. Pollock, D.; Peters, M.D.; Khalil, H.; McInerney, P.; Alexander, L.; Tricco, A.C.; Evans, C.; de Moraes, B.; Godfrey, C.M.; Pieper, D.; et al. Recommendations for the extraction, analysis, and presentation of results in scoping reviews. *JBI Évid. Synth.* **2023**, *21*, 520–532. [CrossRef] [PubMed]
63. Narayana, S.; Franklin, C.; Peterson, E.; Hunter, E.J.; Robin, D.A.; Halpern, A.; Spielman, J.; Fox, P.T.; Ramig, L.O. Immediate and long-term effects of speech treatment targets and intensive dosage on Parkinson's disease dysphonia and the speech motor network: Randomized controlled trial. *Hum. Brain Mapp.* **2022**, *43*, 2328–2347. [CrossRef] [PubMed]
64. Levy, E.S.; Ramig, L.O.; Camarata, S.M. The Effects of Two Speech Interventions on Speech Function in Pediatric Dysarthria. *J. Med. Speech-Lang. Pathol.* **2013**, *20*, 82–87. Available online: www.cengage.com/community/JMSLP (accessed on 22 August 2024).
65. Watts, C.R. Preliminary experimental evidence supports the need for further research into the effects of LSVT LOUD on voice and speech function in children with spastic cerebral palsy. *Evid.-Based Commun. Assess. Interv.* **2013**, *7*, 139–144. [CrossRef]
66. Fortin, A.J.; Hamel, A.; Asselin-Giguère, F.; Poulin, S.; McFarland, D.H. Report on the Impact of LSVT LOUD in Improving Communication of a Preschool Child and a Young Adult With Cerebral Palsy Rapport clinique de l'impact du protocole LSVT LOUD pour améliorer la communication d'un enfant d'âge préscolaire et d'un jeune adulte ayant une paralysie cérébrale. *Can. J. Speech-Lang. Pathol. Audiol.* **2023**, *47*, 125–140. Available online: www.cjslpa.ca (accessed on 22 August 2024).
67. Bakhtiari, R.; Cummine, J.; Reed, A.; Fox, C.M.; Chouinard, B.; Cribben, I.; Boliek, C.A. Changes in brain activity following intensive voice treatment in children with cerebral palsy. *Hum. Brain Mapp.* **2017**, *38*, 4413–4429. [CrossRef]
68. Levy, E.S. Implementing two treatment approaches to childhood dysarthria. *Int. J. Speech-Lang. Pathol.* **2014**, *16*, 344–354. [CrossRef]
69. Boliek, C.A.; Fox, C.M. Individual and environmental contributions to treatment outcomes following a neuroplasticity-principled speech treatment (LSVT LOUD) in children with dysarthria secondary to cerebral palsy: A case study review. *Int. J. Speech-Lang. Pathol.* **2014**, *16*, 372–385. [CrossRef]
70. Boliek, C.A.; Fox, C.M. Therapeutic effects of intensive voice treatment (LSVT LOUD®) for children with spastic cerebral palsy and dysarthria: A phase I treatment validation study. *Int. J. Speech-Lang. Pathol.* **2017**, *19*, 601–615. [CrossRef]
71. Arksey, H.; O'Malley, L. Scoping studies: Towards a methodological framework. *Int. J. Soc. Res. Methodol.* **2005**, *8*, 19–32. [CrossRef]
72. Levac, D.; Colquhoun, H.; O'Brien, K.K. Scoping studies: Advancing the methodology. *Implement. Sci.* **2010**, *5*, 69. [CrossRef] [PubMed]
73. Peters, M.D.J.; Marnie, C.; Tricco, A.C.; Pollock, D.; Munn, Z.; Alexander, L.; McInerney, P.; Godfrey, C.M.; Khalil, H. Updated methodological guidance for the conduct of scoping reviews. *JBI Évid. Synth.* **2020**, *18*, 2119–2126. [CrossRef] [PubMed]
74. López-Nieto, L.; Compañ-Gabucio, L.M.; Torres-Collado, L.; la Hera, M.G.-D. Scoping Review on Play-Based Interventions in Autism Spectrum Disorder. *Children* **2022**, *9*, 1355. [CrossRef]
75. Papadopoulos, A.; Prentza, A.; Voniati, L.; Tafiadis, D.; Trimmis, N.; Plotas, P. Assessing the Impact of Bilingualism on the Linguistic Skills of Children with Autism Spectrum Disorder (ASD) in Greece: A Scoping Review. *Medicina* **2024**, *60*, 894. Available online: https://www.mdpi.com/1648-9144/60/6/894 (accessed on 22 August 2024). [CrossRef] [PubMed]
76. Higgins, J.P. Cochrane Handbook for Systematic Reviews of Interventions. Version 5.1. 0 [updated March 2011]. The Cochrane Collaboration. 2011. Available online: www.cochrane-handbook.org (accessed on 22 August 2024).
77. Wilson, E.M.; Abbeduto, L.; Camarata, S.M.; Shriberg, L.D. Estimates of the prevalence of speech and motor speech disorders in adolescents with Down syndrome. *Clin. Linguist. Phon.* **2019**, *33*, 772–789. [CrossRef] [PubMed]
78. Gilhuber, C.S.; Raulston, T.J.; Galley, K. Language and communication skills in multilingual children on the autism spectrum: A systematic review. *Autism* **2023**, *27*, 1516–1531. [CrossRef] [PubMed]
79. Fox, C.; Ebersbach, G.; Ramig, L.; Sapir, S. LSVT LOUD and LSVT BIG: Behavioral treatment programs for speech and body movement in Parkinson disease. *Park. Dis.* **2012**, *2012*, 391946. [CrossRef] [PubMed]
80. Martel Sauvageau, V.; Roy, J.P.; Langlois, M.; Macoir, J. Impact of the LSVT on vowel articulation and coarticulation in Parkinson's disease. *Clin. Linguist. Phon.* **2015**, *29*, 424–440. [CrossRef]

81. Levy, E.S.; Leone, D.; Moya-Gale, G.; Hsu, S.C.; Chen, W.; Ramig, L.O. Vowel intelligibility in children with and without dysarthria: An exploratory study. *Commun. Disord. Q.* **2016**, *37*, 171–179. [CrossRef]
82. Boyd, E.V. The Effects of LSVT® LOUD and LSVT® CLEAR on Vowel Production in STN-DBS Subjects with Idiopathic Parkinson's Disease. Master's Thesis, University of Colorado at Boulder, Boulder, CO, USA, 2006.
83. Narayana, S.; Fox, P.T.; Zhang, W.; Franklin, C.; Robin, D.A.; Vogel, D.; Ramig, L.O. Neural correlates of efficacy of voice therapy in Parkinson's disease identified by performance–correlation analysis. *Hum. Brain Mapp.* **2010**, *31*, 222–236. [CrossRef]
84. Kleim, J.A.; Hogg, T.M.; VandenBerg, P.M.; Cooper, N.R.; Bruneau, R.; Remple, M. Cortical synaptogenesis and motor map reorganization occur during late, but not early, phase of motor skill learning. *J. Neurosci.* **2004**, *24*, 628–633. [CrossRef] [PubMed]
85. Schertz, M.; Gordon, A.M. Changing the model: A call for a re-examination of intervention approaches and translational research in children with developmental disabilities. *Dev. Med. Child Neurol.* **2009**, *51*, 6–7. [CrossRef] [PubMed]
86. Adkins, D.L.; Boychuk, J.; Remple, M.S.; Kleim, J.A. Motor training induces experience-specific patterns of plasticity across motor cortex and spinal cord. *J. Appl. Physiol.* **2006**, *101*, 1776–1782. [CrossRef] [PubMed]
87. Braza, M.D.; Sakash, A.; Natzke, P.; Hustad, K.C. Longitudinal change in speech rate and intelligibility between 5 and 7 years in children with cerebral palsy. *Am. J. Speech Lang. Pathol.* **2019**, *28*, 1139–1151. [CrossRef]
88. Mahler, L.A.; Jones, H.N. Intensive treatment of dysarthria in two adults with Down syndrome. *Dev. Neurorehabilit.* **2012**, *15*, 44–53. [CrossRef]

brain sciences

MDPI

Article

Multilingual Prediction of Cognitive Impairment with Large Language Models and Speech Analysis

Felix Agbavor and Hualou Liang *

School of Biomedical Engineering, Science and Health Systems, Drexel University, Philadelphia, PA 19104, USA; fa424@drexel.edu

* Correspondence: hualou.liang@drexel.edu

Abstract: Background: Cognitive impairment poses a significant global health challenge, emphasizing the critical need for early detection and intervention. Traditional diagnostics like neuroimaging and clinical evaluations are often subjective, costly, and inaccessible, especially in resource-poor settings. Previous research has focused on speech analysis primarily conducted using English data, leaving multilingual settings unexplored. Methods: In this study, we present our results from the INTERSPEECH 2024 TAUKADIAL Challenge, where we aimed to automatically detect mild cognitive impairment (MCI) and predict cognitive scores for English and Chinese speakers (169 in total). Our approach leverages Whisper, a speech foundation model, to extract language-agnostic speech embeddings. We then utilize ensemble models to incorporate task-specific information. Results: Our model achieved unweighted average recall of 81.83% in an MCI classification task, and root mean squared error of 1.196 in cognitive score prediction task, which placed the model at the second and the first position, respectively, in the ranking for each task. Comparison between language-agnostic and language-specific models reveals the importance of capturing language-specific nuances for accurate cognitive impairment prediction. Conclusions: This study demonstrates the effectiveness of language-specific ensemble modeling with Whisper embeddings in enabling scalable, non-invasive cognitive health assessments of Alzheimer's disease, achieving state-of-the-art results in multilingual settings.

Keywords: large language model; mild cognitive impairment; multilingual processing; speech analysis; whisper; dementia detection

1. Introduction

Cognitive impairment, encompassing conditions such as Alzheimer's disease (AD), poses a significant challenge affecting millions of individuals worldwide, without a cure, currently [1]. Early detection and intervention are critical for managing these conditions effectively, yet conventional diagnostic approaches largely rely on subjective assessments and expensive neuroimaging techniques [2].

Currently, the diagnosis of Alzheimer's disease and related dementias often involves a combination of clinical evaluation, cognitive score testing, such as the Mini Mental State Examination (MMSE) [3], and neuroimaging, etc. In addition to brain scans and cognitive tests, speech analysis has emerged as a promising tool for diagnosing AD. Changes in speech patterns, including disruptions in fluency, vocabulary, and syntactic complexity, have been observed in individuals with cognitive impairment, providing valuable insights into disease progression [4–7].

The utilization of speech in mild cognitive impairment (MCI) detection offers several advantages. It allows for easy and non-invasive sample collection, streamlining the assessment process. Additionally, the cost-effectiveness and ability to continuously monitor MCI progression make it a valuable tool for large-scale screening. The existing work on utilizing speech for AD detection has primarily been conducted in English [2,8–10]. However, the applicability of these findings to multilingual settings, such as English and Chinese, remains unclear.

In recent years, the importance of early detection of cognitive impairment has become increasingly recognized, with growing evidence suggesting that early intervention can delay the progression of conditions like Alzheimer's disease [11–13]. This has fueled a surge in research focused on developing cost-effective, scalable diagnostic tools that can be applied across different populations. As neurodegenerative diseases continue to affect aging populations globally, particularly in countries with limited access to advanced neuroimaging facilities, the need for innovative and accessible diagnostic methods is more urgent than ever [14,15]. Speech analysis offers an excellent opportunity to bridge the gap by providing clinicians with a non-invasive, efficient means for detecting cognitive decline at its early stages. Advances in Artificial Intelligence (AI), particularly large language models (LLMs), have further accelerated progress in this area. These AI-powered technologies allow the analysis of vast amounts of speech data, thus enabling researchers to uncover subtle patterns in language and communication that may be indicative of cognitive decline. Such innovations are paving the way for the deployment of speech-based screening tools in both clinical and community-based settings, enabling earlier identification and subsequent interventions for at-risk populations.

LLMs, such as Whisper [16], have become indispensable tools for speech analysis. Whisper, trained on a massive dataset of multilingual and multitask data, is particularly powerful as it can accurately transcribe speech in various languages and accents, even in noisy environments. Previous approaches [17,18] have used LLMs such as data2vec [19] and wav2vec2 [20] for acoustic analysis of speech data. Other studies [9,21,22] have utilized linguistic features from LLMs such as Bidirectional Encoder Representations from Transformers (BERT) [23] and Generative Pre-trained Transformer 3 (GPT-3) [24] for dementia detection. However, these studies are conducted using only English speech data. In the current study, we use data from the INTERSPEECH 2024 TAUKADIAL Challenge, which consists of speech samples and clinical information for speakers of Mandarin Chinese and English with different levels of cognitive impairment as well as individuals with normal cognition (NC) [25]. Given the global rise in dementia cases [26], the ability to conduct multilingual cognitive assessment has become increasingly important in healthcare.

In this work, we focus on spontaneous speech analysis for automatically detecting people with mild cognitive impairment and predicting a cognitive score on English and Chinese speakers. We explore two primary tasks:

- MCI Classification: This is a classification task, with a goal of distinguishing healthy control speech from mild cognitive impairment (MCI) speech.
- MMSE Score Prediction: This is a regression task, with a goal to infer the subject's Mini Mental Status Examination (MMSE) scores based on spontaneous speech data.

Our main contributions are as follows. First, we leverage Whisper, a speech foundation model, to extract language-agnostic speech embeddings. Second, we propose a novel ensemble model by incorporating task-specific information, with improvement over competition baseline results [25]. Third, we demonstrate the superiority of language-specific pipelines over language-agnostic methods. Lastly, we evaluate the effectiveness of within-language prediction for detecting MCI as compared to between-language prediction. The integration of majority voting in the ensemble model and task-specific information in our approach represents an innovative strategy to enhance model robustness and accuracy. Majority voting helps to mitigate the risks of overfitting and biases that could arise from individual tasks, ensuring that the final predictions are well-rounded and reflective of the broader cognitive patterns across different task types.

2. Materials and Methods

2.1. Dataset Description

The dataset used in this work was provided by the organizers of the INTERSPEECH 2024 TAUKADIAL Challenge [23], which consists of Chinese and English speech samples collected while the speakers participated in picture description tasks conducted as cognitive assessments in clinical settings. In the English language dataset, each participant was asked to describe three different pictures: Cookie Theft, Cat Rescue, and Coming and Going picture description tasks. In the Chinese language dataset, participants similarly were asked to provide a detailed description of three pictures depicting Taiwanese culture. The full dataset (English and Chinese) was balanced across age and gender groups to mitigate biases in modeling. It is important to note that the pictures used for descriptions in each language were different.

The training dataset comprises 387 voice recordings from 129 participants from MCI patients and normal cognition (NC) subjects. The test dataset consists of 120 recordings from 40 participants. Within the training set, 222 samples belong to the MCI group, while the remaining 165 samples are healthy control. The dataset is curated to ensure a balanced representation of age and sex among participants, thereby mitigating potential confounding factors and bias using a propensity score matching approach [25,26]. Table 1 shows the basic data description of the participants for age, sex and MMSE scores in the training dataset.

Table 1. Basic characteristics of participants for age, sex, and mean MMSE scores with their corresponding standard deviation (sd) in the training data.

Age	MCI			NC		
	M	F	MMSE (SD)	M	F	MMSE (SD)
[60, 65)	12	6	28.0 (1.3)	12	3	29.6 (0.8)
[65, 70)	3	42	26.9 (3.1)	9	57	29.2 (0.9)
[70, 75)	24	51	25.6 (3.9)	15	12	28.1 (1.4)
[75, 80)	30	15	25.4 (4.1)	18	9	29.7 (0.5)
[80, 85)	12	21	25.0 (3.3)	6	21	29.0 (1.0)
[85, 90)	6	0	22.5 (6.0)	3	0	27.0 (0.0)
Total	87	135	25.8 (3.7)	63	102	29.1 (1.1)

2.2. Extracting Linguistic and Acoustic Features from Speech

Acoustic feature extraction is a critical step in analyzing spontaneous speech for cognitive impairment prediction [27,28]. In this study, we employ the Whisper model to extract embeddings representing acoustic features from both Chinese and English datasets.

Whisper is a simple, yet effective, end-to-end approach implemented as an encoder–decoder model. It was trained on a massive dataset of multilingual and multitask supervised data collected from various sources on the web [16]. The extensive training data allows Whisper to effectively capture a wide range of speech patterns and nuances across different languages, resulting in robust acoustic feature extraction for predicting cognitive impairment. In this study, we utilize the *whisper-large-v3* model for acoustic feature extraction. This version of the Whisper model offers embeddings of 1280 dimensions, providing a rich representation of the input speech signals. Notably, the *whisper-large-v3* model represents the latest open-source iteration of the Whisper model available at the time of writing.

The encoder component of the Whisper model generates embeddings, which serve as representations of the acoustic features extracted from the input speech. These embeddings encapsulate relevant acoustic information, facilitating subsequent analysis and prediction tasks.

To represent linguistic features, the speech data underwent transcriptions, respectively, to convert English audio into English text and Chinese audio into Chinese text using Whisper. This step ensures that the linguistic nuances in both English and Chinese speech samples were accurately captured. Following transcriptions, distinct LLMs were applied to obtain text embeddings for each language. For English linguistic features, the *Voyage-large-v2 model* was utilized, which generated text embeddings capable of capturing linguistic information within the English speech data. Similarly, for Chinese linguistic feature representation, the FLAG embedding model, specifically the *FLAG-bge-large-zh-v1.5* variant tailored for Chinese language processing, was employed. We chose these models largely due to their top performance in the Massive Text Embedding Benchmark (MTEB) Leaderboard. Comparisons of linguistic features with other multilingual embedding models did not yield improved performance. Therefore, the results reported in this work are based only on acoustic features.

2.3. Model Pipelines

2.3.1. Language-Agnostic Modeling

We started with a language-agnostic model which serves as a baseline for our analysis. This model, depicted in Figure 1A, was designed to be independent of specific languages and aimed to capture general patterns present in the speech data. The baseline model involved extracting embeddings directly from the encoder of the Whisper. These embeddings were further normalized by using z-score normalization, which were then utilized as inputs for classifiers, such as neural networks (NNs). Note that this approach does not incorporate any language identification mechanisms, highlighting the language-agnostic nature of the pipeline.

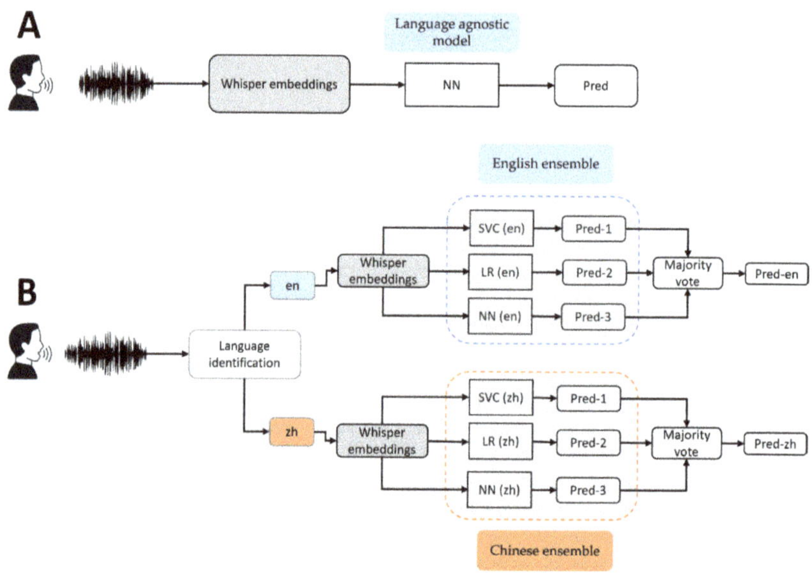

Figure 1. Diagram of language-agnostic (**A**) and language-specific ensemble (**B**) model pipelines.

2.3.2. Language-Specific Ensemble Modeling

For language-specific ensemble modeling, we introduced our approach, as depicted in Figure 1B. This method aimed to leverage the information encoded in three different picture description tasks using an ensemble model. The process began with identifying the language of the speech data, followed by feature extraction using the Whisper model. This step ensures that subsequent modeling processes were tailored to specific language.

For each language, we implemented an ensemble model using a modified stacking technique. This involved training three different machine learning models, each using two of the three picture-description tasks for training and the third task for testing. This setup resulted in three distinct predictions for each subject, based on different tasks. To determine the final prediction, we applied a majority voting strategy, where the most frequent prediction across the three models was selected. This approach leveraged complementary information from the different tasks, ensuring a more robust prediction. Finally, we manually checked the predictions to ensure consistency across the tasks for each subject before submission to the competition organizer.

2.4. Experiments

2.4.1. MCI Classification Task

The first task focused on MCI classification, aimed at detecting the presence or absence of MCI based on spontaneous speech recordings. We utilized three machine learning models for this task: a Support Vector Classifier (SVC), Logistic Regression (LR) and neural network (NN). We selected the SVC, LR, and NN, given their proven performance and widespread use in machine learning, as evidenced by our previous research [9,17]. Specifically, the SVC was chosen for its ability to handle high-dimensional data and capture nonlinear relationships through kernel functions, aligning with the complex Whisper embedding space. LR was selected for its simplicity, interpretability, and strong performance in binary classification, providing a baseline. The NN was included to model intricate nonlinear relationships and fully leverage the richness of the Whisper embeddings. By combining these models in an ensemble approach and applying majority voting, we improved the robustness and accuracy of our predictions.

For the classification task, we used two approaches: a language-agnostic model and language-specific model. For language agnostic modeling, we directly used the embeddings from Whisper as input into a neural network, regardless of language. This approach was used as a baseline to compare with the language-specific model. In a language-specific modeling approach, for each language identified and its Whisper embedding, we built an ensemble model where we trained three different machine learning models. Specifically, we trained LR, the SVC, and the NN independently, and combined their predictions using majority voting for a final prediction such that each subject, regardless of the picture description task, should have the same prediction. We additionally implemented the conventional stacking approach using the NN and SVC as the base estimators and LR as the final estimator to compare our task-specific ensemble to the conventional stacking method.

2.4.2. MMSE Score Prediction Task

This is a regression task aimed at predicting MMSE scores based on spontaneous speech recordings. We leveraged the framework developed for the MCI classification for this prediction. In this task, our objective is to predict MMSE cognitive scores using regression models such as Ridge Regression and a neural network. The MMSE scores normally range from 0 to 30, with scores of 26 or higher being considered normal [3]. A score of 20 to 24 suggests mild dementia, 13 to 20 suggests moderate dementia, and less than 12 indicates severe dementia. As such, the prediction is clipped to a range between 0 and 30.

The key difference from the MCI classification lies in the use of regression models tailored for predicting continuous MMSE scores, rather than classification models used for detecting the presence or absence of MCI. As such, instead of using majority voting in classification, we use the average of the predicted MMSE scores.

2.4.3. Between-Language vs. Within-Language Analysis

So far, our ensemble approach has focused on within-language analysis, wherein inferences are performed for tasks with the same language, either English or Chinese. Here, we seek to explore between-language analysis, where we aim to evaluate to what extent transfer learning between languages can be used for MCI detection. Briefly, we first train a machine learning model on the English (Chinese) dataset. Then, we make inferences for a picture description task on the Chinese (English) dataset. To permit direct comparison with the within-language analysis, we use two picture description tasks in the same language (e.g., English) to train the model, and test it with a task in different language (e.g., Chinese). This process allows us to assess the effectiveness of transferring knowledge and model performance across languages.

2.5. Performance Evaluation

For the MCI classification task, we assess model performance using the following metrics from the challenge: unweighted average recall (UAR, also called balanced accuracy), specificity, sensitivity, and F1 Score. For the MMSE prediction task, our evaluation metrics include Root Mean Square Error (RMSE) score and Pearson correlation coefficient (i.e., R-squared).

To determine optimal hyperparameters for our models, we adopt a methodology that involves dividing the training data into an 80/20 split for training and validation sets. This split ensures adequate data for both training and validation of the models. Additionally, we use a stratified approach based on both the detected language and diagnosis when dividing the data, ensuring a balanced distribution of classes in the training and validation sets. Such an approach facilitates the effective tuning of hyperparameters and the robust evaluation of model performance, hence enhancing the reliability and generalizability of our findings on the unseen test set.

For the SVC, we tune the regularization parameter and perform a search across various kernel functions (radial basis function, sigmoid function and linear function). The NN model is a feedforward neural network with one hidden layer of size 100. The Adam optimizer is used with a learning rate of 0.001.

3. Results

3.1. MCI Classification Task

The results of within-language analysis for the English dataset are presented in Table 2, showing the performance metrics of our ensemble approach obtained for inference on each picture description task after model trained on the other two tasks. The results indicate that, in general, majority voting performs better across all the metrics.

Table 2. Within-language results on English validation set for SVC, LR and NN. Inferences are made on picture description tasks 1, 2 and 3, respectively. Majority: majority voting. Bold indicates best performance for the metric.

	Inference	Balanced Accuracy	Specificity	Sensitivity	F1
SVC	1	0.761	0.619	0.902	0.860
	2	0.701	0.524	0.878	0.828
	3	0.663	0.619	0.707	0.744
	Majority	**0.797**	**0.667**	**0.927**	**0.883**
LR	1	0.749	0.619	0.878	0.847
	2	0.712	0.619	0.805	0.805
	3	0.722	**0.762**	0.683	0.757
	Majority	**0.761**	0.619	**0.902**	**0.860**
NN	1	0.796	0.714	0.878	0.867
	2	0.712	0.571	0.854	0.826
	3	0.735	0.714	0.756	0.795
	Majority	**0.808**	**0.714**	**0.902**	**0.881**

Table 3 presents the within-language results for the Chinese dataset, depicting the performance metrics obtained for inferences on each picture description task after the model was trained on the other two tasks. The findings indicate that, overall, majority voting utilizing task-specific information outperforms inferences made using individual picture description tasks.

Table 3. Within-language results on Chinese validation set for SVC, LR and NN. Inferences are made on picture description tasks 1, 2 and 3, respectively. Majority: majority voting. Bold indicates the best performance for the metric.

	Inference	Balanced Accuracy	Specificity	Sensitivity	F1
SVC	1	0.882	0.794	0.970	0.889
	2	0.896	**0.882**	0.910	0.896
	3	0.897	0.824	0.970	0.901
	Majority	**0.911**	0.853	**0.970**	**0.914**
LR	1	0.926	0.853	**1.000**	0.930
	2	0.836	0.853	0.818	0.831
	3	0.926	0.912	0.939	0.925
	Majority	**0.955**	**0.941**	0.970	**0.955**
NN	1	0.941	0.882	**1.000**	0.943
	2	0.850	0.882	0.818	0.844
	3	0.926	0.882	0.970	0.928
	Majority	**0.955**	**0.941**	0.970	**0.955**

In Figure 2, the receiver operating characteristic (ROC) curves for both the English and Chinese datasets are presented to demonstrate the model classification performance. The area under the ROC curve (AUC) values of 0.87 and 0.99 are achieved, respectively, for English and Chinese datasets. The near-perfect AUC for the Chinese dataset suggests that the model is particularly well-suited to the linguistic and acoustic characteristics of the Chinese language, achieving high accuracy in distinguishing cognitive impairment. These ROC curves provide a clear indication of the model's robust classification capabilities.

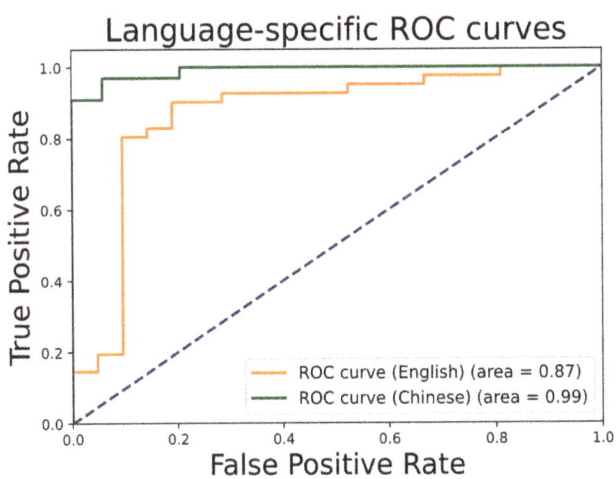

Figure 2. ROC curves of our language-specific model for Chinese and English datasets.

We compare the results of our task-specific approach to the conventional stacking method in Table 4 for both the English and Chinese dataset. It can be seen that for the English data, our task-specific approach achieves a higher balanced accuracy and specificity while having comparable sensitivity and F1 scores. For the Chinese dataset results, our task-specific approach significantly outperforms the conventional stacking method, showing a clear advantage over the conventional method.

Table 4. Comparison of our task-specific ensemble approach (Task-Specific) to the conventional stacking method (Stacking). Bold indicates the best performance for the metric.

Dataset	Method	Balanced Accuracy	Specificity	Sensitivity	F1
English	Task-Specific	**0.808**	**0.714**	0.902	0.881
	Stacking	0.787	0.615	**0.960**	**0.888**
Chinese	Task-Specific	**0.955**	**0.941**	**0.970**	**0.955**
	Stacking	0.829	0.809	0.850	0.829

Table 5 presents the between-language results obtained from training on the English dataset and inferring on different tasks of the Chinese dataset. The results indicate difficulties in transferring task information from English to Chinese. These results indicate that, despite utilizing task-specific information from the English dataset, the performance of the transfer learning approach in inferring on the Chinese dataset is subpar compared to the within-language results. Similar findings are observed for transfer learning from Chinese to English.

Table 5. Between-language results from training on English tasks and inference on Chinese picture description tasks 1, 2, and 3 using SVC, LR, and NN.

Model	Dataset	Balanced Accuracy	Specificity	Sensitivity	F1
SVC	1	0.491	0.588	0.394	0.433
	2	0.515	0.029	1.000	0.667
	3	0.506	0.588	0.424	0.459
LR	1	0.418	0.382	0.454	0.435
	2	0.431	0.529	0.333	0.367
	3	0.521	0.618	0.424	0.467
NN	1	0.529	0.088	0.970	0.667
	2	0.529	0.088	0.970	0.667
	3	0.445	0.647	0.242	0.302

In Table 6, we show the results of the MCI classification task on the unseen test set. Specifically, we compare the performance of the language-agnostic model with the language-specific ensemble modeling approach (Method 2). Results on the unseen test set reveal that the ensemble approach outperforms the language-agnostic model (Method 1) across all the metrics. The language-specific ensemble modeling approach has much better performance in accurately detecting the presence or absence of MCI, with a balanced accuracy of 0.818 compared to 0.612 for language-agnostic methods.

Table 6. Classification results on unseen test set for balanced accuracy, specificity, sensitivity and F1 score for language-agnostic and language-specific pipelines using ensemble approach.

Method	Balanced Accuracy	Specificity	Sensitivity	F1
Language-agnostic	0.612	0.620	0.612	0.566
Language-specific	0.818	0.684	0.952	0.851

3.2. MMSE Score Prediction Task

We present the results of the MMSE cognitive score prediction task in Table 7, where we compare the performance of the language-agnostic model with the language-specific ensemble model. The MMSE prediction results are reported in terms of RMSE scores (lower is better) and the R-squared method (higher is better). These results indicate that the language-specific ensemble modeling approach outperforms the language-agnostic model in terms of both RMSE scores and R-squared.

Table 7. RMSE and R-squared values of MMSE prediction task on unseen test set for language-agnostic and language-specific ensemble approaches.

Method	RMSE	R_Squared
Language-agnostic	2.582	0.496
Language-specific	1.196	0.920

3.3. Comparison with Other Competition Teams

We compare the performance of our approach with the other participants in the challenge. The competition rankings were based on UAR (i.e., balanced accuracy) for the MCI classification task and root mean squared error (RMSE) for the MMSE cognitive score prediction task. Figure 3 presents the competition results where only the top ten teams are provided for each task, with the baseline also shown. Refer to the full list of results here: (https://taukadial-luzs-69e3bf4b9878b99a6f03aea43776344580b77b9fe54725f4.gitlab.io/ranking.html, accessed on 29 September 2024). We can see from Figure 3 that our model performs remarkably well. We ranked second for MCI detection (Figure 3A), and first for MMSE prediction (Figure 3B).

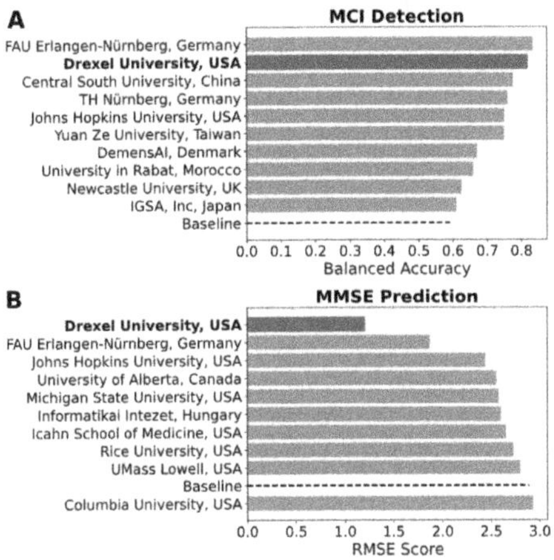

Figure 3. Rankings of the top ten teams and the baseline method in the challenge, assessed using balanced accuracy for the MCI classification task (**A**) where our model ranked second (dark gray), and the RMSE for the MMSE cognitive score prediction task (**B**) where our model ranked first (dark gray). Note that the baseline is represented by black dashed lines.

4. Discussion

In this paper, we present the findings of our participation in the INTERSPEECH 2024 TAUKADIAL Challenge, which focused on the automated detection of mild cognitive impairment and the prediction of cognitive scores for both English and Chinese speakers. Our approach harnesses the power of Whisper, a state-of-the-art speech foundation model, to extract language-independent speech embeddings. These embeddings are then integrated into ensemble models to capture task-specific information. Our model achieved an unweighted average recall of 81.83% and a root mean squared error of 1.196 on the MCI classification and cognitive score prediction tasks, respectively. This performance placed our model in the second and first positions in the overall rankings for the respective tasks. Our method relies only on acoustic features extracted from spontaneous speech recordings using Whisper encoder embeddings [14], demonstrating the potential in identifying cognitive impairment based purely on speech. These results highlight the growing potential of speech-based biomarkers in clinical applications, as they offer a non-invasive, scalable, and cost-effective solution for the early detection of cognitive decline.

We performed a comparative analysis of the language-agnostic and language-specific ensemble modeling techniques. Our findings showed a clear advantage of language-specific models in both the MCI classification and MMSE prediction tasks. These models are able to leverage subtle but important language-specific patterns, which play a crucial role in the accurate prediction of cognitive impairment. We additionally utilized task-specific information from the language-specific models by training models on two out of three picture description tasks and testing them on the remaining third task. This process was repeated across all possible combinations of the picture tasks, resulting in three predictions for each language. The final ensemble prediction was obtained through majority voting. Majority voting, a simple yet powerful ensemble technique, allowed us to aggregate three predictions and make a final decision for the same subject in the classification task. In doing so, we ensured that the final prediction was not biased toward any single task but rather reflected the consensus across all task-specific models.

We also conducted an analysis of cross-language predictions, where models trained on data in one language were evaluated on speech data from a different language. This analysis provided insights into the limitations of between-language prediction. We observed a significant reduction in performance when models were applied to different languages from that they were trained on. This drop in accuracy and MMSE prediction underscores the challenge of building generalized models that work across various languages without specific adaptations. This suggests that incorporating language-specific information, even across similar languages, can lead to better predictions than models trained on diverse language datasets without any language-specific adaptations. Our findings highlight the importance of tailoring models to specific linguistic and cultural contexts, especially when dealing with cognitive impairment, where subtle differences in speech can carry diagnostic significance.

MMSE prediction is an important task in clinical settings, as the MMSE is a widely used metric to assess cognitive function. Our results further validate the efficacy of language-specific ensemble modeling techniques for MMSE prediction tasks. In addition to yielding a lower and better RMSE score for the prediction, the use of these models resulted in a higher Pearson correlation coefficient, indicating their effectiveness in accurately predicting MMSE scores from acoustic features. The ability to accurately predict MMSE scores from speech recordings provides a non-invasive and accessible method for monitoring cognitive health, particularly in large-scale screening programs. Moreover, our results indicate that language-specific models are better suited to capturing the relationship between speech patterns and cognitive function. This is consistent with the idea that certain cognitive impairments can manifest differently across languages due to variations in grammar, phonetics, and syntax. By leveraging language-specific features, our ensemble models are able to effectively capture these variations, resulting in accurate MMSE predictions.

One strength of our approach is its reliance solely on acoustic features derived from spontaneous speech. By focusing on acoustic data, we avoid the need for transcribed or labeled text, which can be resource-intensive and time-consuming to obtain, particularly in a clinical setting. Additionally, this approach demonstrates the potential of using spontaneous speech recordings to detect cognitive impairment in a non-invasive and accessible way, which can have far-reaching applications in telemedicine and large-scale screening programs. However, there are some limitations to our study that must be acknowledged. First, while our language-specific models outperformed language-agnostic approaches, we only tested in English and Chinese. Hence, the generalizability of our findings to other languages is unclear. Future work should explore the inclusion of a more diverse set of languages to validate the effectiveness of language-specific modeling across a wider range of linguistic contexts. Second, our focus on acoustic features, while effective, may have missed important linguistic or semantic cues that could further enhance prediction accuracy. This motivates our future work, in which we aim to incorporate text embeddings, as performed in our previous work [9], in addition to using acoustic features to capture a more holistic view of speech and cognitive function. Third, one limitation of this study is the reliance on the MMSE as the primary assessment tool. While widely used, the MMSE is a relatively basic screening instrument with notable shortcomings [29,30], such as limited sensitivity to mild cognitive impairment, susceptibility to ceiling and floor effects, and biases related to education and culture. While our results demonstrate the promise of speech-based biomarkers as complementary tools, the limitations of the MMSE underscore the need for future research to integrate more sensitive and inclusive cognitive assessment instruments to ensure robust and equitable evaluations across diverse populations. Fourth, one of the significant challenges in working with LLMs is the inherent black-box nature of their embeddings. While these embeddings are incredibly powerful in capturing semantic and syntactic meaning, they often lack interpretability. This lack of transparency can hinder our ability to understand why specific predictions are made. As LLM technology continues to advance, it is essential to prioritize interpretability.

While our study demonstrates strong technical performance, it is essential to consider how these findings might translate into practical clinical utility. While our current approach leverages speech-based biomarkers to detect cognitive impairment, its integration into clinical settings would require additional validation with more comprehensive neuropsychological assessments. Future work should explore the application of our model alongside well-established cognitive tests, such as the Montreal Cognitive Assessment (MoCA), or other domain-specific cognitive tests, to evaluate its efficacy in detecting subtle cognitive changes that may not be captured by current screening tools like the MMSE. Furthermore, for our approach to be clinically applicable, we must address key questions regarding thresholds for accurate classification, sensitivity, and specificity in real-world scenarios. This includes the determination of optimal cut-off values for the classification of cognitive impairment across diverse populations, ensuring that the approach maintains an appropriate balance between sensitivity and specificity in all settings. Moreover, the integration of speech-based assessments into existing clinical workflows will require careful consideration of factors such as ease of use, clinician training, and the potential for automating the scoring process.

5. Conclusions

This paper presents our results from the INTERSPEECH 2024 TAUKADIAL Challenge, focused on automated detection of mild cognitive impairment (MCI) and cognitive score prediction for English and Chinese speakers. Using Whisper, a leading speech foundation model, we extracted language-independent speech embeddings and incorporated them into ensemble models. Our approach achieved an unweighted average recall of 81.83% for MCI classification (second place) and a root mean squared error of 1.196 for cognitive score prediction (first place). Our method demonstrates the potential of speech-based biomarkers for non-invasive, scalable, and cost-effective early detection of cognitive decline.

Looking ahead, several avenues for future research arise from this study. First, we plan to incorporate text embeddings alongside acoustic features to enhance the model's ability to capture both the acoustic and linguistic dimensions of speech. By integrating text-based features, we can explore how semantic information and language structure contribute to the detection of cognitive impairment. This multimodal approach could potentially lead to even more accurate predictions of both MCI classification and MMSE scores.

Second, a more detailed comparison between acoustic-based and linguistic-based approaches is warranted. This would provide further insights into how these two types of features interact and whether they offer complementary information that could be leveraged in ensemble modeling. A deeper understanding of these relationships could inform the design of more sophisticated models that make use of both acoustic and linguistic cues to predict cognitive decline.

Finally, we aim to investigate the impact of incorporating additional non-verbal features, such as facial expressions or gestures, which may carry important information about cognitive functions. Multimodal models that combine speech with other behavioral cues could provide a richer picture of cognitive health, offering new opportunities for non-invasive cognitive assessment in diverse populations.

Author Contributions: Conceptualization: F.A. and H.L.; Methodology: F.A. and H.L.; Formal analysis and investigation: F.A.; Writing—Original Draft, F.A. and H.L.; Writing—Review and Editing, F.A. and H.L.; Supervision: H.L. All authors have read and agreed to the published version of the manuscript.

Funding: This work was partly supported by NIH/NIA 5P30AG073105.

Institutional Review Board Statement: Not applicable.

Informed Consent Statement: Not applicable.

Data Availability Statement: All the data are available at https://dementia.talkbank.org, accessed on 29 September 2024.

Acknowledgments: We thank the TAUKADIAL Challenge data that was available via DementiaBank. Work reported here was run on hardware supported by Drexel's University Research Computing Facility.

Conflicts of Interest: The authors declare no conflicts of interest.

References

1. Yiannopoulou, K.G.; Papageorgiou, S.G. Current and future treatments in Alzheimer disease: An update. *J. Cent. Nerv. Syst. Dis.* **2020**, *12*, 1179573520907397. [CrossRef] [PubMed]
2. Luz, S.; Haider, F.; De La Fuente, S.; Fromm, D.; MacWhinney, B. Detecting cognitive decline using speech only: The ADReSSO challenge. *arXiv* **2021**, arXiv:2104.09356. [CrossRef]
3. Folstein, M.F.; Folstein, S.E.; McHugh, P.R. "Mini-mental state": A practical method for grading the cognitive state of patients for the clinician. *J. Psychiatr. Res.* **1975**, *12*, 189–198. [CrossRef]
4. Kong, W.; Jang, H.; Carenini, G.; Field, T.S. Exploring neural models for predicting dementia from language. *Comput. Speech Lang.* **2021**, *68*, 101181. [CrossRef]
5. Östberg, P.; Fernaeus, S.-E.; Hellström, Å.; Bogdanović, N.; Wahlund, L.-O. Impaired verb fluency: A sign of mild cognitive impairment. *Brain Lang.* **2005**, *95*, 273–279. [CrossRef]
6. McDonnell, M.; Dill, L.; Panos, S.; Amano, S.; Brown, W.; Giurgius, S.; Small, G.; Miller, K. Verbal fluency as a screening tool for mild cognitive impairment. *Int. Psychogeriatr.* **2020**, *32*, 1055–1062. [CrossRef] [PubMed]
7. Zhao, Q.; Guo, Q.; Hong, Z. Clustering and switching during a semantic verbal fluency test contribute to differential diagnosis of cognitive impairment. *Neurosci. Bull.* **2013**, *29*, 75–82. [CrossRef] [PubMed]
8. Luz, S.; Haider, F.; de la Fuente, S.; Fromm, D.; MacWhinney, B. Alzheimer's Dementia Recognition through Spontaneous Speech: The ADReSS Challenge. *arXiv* **2020**, arXiv:2004.06833. [CrossRef]
9. Agbavor, F.; Liang, H. Predicting Dementia from Spontaneous Speech using Large Language Models. *PLoS Digit. Health* **2022**, *1*, e0000168. [CrossRef] [PubMed]
10. Balagopalan, A.; Eyre, B.; Rudzicz, F.; Novikova, J. To BERT or Not To BERT: Comparing Speech and Language-based Approaches for Alzheimer's Disease Detection. *arXiv* **2020**, arXiv:2008.01551. [CrossRef]
11. Rasmussen, J.; Langerman, H. Alzheimer's Disease—Why We Need Early Diagnosis. *Degener. Neurol. Neuromuscul. Dis.* **2019**, *9*, 123–130. [CrossRef] [PubMed]
12. Leifer, B.P. Early Diagnosis of Alzheimer's Disease: Clinical and Economic Benefits. *J. Am. Geriatr. Soc.* **2003**, *51*, S281–S288. [CrossRef]
13. Porsteinsson, A.P.; Isaacson, R.S.; Knox, S.; Sabbagh, M.N.; Rubino, I. Diagnosis of Early Alzheimer's Disease: Clinical Practice in 2021. *J. Prev. Alzheimers Dis.* **2021**, *8*, 371–386. [CrossRef]
14. Better, M.A. Alzheimer's disease facts and figures. *Alzheimers Dement.* **2023**, *19*, 1598–1695.
15. Alzheimer's Disease International; Guerchet, M.; Prince, M.; Prina, M. Numbers of People with Dementia Worldwide: An Update to the Estimates in the World Alzheimer Report 2015. November 2020. Available online: https://www.alzint.org/resource/numbers-of-people-with-dementia-worldwide/ (accessed on 10 December 2024).
16. Radford, A.; Kim, J.W.; Xu, T.; Brockman, G.; McLeavey, C.; Sutskever, I. Robust Speech Recognition via Large-Scale Weak Supervision. *arXiv* **2022**, arXiv:2212.04356. [CrossRef]
17. Agbavor, F.; Liang, H. Artificial Intelligence-Enabled End-To-End Detection and Assessment of Alzheimer's Disease Using Voice. *Brain Sci.* **2022**, *13*, 28. [CrossRef] [PubMed]
18. Balagopalan, A.; Novikova, J. Comparing Acoustic-based Approaches for Alzheimer's Disease Detection. *arXiv* **2021**, arXiv:2106.01555. [CrossRef]
19. Baevski, A.; Hsu, W.-N.; Xu, Q.; Babu, A.; Gu, J.; Auli, M. data2vec: A General Framework for Self-supervised Learning in Speech, Vision and Language. *arXiv* **2022**, arXiv:2202.03555. [CrossRef]

20. Baevski, A.; Zhou, Y.; Mohamed, A.; Auli, M. wav2vec 2.0: A Framework for Self-Supervised Learning of Speech Representations. In *Advances in Neural Information Processing Systems*; Curran Associates, Inc.: Red Hook, NY, USA, 2020; pp. 12449–12460. Available online: https://proceedings.neurips.cc/paper/2020/hash/92d1e1eb1cd6f9fba3227870bb6d7f07-Abstract.html (accessed on 14 July 2022).
21. Orimaye, S.O.; Wong, J.S.; Golden, K.J.; Wong, C.P.; Soyiri, I.N. Predicting probable Alzheimer's disease using linguistic deficits and biomarkers. *BMC Bioinform.* **2017**, *18*, 34. [CrossRef]
22. Fraser, K.C.; Meltzer, J.A.; Rudzicz, F. Linguistic Features Identify Alzheimer's Disease in Narrative Speech. *J. Alzheimers Dis.* **2016**, *49*, 407–422. [CrossRef]
23. Devlin, J.; Chang, M.-W.; Lee, K.; Toutanova, K. BERT: Pre-training of Deep Bidirectional Transformers for Language Understanding. *arXiv* **2019**, arXiv:1810.04805. [CrossRef]
24. Brown, T.; Mann, B.; Ryder, N.; Subbiah, M.; Kaplan, J.D.; Dhariwal, P.; Neelakantan, A.; Shyam, P.; Sastry, G.; Askell, A.; et al. Language Models are Few-Shot Learners. In *Advances in Neural Information Processing Systems*; Curran Associates, Inc.: Red Hook, NY, USA, 2020; pp. 1877–1901. Available online: https://proceedings.neurips.cc/paper/2020/hash/1457c0d6bfcb4967418bfb8ac142f64a-Abstract.html (accessed on 14 July 2022).
25. Luz, S.; Garcia, S.D.L.F.; Haider, F.; Fromm, D.; MacWhinney, B.; Lanzi, A.; Chang, Y.-N.; Chou, C.-J.; Liu, Y.-C. Connected Speech-Based Cognitive Assessment in Chinese and English. *arXiv* **2024**, arXiv:2406.10272. [CrossRef]
26. Hebert, L.E.; Beckett, L.A.; Scherr, P.A.; Evans, D.A. Annual Incidence of Alzheimer Disease in the United States Projected to the Years 2000 Through 2050. *Alzheimer Dis. Assoc. Disord.* **2001**, *15*, 169–173. [CrossRef] [PubMed]
27. Al-Hameed, S.; Benaissa, M.; Christensen, H. Detecting and Predicting Alzheimer's Disease Severity in Longitudinal Acoustic Data. In Proceedings of the 4th International Conference on Bioinformatics Research and Applications (ICBRA '17), Barcelona, Spain, 8–10 December 2017; Association for Computing Machinery: New York, NY, USA, 2017; pp. 57–61. [CrossRef]
28. Pan, Y.; Mirheidari, B.; Tu, Z.; O'Malley, R.; Walker, T.; Venneri, A.; Reuber, M.; Blackburn, D.; Christensen, H. Acoustic feature extraction with interpretable deep neural network for neurodegenerative related disorder classification. In Proceedings of the Interspeech 2020, Shanghai, China, 25–29 October 2020. Available online: https://eprints.whiterose.ac.uk/178305/ (accessed on 2 March 2024).
29. EDevenney, E.; Hodges, J.R. The Mini-Mental State Examination: Pitfalls and limitations. *Pract. Neurol.* **2017**, *17*, 79–80. [CrossRef]
30. Franco-Marina, F.; García-González, J.J.; Wagner-Echeagaray, F.; Gallo, J.; Ugalde, O.; Sánchez-García, S.; Espinel-Bermúdez, C.; Juárez-Cedillo, T.; Rodríguez, M.Á.V.; García-Peña, C. The Mini-mental State Examination revisited: Ceiling and floor effects after score adjustment for educational level in an aging Mexican population. *Int. Psychogeriatr.* **2010**, *22*, 72–81. [CrossRef] [PubMed]

brain sciences

MDPI

Article

Exploring Narrative Ability in Greek-Speaking Children with High-Functioning ASD: Associations with Memory and Attention

Vasiliki Zarokanellou [1], Alexandros Gryparis [1,*] and Katerina Papanikolaou [2]

[1] Department of Speech and Language Therapy, University of Ioannina, 45500 Ioannina, Greece; vzarokanellou@uoi.gr

[2] Department of Child Psychiatry, Agia Sophia Children's Hospital, Medical School, National and Kapodistrian University of Athens, 11527 Athina, Greece; katpapan@med.uoa.gr

* Correspondence: alexandros@uoi.gr

Abstract: Background/Objectives: Narration is a sensitive tool for the assessment of language in children with high-functioning autism spectrum disorder (HF-ASD) since mild language deficits beyond the sentential level are not always noticeable through the administration of standardized language tests targeting the lexical or sentential level. This study investigated the narrative ability of monolingual Greek-speaking HF-ASD children in comparison to that of their typically developing (TD) peers and explored the associations between narrative variables, ADHD symptomatology, and memory skills in the participants on the autistic spectrum. **Methods**: The participants were 39 children aged 7 to 12 years, 19 with HF-ASD and 20 age-matched, vocabulary-matched, and cognitively matched TD peers. **Results**: The two groups were similar in most microstructural and macrostructural variables but differed significantly in syntactic complexity ($p = 0.024$; $d = 0.754$) and subordination ($p < 0.001$; $d = -1.576$) indices, implying that the HF-ASD group presented syntactic delay in comparison to their TD peers. The HF-ASD participants showed significantly higher heterogeneity in the amount of information generated for the story's main character ($p = 0.004$; $d = -0.093$) in comparison to their TD peers. Significant associations were observed between verbal and visual memory, complex syntactic structures, and Theory of Mind-related internal state terms. ADHD symptomatology was negatively correlated with the generation of simple and coordinated clauses. Finally, complex syntax and delayed vSTM were correlated with retelling total scores, indicating that language ability and verbal memory compensate for narrative competence in HF-ASD children. **Conclusions**: The findings highlight the impact that language skills, memory ability, and ADHD symptomatology have on narrative competence in children with HF-ASD, as well as the importance of narrative use for assessing the language skills in populations with mild language impairment.

Keywords: narratives; narrative competence; microstructure; macrostructure; memory; ADHD

1. Introduction

Autism spectrum disorder (ASD) is a complex neurodevelopmental disorder that is associated with persistent impairment in social interaction and social communication across various environments, as well as restricted and repetitive patterns of behaviors, interests, and activities [1]. According to the Centers for Disease Control and Prevention [2],

https://doi.org/10.3390/brainsci15010073

most children on the Autism spectrum present normal (41.7%) or borderline (23.1%) intellectual ability and are characterized as children with high-functioning ASD (HF-ASD). Even though these children have a normal range of intelligence, they commonly present significant language deficits. Findings from several studies propose high comorbidity between HF-ASD and Developmental Language Disorder (DLD) [3–5]. Children with ASD and normal intelligence form two distinct subgroups, one with language skills that fall in the normal range and one with concurrent DLD [5]. Relevant research suggests that children with HF-ASD present an uneven language profile with more profound language deficits in pragmatics, specifically in narrative skills, while language core abilities (syntax and semantics) can be relatively intact [6,7].

1.1. Narrative Skills in Children with HF-ASD

Narrative is described as a common type of discourse, involving the relation of a sequence of events [8]. The ability to narrate is significant for sharing experiences and communicating memories leading to the development of social and emotional relationships [9]. Narrative skills also play a major role in academic performance and lay the foundation for written production [9]. Narration is a complex task that incorporates language, cognitive, and social skills, starting to develop early, even before the age of 3, in typically developing (TD) children [10]. Research regarding narrative ability in children with HF-ASD reports inconsistent findings, which may be due to the wide age range of HF-ASD participants and the divergent matching criteria with the control groups used, as well as the different methodologies administered for the elicitation of narrations [8,9]. A meta-analysis including 24 studies reported significant mean differences in most microstructural and macrostructural variables measured in children with HF-ASD in comparison to their TD peers, but these findings presented moderate to high heterogeneity [9]. Regarding microstructure, several studies did not find significant differences in story length produced by children with HF-ASD and control groups [11–16], but some reported the opposite [7,9,17,18]. Studies examining the syntactic complexity of generated narratives suggest that the language ability of HF-ASD children significantly affects grammatical performance in narrative tasks [9,18]. Peristeri et al. [16] reported that HF-ASD children with language ability in the lower end of the normal range produced simpler and syntactically less complex sentences than their TD peers, while HF-ASD children with language ability in the upper end of the normal range performed syntactically similarly to the control group. In line with the above, several studies did not disclose significant differences in the mean length of utterances or other syntactic complexity measures between children with HF-ASD and their TD age-matched and/or language-matched peers [7,12–15,19], even though some did [17,18,20]. In contrast to the findings about microstructure, most studies referring to a macrostructure analysis (overall content of story and hierarchical sequencing of events) report significant difficulties in children with HF-ASD [9,11,17,18,21] in comparison to the TD language-matched and age-matched controls. These difficulties include a lower number of features described, fewer causal relations between events, more incoherent narratives containing more irrelevant comments, fewer inferences from the story events, and a less frequent reference to the cognitive states and emotions of the characters of the story [7,9,17].

1.2. Links Between Narrative Performance and Executive Functions in HF-ASD Children

"Executive functions" (EFs) include a broad range of purposeful upper cognitive processing skills serving for goal-directed behavior, abstract reasoning, decision-making, and social regulation [22,23]. A recent model of EFs divides them into a. cold (i.e., memory, inhibition, attention, cognitive flexibility, and monitoring), serving to manipulate abstract concepts, letters, and numbers, and b. hot, participating in emotional regulation, empathy, and the Theory of Mind (ToM) [24]. Deficits in EFs have a significant impact on generated narratives, and difficulties in narration in children with HF-ASD may be explained with reference to their cognitive differences [9]. HF-ASD children present important deficits in planning, cognitive flexibility, and attention, which play an instrumental role in the encoding of new information into episodes, the hierarchical organization of the episodes in a story, and the understanding of causal relationships between the narrated events [9]. Moreover, narration requires an intact working memory system since the speaker must remember what they narrated while planning the next utterance. Relevant research reveals that the narration of future episodes attests to supplementary working memory resources compared to the narration of past events from episodic memory, making the narration of future episodes a more difficult task [25]. Tsimpli et al. [26] suggest that higher performance in working memory (WM) is associated with better narrative competence in TD children. Along the same line, research in children with ASD shows that lower WM capacity is associated with lower syntactic complexity and less coherent and cohesive narrations [27–29]. Additionally, children with HF-ASD present a local over a global style of information processing, missing the gist of a story [12]. This deficit in central coherence makes the integration of new information into meaningful coherent events and the use of context to infer meaning more difficult, accounting for more superficial narratives with less causal relationships between events [7–9]. Finally, HF-ASD children have difficulties in empathizing; in understanding complex feelings such as jealousy, grief, regret, etc.; and in attributing mental states in themselves and others, and they often show a limited range of emotions [7]. These abilities are essential for the successful generation of a story because the speaker must "transmit" the feelings, thoughts, and conflicts of the main character of the story and not just describe the events. Moreover, the narrator has to take into consideration the reaction of the audience toward the story to adapt the narration accordingly [7,9]. Previous studies provide contradictory results with some proposing that children with HF-ASD produce fewer internal mental state terms (ISTs) [7], emotions [11], and causal explanations about the behavior of the main character of the story [11] than controls, while others did not report a difference between the two groups [10].

1.3. Purpose of This Study

The current study aims to investigate narrative competence in a group of monolingual Greek-speaking children with HF-ASD and to compare their narrative ability to that of a group of age-matched, vocabulary-matched, and cognitively matched TD peers. Furthermore, this study investigates the relationships between the microstructural and macrostructural indices on narrative tasks, ADHD symptomatology, and memory skills since previous research suggests that executive functions, specifically attention and memory skills, significantly affect narration in children with HF-ASD [27–29]. The research questions that the present study poses are the following:

1. Does the narrative microstructure (productivity and grammar) differ significantly in children with HF-ASD compared to their TD peers?
2. Does the narrative macrostructure differ significantly in children with HF-ASD compared to their TD peers?
3. Does ISL produced on narratives differ significantly in children with HF-ASD compared to their TD peers?

2. Materials and Methods

2.1. Participants

The participants included nineteen children with ASD and twenty typically developing (TD) peers, aged 7 to 12 years old. All the participants were monolingual Greek speakers and attended general elementary schools, each with a performance and verbal IQ over 70, as measured by Raven's Educational [30]. The ASD children were recruited from a special ASD clinic in Athens and had taken a formal diagnosis of ASD without accompanying intellectual impairment (Level 1) according to the criteria of the Diagnostic and Statistical Manual of Mental Disorders (DSM)-5 [1] by an experienced child-psychiatrist specialized in ASD. The TD participants were recruited from different general elementary schools in two regions of Greece. They had free medical and educational history records and were typically developing according to their teacher and parent reports. The TD participants had scores below 11 on the Greek version of the Social Communication Questionnaire [31] and of 93% on the Greek Assessment Scale for ADHD-IV [32], as completed by their primary caregivers. The two groups were similar in terms of age, gender, IQ, naming ability, and verbal memory skills but presented significant differences in SCQ and ADHD total scores as expected (Table 1).

Table 1. Characteristics of participants.

	Group			
	TD (*n*1 = 20)	**ASD (*n*2 = 19)**		
	Mean (SD)	**Mean (SD)**	*p*	*Cohen's d*
Age (months)	118.2 (17.7)	118.0 (17.6)	0.946	0.014
Gender, N (%)				
Boys	15 (75%)	15 (78.9%)	0.770	
Girls	5 (25%)	4 (21.1%)		
Total score SCQ	5.2 (2.5)	13.0 (7.4)	0.002	−1.439
Total score ADHD	8.4 (5.1)	22.3 (12.5)	<0.001	−1.468
EOWPVT-R (RS)	69.9 (9.6)	68.4 (8.9)	0.864	0.160
Raven's Educational				
Performance IQ	108.8 (15.8)	109.2 (14.5)	0.764	−0.030
Verbal IQ	96.5 (12.2)	89.0 (12.4)	0.864	0.615
Memory Assessment Battery				
Verbal short-term memory (RS)	27.3 (4.6)	28.4 (6.0)	0.919	−0.201
Delayed verbal short-term memory (RS)	6.1 (1.1)	5.8 (1.9)	0.076	0.169
Immediate visual memory (RS)	31.6 (4.4)	30.5 (5.6)	0.091	0.215
Delayed visual memory (RS)	6.4 (0.9)	5.7 (2.0)	0.012	0.473

2.2. Materials and Measurements

2.2.1. Screening Tools

1. *Raven's Educational* [30]. Raven's Educational is a screening norm-referenced instrument of general cognitive ability for monolingual Greek children aged 4;0 to 11;11 years old. It includes two separate scales, the Raven's Colored Progressive Matrices (CPM), which assess performance intelligence (PIQ), and the Crichton Vocabulary Scales (CVS), which assess the production of definitions (verbal IQ). In the current study, it was employed as a criterion assessment instrument to ensure that all the participants had performance and verbal IQs over 70.
2. *Greek Assessment Scale for Attention Deficit Hyperactivity Disorder-IV* [32]. It is a screening standardized questionnaire of attention deficit hyperactivity disorder (ADHD) for children and adolescents aged 5 to 19 years old. The questionnaire has two forms, one for caregivers and one for the teacher of the class. In the current study, the caregiver's form was used to exclude participants with possible ADHD in the control group and to assess ADHD symptomatology in the ASD group. For the parents'

form, a cut-off of 93% is recommended as the optimum point for possible ADHD with 74% sensitivity, 70% specificity, 72% positive predictive value, and 72% negative predictive value.

3. *The Greek version of the Social Communication Questionnaire (SCQ, Lifetime Form)* [31]. It is a screening instrument for ASD symptomatology, which is easily completed by the primary caregivers of the child in less than 10 min. It is appropriate for individuals over the age of 4 years old with cognitive functioning above the age of 2 and it investigates the full developmental history of each examinee. In this study, it was used to exclude participants with possible ASD in the control group and to assess ASD severity in the ASD group. An optimum cut-off of 15 points is recommended by the authors as positive for autism, while a score above 11 is recommended for HF-ASD in a Greek sample of 7–10-year-old children with high sensitivity and specificity (autism vs. TD children: 96.3% sensitivity and 98.7% specificity; HF-ASD vs. TD group: 90.6% sensitivity and 89.6% specificity).

2.2.2. Naming Skills

The Greek version of the Expressive One-Word Picture Vocabulary Test-Revised (EOWPVT-R) [33]. It is a naming test of 100 words presented with colorful line drawings, evaluating expressive vocabulary in children aged 2 to 12 years old. In this task, the child must name the picture that the examiner shows. The target words comprise mainly high-frequency and low-frequency nouns and a few verbs, (e.g.,"watch" and "sew"). In this study, a Greek-translated and adapted version was used since in Greek there is no standardized naming test for children over the age of eight. The administration of the test always started from the beginning of the tool regardless of the age of the participant and terminated when the participant made six consecutive mistakes. The children were awarded one point for each correct answer, while no points were given for incorrect answers, leading to a possible maximum raw score of 100 points. The raw scores were used for between-group comparisons.

2.2.3. Memory Skills

The Screening Scale of Visual and Verbal Memory from the Memory Assessment Battery for preschool and school-aged children [34] was implemented. It is a standardized instrument, which evaluates verbal and visual short-term and delayed short-term memory skills for children aged 5 to 12 years old.

For the assessment of verbal memory, participants have to recall a list of five or seven unrelated words depending on their chronological age once they hear them from the examiner. The words are produced by the examiner at a steady speaking rate of one word per second. Each examinee has five trials to succeed in this goal. Every correctly recalled word gives one point, and once the examinee has recalled all the words accurately in one attempt, they get five points for all remaining trials. Afterward, independently of the number of correctly recalled words from each examinee, the examiner asks the participant once again to recall the list of words, giving a distinctive cue for each of them, e.g., for the word trousers, the examiner says the following to the child: "We wear it". This is done to ensure that participants have adequately encoded the information. After a break of 5 to 10 min, the examiner asks the examinee to reproduce the same list of words once again.

For the assessment of visual memory, participants have to recall the positions of five or seven monochrome tokens, depending on their chronological age, once they have seen the positions they are placed in by the examiner on a white table board with squares. Similar to the verbal memory task, each examinee has five trials to succeed in this goal. Every correctly recalled position of a token gets one point, and once the examinee has recalled all the positions of the tokens correctly in one attempt, they get five points for all remaining trials. After a short interval of 5 to 10 min, where another verbal task can be administered, the examiner asks the examinee to reproduce the same positions of the tokens once again.

2.2.4. Narrative Ability

The participants' oral retellings were elicited by using the two short stories of the "Storytelling Scale" from the Memory Assessment Battery for preschool and school-aged children [34]. The stories were presented orally and without visual support for retelling, according to the manual of the test. These stories were chosen as they are culturally and socially neutral. Furthermore, their themes are carefully selected to be familiar to the children, triggering their interest, and they are also compatible with the age of the participants. For each story, to assess the microstructure, the following linguistic indicators were measured as proposed by the relevant literature [9,16]:

(1) Total number of words: The total number of words produced in the two retelling tasks was compared between the children with HF-ASD and their TD peers.
(2) Lexical diversity: The lexical diversity index is calculated as the number of different content words divided by the total number of content words. The lexical diversity indices were compared between the two study groups.
(3) Syntactic complexity: The syntactic complexity index is measured as the number of complex sentences, which include subordinate and coordinate clauses, divided by the total number of simple and complex clauses.
(4) Subordination index: It is calculated as the number of subordinate clauses divided by the total number of complex sentences
(5) Types of subordination: According to Peristeri et al.'s [16] study, the types of subordination were investigated, including the total number of verb-complement clauses and modifiers, such as adverbial and relative clauses. In complement clauses, the type of complementizers (subjective na vs. indicative oti/pos and factive pu vs. non-factive oti/pos) were further examined. In Greek, the complementizer na is the least complex and the first to be acquired by native Greek-speaking children [16].

For the assessment of the macrostructure, the following information units were evaluated, as the relevant literature proposes [9,16]: a. the information given about the place and the time that the story occurred, b. the information given about the main characters of the story, c. the information given about the event content of the story, d. the information given about the conflict and the attempts to resolve the main problem, and e. the information given about the final result and the conclusions of the story. Each component of macro narrative skills was assessed with 0, 1, 2, or 3 points, giving a total possible score of 15 points for each narrative task. Moreover, the following were calculated: the total number of +ToM-related ISTs (emotion and mental verb ISTs) and the total number of −ToM-related ISTs, including the number of perceptual, physiological, and linguistic action verbs [7,16].

Finally, the transcription of each narrative was made by this study's first author, an SLP with more than 15 years of clinical experience. An independent rater with a background in linguistics analyzed all the transcripts for grammatical microstructure variables and ±ToM ISTs. Subsequently, the first author of this study re-evaluated all the transcripts for microstructure and macrostructure variables. Both raters were blind to the status of the child that produced each narrative. Disagreements between the two raters were resolved through discussion. Spearman's correlation coefficients between the first and the second rater were excellent. Specifically, Spearman's correlation coefficients for attested grammatical microstructure variables exceeded 0.95 ($p < 0.001$), while for ±ToM ISTs, they exceeded 0.97 (+ToM ISTs: $rho = 0.971$ and $p < 0.001$; −ToM ISTs: $rho = 0.984$ and $p < 0.001$).

2.3. Procedure

Before the beginning of this study, approval was received by the Ethics Committee of the Greek Institution of Educational Policy (Φ15/176437/204757/Δ1) and the Ethics Committee of the Medical School. According to the research protocol, all the caregivers provided a written consent form and completed the Greek version of the SCQ, Lifetime Form [31], as well as the Greek Assessment Scale for Attention Deficit Hyperactivity Disorder-IV [31], before the individual assessment of each child. For the cognitive evaluation of the participants, the Greek edition of Raven's Educational [30] was administered, followed by the Screening Scale of Visual and Verbal Memory [34]. Lastly, the EOWPVT-R [33] was given followed by the two narrative tasks. The tests were administered in the same order to all the participants. Each participant was evaluated individually in a quiet room free of distraction. The participants with ASD were assessed in the clinic, from where they were recruited, while their TD peers were assessed in either their home environment or in their school setting. All the answers of the participants were audio-recorded and then transferred to their corresponding answer sheets. Overall, the procedure followed for the assessment of the participants is pictured in the following diagram (Figure 1):

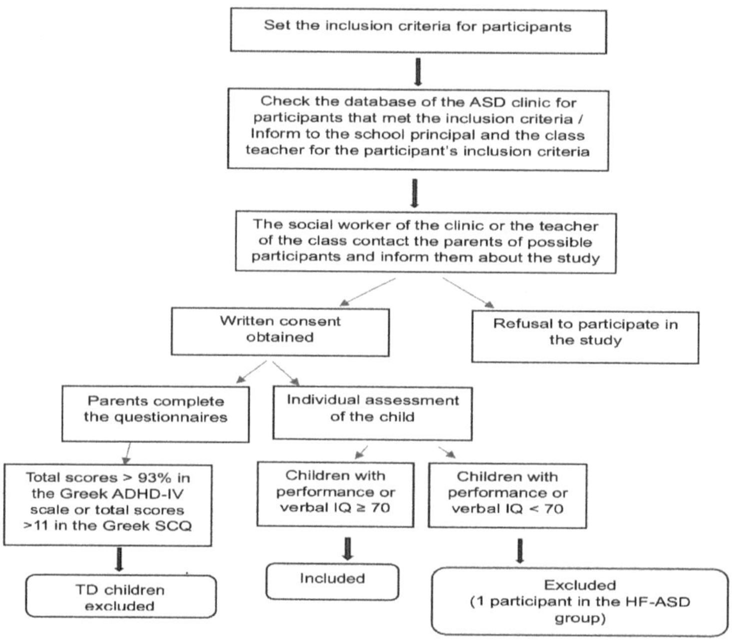

Figure 1. Procedure flow diagram.

2.4. Statistical Analysis

The quantitative variables were expressed as the mean and standard deviation. The qualitative variables were expressed as absolute and relative (%) frequencies. For the comparison of proportions, the chi-square test was used. Student's *t*-test was used for the comparison of the continuous variables between two groups. Spearman's correlation coefficient was used to investigate the linear relationship between the micro- and macrostructure indices and cognitive variables in the ASD group. The effect sizes were calculated using Cohen's d coefficient. Effect sizes up to 0.2 are considered small, between 0.2 and 0.5 are moderate, sizes between 0.5 and 0.8 are considered large, and sizes over 0.8 are considered very large [35]. Finally, the inter-rater reliability of the narrative measures between two independent raters was calculated using Spearman's correlation coefficient. All the reported p-values are two-tailed. Statistical significance was set at $p < 0.05$ and the analyses were conducted using IBM SPSS v29 [36].

3. Results

3.1. Microstructure Measures

The two groups were similar in terms of vocabulary productivity since they performed alike in the lexical diversity index ($p = 0.079$; $d = -0.002$) and in the total number of words ($p = 0.976$; $d = -0.0383$) they generated during the narration of the two stories (Table 2). Also, the two groups used an equivalent number of coordinate ($p = 0.563$; $d = 0.504$), subordinate ($p = 0.554$; $d = -0.050$), verb-complement ($p = 0.887$; $d = 0.064$), and modifier clauses ($p = 0.775$; $d = 0.291$) in their narrations (Table 2). However, the TD peers produced significantly more simple clauses than the ASD participants ($p < 0.001$; $d = 2.205$) and they used more coordinate clauses in their retellings than the ASD children ($p = 0.563$; $d = 0.504$) even though this difference did not reach statistical significance. Finally, the TD participants showed higher syntactic complexity ($p = 0.024$; $d = 0.754$) and subordination ($p < 0.001$; $d = -1.576$) indices in relationship to their ASD peers, indicating that the ASD group presented syntactic delay in comparison to the TD group. The above findings are presented in the following table (Table 2).

Table 2. Group means (and SDs) of total scores for microstructure measures.

Total Scores of Microstructural Measures	TD Group ($n1 = 20$)	ASD Group ($n2 = 19$)	p-Value	Effect Size *Cohen's d*
Total number of words	139.5 (31.6)	151.6 (31.9)	0.976	−0.383
Lexical diversity (%)	0.8 (0.04)	0.8 (0.05)	0.079	−0.002
Simple clauses	12.6 (3.1)	5.7 (3.1)	<0.001	2.205
Coordinate clauses	7.1 (2.7)	5.8 (2.2)	0.563	0.504
Subordinate clauses	8.7 (4.1)	8.9 (3.6)	0.554	−0.050
Syntactic complexity (%)	0.7 (0.2)	0.6 (0.1)	0.024	0.754
Subordination index (%)	0.6 (0.1)	0.4 (0.1)	<0.001	−1.576
Na complementizer	3.4 (1.7)	3.6 (1.8)	0.578	−0.102
Oti/pu complementizers	0.9 (1.0)	0.5 (0.7)	0.110	0.364
Verb-complement clauses	4.3 (2.2)	4.2 (2.3)	0.887	0.064
Modifier clauses	4.9 (2.0)	4.3 (2.0)	0.775	0.291

3.2. Macrostructure Measure

The two groups performed similarly in all the macrostructure measures except the information describing the story's main character ($p = 0.004$; $d = -0.093$), in which the participants with ASD presented significantly higher heterogeneity in the amount of information generated in comparison to their TD peers (TD: mean = 4.8 and $SD = 0.6$ vs. ASD: mean = 4.8 and $SD = 1.3$). The results are presented in Table 3.

Table 3. Group means (and SDs) of total scores for macrostructure measures.

Total Raw Scores of Macrostructural Measures	TD Group ($n1 = 20$)	ASD Group ($n2 = 19$)	p-Value	Effect Size *Cohen's d*
Place/Time	2.5 (1.7)	2.5 (1.7)	0.874	−0.014
Character/s	4.8 (0.6)	4.8 (1.3)	0.004	−0.093
Event Structure	4.3 (2.0)	3.2 (2.3)	0.230	0.487
Conflict–Solution	3.8 (1.2)	3.6 (1.2)	0.663	0.145
Conclusion	4.8 (1.1)	4.4 (1.5)	0.111	0.249
Total Scores	20.0 (4.2)	18.6 (5.1)	0.271	0.304
Internal State Language				
+ToM ISTs	7.0 (3.1)	6.6 (2.9)	0.438	0.140
−ToM ISTs	6.8 (2.7)	7.9 (2.8)	0.752	−0.400

3.3. Correlations Between Microstructure and Macrostructure Measures in the ASD Group

To investigate the extent to which the core language abilities (syntax and vocabulary) impact retelling skills in the ASD group, we examined the correlation between the microstructural and macrostructural measures. The statistical analysis revealed significant correlations between the coordinate clauses and story total scores ($rho = 0.5$; $p < 0.05$), subordinate clauses and story total scores ($rho = 0.5$; $p \leq 0.01$), and complex clauses and story total scores ($rho = 0.6$; $p \leq 0.01$). Finally, the statistical analysis showed significant correlations between the +ToM ISTs and subordinate clauses ($rho = 0.4$; $p < 0.05$), +ToM ISTs and complex clauses ($rho = 0.4$; $p \leq 0.01$), −ToM ISTs and subordinate clauses ($rho = 0.6$; $p \leq 0.01$), and −ToM ISTs and complex clauses ($rho = 0.5$; $p \leq 0.01$). The results are presented in Table 4.

Table 4. Spearman's correlation coefficients between the microstructural and macrostructural variables in the ASD group.

Macrostructure Variables	Microstructure Variables						
	Syntactic Complexity	Simple Clauses	Coordinate Clauses	Subordinate Clauses	Complex Clauses	Verb-Complement	Modifier
+ToM ISTs	0.1	0.2	0.3	0.4 *	0.4 **	0.1	0.2
−ToM ISTs	−0.3	−0.0	0.3	0.6 **	0.5 **	−0.1	−0.1
Story total scores	0.2	0.0	0.5 **	0.5 **	0.6 **	0.1	−0.0

* $p < 0.05$; ** $p \leq 0.01$.

3.4. Correlations Between Language and Executive Function Measures in the ASD Group

To begin with, ADHD symptomatology was associated significantly and negatively with the generation of simple ($rho = -0.3$; $p < 0.05$) and coordinate ($rho = -0.3$; $p < 0.05$) clauses. As it concerns the syntactic complexity index, it was not related to any executive function measure. However, the generation of complex clauses was associated with immediate verbal ($rho = 0.4$; $p < 0.05$) and visual ($rho = 0.4$; $p < 0.05$) memory skills, as well as delayed verbal ($rho = 0.4$; $p \leq 0.01$) and visual ($rho = 0.5$; $p \leq 0.01$) memory skills. Similarly, subordinate clauses presented significant associations with all the memory variables. Finally, modifier clauses were associated with all the memory tasks, except delayed visual memory, while verb-complement clauses were related only to vSTM ability ($rho = 0.3$; $p < 0.05$). Regarding the macrostructure language measures, the retelling total scores were associated only with delayed vSTM ($rho = 0.3$; $p < 0.05$), while the production of emotion words and mental state verbs (+ToM ISTs) was related to all the memory tasks. Lastly, the generation of perceptual, physiological, and linguistic action verbs (−ToM ISTs) was associated only with vSTM ability ($rho = 0.3$; $p < 0.05$). The correlational analysis is presented in Table 5.

Table 5. Spearman's correlations between cognitive, microstructural, and macrostructural variables in the ASD group.

Variables	RS ADHD	vSTM	Delayed vSTM	Immediate Visual Memory	Delayed Visual Memory
Microstructure					
Syntactic complexity	−0.2	0.0	0.0	0.1	−0.0
Simple clauses	−0.3 *	−0.0	0.0	0.2	0.3
Complex clauses	−0.3	0.4 *	0.4 **	0.4 *	0.5 **
Coordinate clauses	−0.3 *	0.0	0.1	0.2	0.4 *
Subordinate clauses	−0.1	0.4 *	0.3 *	0.4 **	0.4 **
Verb-complement	0.2	0.3 *	0.3	0.3	0.1
Modifier	−0.0	0.3 *	0.4 **	0.4 **	0.3
Macrostructure					
Retelling total scores	−0.1	0.3	0.3 *	0.2	0.1
+ToM IST	0.1	0.4 **	0.6 **	0.5 **	0.6 **
−ToM IST	0.1	0.3 *	0.2	0.2	0.2

* $p < 0.05$; ** $p \leq 0.01$.

4. Discussion

The present study a. compared the syntactic and story structure complexity of narratives of monolingual Greek-speaking children with HF-ASD and neurotypical controls matched in age, gender, and cognitive abilities (IQ scores and memory skills), expanding on findings by previous research in Greek [15,16,19], and b. investigated the associations between the linguistic indicators of narratives, ADHD symptomatology, and memory abilities in a Greek-speaking sample. Previous research in ASD children supports that narrative competence correlates with language ability [9,16]. Additionally, the existing literature suggests that executive deficits, specifically deficits in attention and memory ability, significantly impact the narrative competence of children on the autistic spectrum [26–28].

Regarding the first question of this study, if children with HF-ASD differ significantly in comparison to their TD peers in narrative microstructure (productivity and grammar), the analysis revealed that the HF-ASD participants had similar performance to their TD peers in most microstructure variables. Analytically, the ASD group did not differ from the control group in story length and lexical diversity, reinforcing similar outcomes from other research in English and Greek [11–16].

Furthermore, the ASD group performed similarly in the production of coordinate and subordinate clauses compared to the control group but presented a significantly lower subordination index in comparison to their TD peers, a finding attributed to the use of more complex clauses by the control group. Moreover, the TD group had higher performance in coordinate sentences than the HF-ASD group and differed also significantly in the syntactic complexity index in relationship to the control group, supporting previous findings in Greek that claim that ASD-HL children produce less complex morpho-syntactic structures than their TD peers [15,20]. Additionally, Peristeri et al. [16] propose that the frequency of the use of coordinate clauses is related to language ability in ASD. Thus, the findings of our analysis align with the above suggestions, indicating that our HF-ASD group presented syntactic delay compared to their neurotypical peers [15,16,20]. A more detailed analysis of the complementizers used by the two groups of participants revealed that the ASD group did not differ from the control group in any variable. These findings contradict the results of Peristeri et al.'s [16] study, in which the ASD-HL group produced significantly fewer modifier clauses compared to the control group. This difference may be attributed to the different narration tasks used in the two studies. In the Edmonton Narrative Norms Instrument (ENNI) [37] used by Peristeri et al. [16], the child first hears a story and then must generate it by looking at its pictures, while in the present study, a retelling story task without pictures was used. Moreover, in Peristeri et al.'s [16] study, modifiers were significantly correlated with overall story structure complexity, a finding also not observed in this study. Previous research has proved that story materials influence the semantic quality and syntactic structures of the narratives generated in TD children [25] and children with ASD [11,38].

Regarding the performance of the two groups on macrostructure variables, the two groups performed alike, except in the amount of information generated about the characters of the story, where the HF-ASD group showed significantly greater variability in the number of features (family relationships and/or causal explanations of the behavior) described but not in the number of terms used to describe the main character's cognitive states and emotions. The above aligns with previous results [15,38,39] and corroborates the Losh and Capps [11] findings, in which ASD participants provided fewer causal explanations about the behavior of the main character of the story than controls. Finally, there was a significant association between the story's total scores and coordinate, subordinate, and complex clauses, revealing that language skills correlate with narrative competence, a result enhancing the outcomes from other studies in Greek [16].

Lastly, as to the third question of this study, the qualitative analysis of the narratives showed that both groups produced an equivalent number of ±ToM ISTs. These results contradict the findings of Peristeri et al. [16], in which the ASD-HL group produced significantly fewer −ToM ISTs, but are in line with the findings of Baldimitsi et al. [15], according to which inferior performance was not detected between the monolingual and bilingual HF-ASD children and their controls in the number of generated ±ToM ISTs. On the contrary, in the Baldimitsi et al. [15] study, the monolingual HF-ASD children generated significantly more −ToM ISTs compared to their monolingual controls. Finally, all the ±ToM ISTs were significantly correlated with subordinate and complex clauses since the use of these terms requires the generation of specific complementizers in Greek, such as oti/pu and na [16].

In addition to the above, significant associations between both verbal and visual memory skills and microstructure and macrostructure variables in the ASD group were observed. Also, a significant negative association between simple and coordinate clauses, and total scores of ADHD symptomatology, was affirmed. Retelling a story requires that the narrator can regenerate the conceptual representation of the story [39]. The production of the macrostructure of a story relies on global processing skills (central coherence and ToM abilities), which play a pivotal role in the understanding of the causal relationships between the different events and in the interpretation of the thoughts and feelings of characters [9,39], as well as on cognitive mechanisms, such as attention and memory [28]. An adequate attention span and good short-term memory skills are required for encoding the new information into episodes, while an intact working memory system is needed to actively hold what the speaker narrated while planning the next utterance [28]. In the current study, significant associations were established between delayed vSTM and retelling total scores, as well as between +ToM ISTs and vSTM and delayed vSTM, a result that supports the bilateral relationship between retelling story ability and memory skills, and ToM skills and working memory abilities. Moreover, there was a significant association between +ToM ISTs and immediate and delayed visual memory, but a similar association between −ToM ISTs and visual memory was not revealed. The above may be accredited to the different psycholinguistic features of the two categories of ±ToM ISTs. Possibly, the +ToM ISTs (emotions and mental verbs) are words presenting higher imageability than −ToM ISTs, and for their acquisition, individuals may activate and rely more on visual memory.

As concerns the associations between microstructure variables and memory skills, previous studies on TD children and children with ASD show a robust link between verbal working memory skills and syntactic capacity [29]. Specifically, working memory seems to get involved more heavily in the processing of complex sentences and not in simple sentences that contain fewer syntactic operations [29]. According to the regeneration of syntax hypothesis, the immediate recall of a sentence is not based on a surface representation of a string of words in memory but is reproduced by a conceptual representation stored in the long-term memory involving recently activated words and a normal mechanism of sentence production, thus enabling the verb of the sentence to determine the syntactic structure of the to-be-recalled sentence [40,41]. In the current study, the statistical analysis revealed significant associations between complex and subordinate clauses and immediate and delayed verbal and visual STM, giving evidence to previous findings, which claim that there is a strong association between verbal memory capacity and syntax in children with ASD [29]. Moreover, the modifier clauses were significantly correlated with immediate and delayed vSTM, as well as immediate visual memory, while verb-complement clauses were only associated with vSTM. A possible explanation for this disparity may relate to the information that carries each type of clause. Adverbial clauses provide temporal or causal information that modifies the event of the main clause [16] and for that reason may correlate also with visuospatial memory. Finally, ADHD symptomatology was negatively associated with simple and coordinate clauses. Another study has found significant associations between inattention and pragmatic skills in children with ASD and our findings are in line with these results [28].

5. Conclusions

Overall, the results of the present study highlight the important role of attention and working memory ability in the regeneration of story retellings in children with HF-ASD. The scarce research that has investigated the link between working memory skills and syntax has found a strong association between working memory capacity and the processing of complex syntactic structures and the current study affirms this link. Also, the results of this study confirm previous findings in Greek attesting to a similar pattern of macrostructure and microstructure narrative performance for children with HF-ASD to that of their neurotypical peers [15,16]. A limitation of the current study is the small sample of participants, which may have limited the statistical power of our findings. As a result, caution should be exercised when interpreting the findings, and further studies with larger sample sizes are warranted to confirm these observations. Another limitation is the non-implementation of a standardized narration task that would permit the direct comparison of the narrative performance of HF-ASD children to the mean performance of a larger sample of their neurotypical peers. Moreover, the language level of the participants was not evaluated with a standardized language tool such as CELF-4 [42], nor were the two study groups matched on standardized language scores. This would have permitted a more detailed examination of the relationship between language ability and narrative competence. In Greek, however, there is not a norm-referenced language test for children older than seven years old, nor is there a standardized assessment tool for narration. Even though this fact has limited the exploration of the associations between language skills and narrative ability, it also underscores the importance of using narratives as a sensitive tool for the assessment of language in children with HF-ASD since mild language deficits beyond the sentential level may not always be notable through the administration of standardized language tests targeting the sentential or lexical level. Future studies should examine to a greater extent the association of narrative indices to executive functions such as working memory and attention in HF-ASD children.

Author Contributions: Conceptualization, V.Z.; methodology, V.Z. and K.P.; formal analysis, A.G.; investigation, V.Z. and K.P.; resources, V.Z. and K.P.; data curation, V.Z. and A.G.; writing—original draft preparation, V.Z.; writing—review and editing, V.Z. and A.G.; visualization, V.Z.; supervision, K.P.; project administration, V.Z. All authors have read and agreed to the published version of the manuscript.

Funding: This research received no external funding.

Institutional Review Board Statement: This study was conducted in accordance with the Declaration of Helsinki and approved by the Ethics Committee of Medical School and the Greek Institution of Educational Policy (protocol code Φ15/176437/204757/Δ1; date of approval 27 November 2018).

Informed Consent Statement: Informed consent was obtained from the caregivers of all the participants involved in this study.

Data Availability Statement: The data that support the findings of this study are available on request from the corresponding author. The data are not publicly available due to privacy or ethical restrictions.

Acknowledgments: The first author of this study wants to express her gratitude to her beloved professor, Angelique Kotsopoulos, who perished suddenly and to whom belonged the idea of the current study. Grateful thanks also to all the children and families who deserve recognition for agreeing to participate in this study.

Conflicts of Interest: The authors declare no conflicts of interest.

References

1. American Psychiatric Association. *Diagnostic and Statistical Manual of Mental Disorders: DSM-5*, 5th ed.; American Psychiatric Association: Arlington, VA, USA, 2013; ISBN 089042554X.
2. Maenner, M.J.; Shaw, K.A.; Bakian, A.V.; Bilder, D.A.; Durkin, M.S.; Esler, A.; Furnier, S.M.; Hallas, L.; Hall-Lande, J.; Hudson, A.; et al. Prevalence and Characteristics of Autism Spectrum Disorder Among Children Aged 8 Years—Autism and Developmental Disabilities Monitoring Network, 11 Sites, United States, 2018. *MMWR Surveill. Summ.* **2021**, *70*, 1–16. [CrossRef]
3. Bishop, D.V. Overlaps between autism and language impairment: Phenomimicry or shared etiology? *Behav. Genet.* **2010**, *40*, 618–629. [CrossRef]
4. Félix, J.; Santos, M.E.; Benitez-Burraco, A. Specific Language Impairment, Autism Spectrum Disorders and Social (Pragmatic) Communication Disorders: Is There Overlap in Language Deficits? A Review. *Rev. J. Autism Dev. Disord.* **2024**, *11*, 86–106. [CrossRef]
5. Georgiou, N.; Spanoudis, G. Developmental Language Disorder and Autism: Commonalities and Differences on Language. *Brain Sci.* **2021**, *11*, 589. [CrossRef] [PubMed]
6. Manolitsi, M.; Botting, N. Language abilities in children with autism and language impairment: Using narrative as a additional source of clinical information. *Child Lang. Teach. Ther.* **2011**, *27*, 39–55. [CrossRef]
7. Rumpf, A.L.; Kamp-Becker, I.; Becker, K.; Kauschke, C. Narrative competence and internal state language of children with Asperger Syndrome and ADHD. *Res. Dev. Disabil.* **2012**, *33*, 1395–1407. [CrossRef]
8. Stirling, L.; Douglas, S.; Leekam, S.; Carey, L. The use of narrative in studying communication in autism spectrum disorders: A review of methodologies and findings. In *Communication in Autism. Trends in Language Acquisition Research*; Arciuli, J., Brock, J., Eds.; Joh Benjamins Publishing Company: Amsterdam, The Netherlands, 2014; pp. 169–216.
9. Baixauli, I.; Colomer, C.; Roselló, B.; Miranda, A. Narratives of children with high-functioning autism spectrum disorder: A meta-analysis. *Res. Dev. Disabil.* **2016**, *59*, 234–254. [CrossRef]
10. Norbury, C.F.; Bishop, D.V. Narrative skills of children with communication impairments. *Int. J. Lang. Commun. Disord.* **2003**, *38*, 287–313. [CrossRef]
11. Losh, M.; Capps, L. Narrative ability in high-functioning children with autism or Asperger's syndrome. *J. Autism Dev. Disord.* **2003**, *33*, 239–251. [CrossRef] [PubMed]
12. Diehl, J.J.; Bennetto, L.; Young, E.C. Story recall and narrative coherence of high-functioning children with autism spectrum disorders. *J. Abnorm. Child Psychol.* **2006**, *34*, 87–102. [CrossRef]
13. Mäkinen, L.; Loukusa, S.; Leinonen, E.; Moilanen, I.; Ebeling, H.; Kunnari, S. Characteristics of narrative language in autism spectrum disorder: Evidence from the Finnish. *Res. Autism Spectr. Disord.* **2014**, *8*, 987–996. [CrossRef]
14. Banney, R.M.; Harper-Hill, K.; Arnott, W.L. The Autism Diagnostic Observation Schedule and narrative assessment: Evidence for specific narrative impairments in autism spectrum disorders. *Int. J. Speech Lang. Pathol.* **2015**, *17*, 159–171. [CrossRef] [PubMed]
15. Baldimtsi, E.; Peristeri, E.; Tsimpli, I.M.; Nicolopoulou, A. Bilingual children with High functioning autism spectrum disorder: Evidence from Oral narratives and Non-verbal executive function tasks. In Proceedings of the 40th Annual Boston University Conference on Language Development; Scott, J., Waughtal, D., Eds.; Cascadilla Press: Somerville, MA, USA, 2016; pp. 18–31.
16. Peristeri, E.; Andreou, M.; Tsimpli, I.M. Syntactic and Story Structure Complexity in the Narratives of High- and Low-Language Ability Children with Autism Spectrum Disorder. *Front. Psychol.* **2017**, *8*, 2027. [CrossRef]
17. Norbury, C.F.; Gemmell, T.; Paul, R. Pragmatics abilities in narrative production: A cross-disorder comparison. *J. Child Lang.* **2014**, *41*, 485–510. [CrossRef]
18. King, D.; Dockrell, J.; Stuart, M. Constructing fictional stories: A study of story narratives by children with autistic spectrum disorder. *Res. Dev. Disabil.* **2014**, *35*, 2438–2449. [CrossRef] [PubMed]
19. Terzi, A.; Marinis, T.; Zafeiri, A.; Francis, K. Subject and Object Pronouns in High-Functioning Children with ASD of a Null-Subject Language. *Front. Psychol.* **2019**, *10*, 1301. [CrossRef] [PubMed]
20. Marinis, T.; Terzi, A.; Kotsopoulou, A.; Francis, K. Pragmatic abilities of high-functioning Greek-speaking children with autism. *Psychol. J. Hell. Psychol. Soc.* **2020**, *20*, 321–337. [CrossRef]
21. Carlsson, E.; Åsberg Johnels, J.; Gillberg, C.; Miniscalco, C. Narrative Skills in Primary School Children with Autism in Relation to Language and Nonverbal Temporal Sequencing. *J. Psycholinguist. Res.* **2020**, *49*, 475–489. [CrossRef]
22. Wallace, G.L.; Yerys, B.E.; Peng, C.; Dlugi, E.; Anthony, L.G.; Kenworthy, L. Assessment and treatment of executive function impairments in Autism Spectrum Disorder: An update. *Int. Rev. Res. Dev. Disabil.* **2016**, *51*, 85–122. [CrossRef]

23. Demetriou, E.A.; Lampit, A.; Quintana, D.S.; Naismith, S.L.; Song, Y.J.C.; Pye, J.E.; Hickie, I.; Guastella, A.J. Autism spectrum disorders: A meta-analysis of executive function. *Mol. Psychiatry* **2018**, *23*, 1198–1204. [CrossRef] [PubMed]
24. Nemeth, D.G.; Chustz, K.M. Understanding "hot and cold" executive functions in children and adolescents. In *Evaluation and Treatment of Neuropsychologically Compromised Children*; Nemeth, D.G., Glozman, J., Eds.; Elsevier Academic Press: Amsterdam, The Netherlands, 2020; pp. 121–130. [CrossRef]
25. Marini, A.; Ferretti, F.; Chiera, A.; Magni, R.; Adornetti, I.; Nicchiarelli, S.; Vicari, S.; Valeri, G. Episodic future thinking and narrative discourse generation in children with Autism Spectrum Disorders. *J. Neurolinguist.* **2019**, *49*, 178–188. [CrossRef]
26. Tsimpli, I.; Papakonstantinou, M.; Nicolopoulou, A. Linguistic Structure and Narrative Quality in Child L1 Production. *Sel. Pap. Theor. Appl. Linguist.* **2017**, *20*, 387–404. [CrossRef]
27. Kuijper, S.J.; Hartman, C.A.; Bogaerds-Hazenberg, S.T.; Hendriks, P. Narrative production in children with autism spectrum disorder (ASD) and children with attention-deficit/hyperactivity disorder (ADHD): Similarities and differences. *J. Abnorm. Psychol.* **2017**, *126*, 63–75. [CrossRef] [PubMed]
28. Baixauli-Fortea, I.; Miranda Casas, A.; Berenguer-Forner, C.; Colomer-Diago, C.; Roselló-Miranda, B. Pragmatic competence of children with autism spectrum disorder. Impact of theory of mind, verbal working memory, ADHD symptoms, and structural language. *Appl. Neuropsychol. Child* **2019**, *8*, 101–112. [CrossRef]
29. Delage, H.; Eigsti, I.M.; Stanford, E.; Durrleman, S. A Preliminary Examination of the Impact of Working Memory Training on Syntax and Processing Speed in Children with ASD. *J. Autism Dev. Disord.* **2022**, *52*, 4233–4251. [CrossRef]
30. Sideridis, G.; Adoniou, F.; Mouzaki, A.; Simos, P. *Raven's Educational CPM/CVS*; Motivo Publishing: Athens, Greece, 2015.
31. Zarokanellou, V.; Kolaitis, G.; Vlassopoulos, M.; Papanikolaou, C. Brief report: A pilot study of the validity and reliability of the Greek version of the Social Communication Questionnaire. *Res. Autism. Spect. Disord.* **2017**, *38*, 1–5. [CrossRef]
32. Kalatzi-Azizi, A.; Ageli, K.; Efstathiou, G. *The Greek Assessment Scale of Attention Deficit Hyperactivity Disorder-IV. Parent's and Teacher's Form*; Pedio: Athens, Greece, 2012.
33. Kotsopoulos, A. (Child and Adolescent Psychiatric Unit of Aetoloakarnania, Messolonghi, Greece). Greek Version of the Expressive One Word Picture Vocabulary Test-Revised. 2003; Unpublished work.
34. Bezevegis, I.; Ikonomou, A.; Milonas, K. *The Memory Assessment Battery for Preschool and School-Aged Children*; Ministry of National Education and Religious Affairs: Athens, Greece, 2007.
35. Cohen, J. *Statistical Power Analysis for the Behavioral Sciences*, 2nd ed.; Lawrence Erlbaum Associates, Publishers: Hillsdale, NJ, USA, 1998; pp. 19–74.
36. IBM Corp. *IBM SPSS Statistics for Windows, Version 29.0*; IBM Corp: Armonk, NY, USA, 2023.
37. Schneider, P.; Hayward, D.; Dubé, R.V. *Edmonton Narrative Norms Instrument (ENNI) [Database Record]*; APA PsycTests: Washington, DC, USA, 2006. [CrossRef]
38. Losh, M.; Gordon, P.C. Quantifying narrative ability in autism spectrum disorder: A computational linguistic analysis of narrative coherence. *J. Autism Dev. Disord.* **2014**, *44*, 3016–3025. [CrossRef] [PubMed]
39. Kimhi, Y.; Kadosh, L.; Tubul-Lavy, G. Oral story retelling after reading or listening: Children with ASD versus typical development. A pilot study. *Prev. Sch. Fail.* **2022**, *68*, 50–59. [CrossRef]
40. Potter, M.C.; Lombardi, L. Regeneration in the short-term recall of sentences. *J. Mem. Lang.* **1990**, *29*, 633–654. [CrossRef]
41. Lombardi, L.; Potter, M.C. The regeneration of syntax in short term memory. *J. Mem. Lang.* **1992**, *31*, 713–733. [CrossRef]
42. Semel, E.; Wiig, E.H.; Compaan, E.; Secord, W.A. *CELF-4: Clinical Evaluation of Language Fundamentals*; The Psychological Corporation/A Harcourt Assessment Company: Toronto, ON, Canada, 2003.

brain sciences

MDPI

Systematic Review

Computer- and Smart-Tablet-Based Self-Administered Treatments in Chronic Post-Stroke Aphasia: A Systematic Review

Célia Ericson [1,*], Alisa Latysheva [2], Sarah-Ève Poirier [3,4] and Marion Fossard [1]

1 Institute of Logopedic Sciences, University of Neuchâtel, rue de la Pierre-à-Mazel 7, 2000 Neuchâtel, Switzerland
2 Clinic La Lignière, La Lignière 5, 1196 Gland, Switzerland; alisa.latysheva@gmail.com
3 Center for Interdisciplinary Research in Rehabilitation and Social Integration (Cirris), CIUSSS de la Capitale-Nationale, Quebec, QC G1M 2S8, Canada; sarah-eve.poirier.1@ulaval.ca
4 School of Rehabilitation Sciences, Faculty of Medicine, Laval University, Quebec, QC G1V 0A6, Canada
* Correspondence: celia.ericson@unine.ch

Abstract: Background: In current clinical practice, resources remain limited and are insufficient to address the needs of people with chronic post-stroke aphasia. To improve access to speech therapy, self-administered therapies using computers or smart tablets are increasingly recommended. In addition to enabling more intensive and prolonged treatment, computer- and smart-tablet-delivered therapies can be highly enjoyable and motivating for patients. Aims: This systematic review aimed to identify computer- and smart-tablet-based self-administered treatments and analyze the proposed interventions in terms of treatment targets, effectiveness (considering specificity, generalization, transfer, and maintenance), and clinician involvement (during and/or prior self-administered therapies). Methods: Terms encompassing three main concepts (rehabilitation, self-administration, and aphasia) were used to search three electronic databases (Scopus, PubMed, and PsycINFO). Two reviewers independently screened titles and abstracts against eligibility criteria. Data extraction of included studies was completed by three reviewers. Results: Thirty-nine studies were included in this study. In terms of treatment targets, anomia is the most treated symptom in published studies ($n = 24$), but the existence of promising studies for other disorders means that the targets can be broadened. Therapies are effective for trained items, and gains are maintained. There is some evidence of transfer effects for treatments targeting the sentence level. Most studies offer training sessions, previous self-administered therapy, and/or observation and monitoring sessions during therapy; more rarely, self-administered therapy is supplemented with face-to-face therapy. Conclusions: This systematic review is the first to focus specifically on self-administered technology-based therapies. It provides important evidence-based information for clinical practice in self-administered therapies via computer or smart tablet.

Keywords: chronic post-stroke aphasia; language therapy; self-administered treatment; computer-based; smart-tablet-based

1. Introduction

Even though the overall ischemic stroke incidence rate has decreased during the first decades of the 21st century, the number of patients living with the long-term effects of stroke is increasing and projected to grow in the coming decades due to population aging and reduced stroke mortality [1,2]. More than one-third of stroke survivors suffer from acquired aphasia, a language disorder that can affect all four language modalities (oral production, auditory comprehension, reading, and writing), and for many individuals, this condition persists well beyond the acute phase [1–3]. Aphasia is associated with higher mortality, disability, and use of health services [4,5] so it remains an important symptom that needs to be considered in stroke care and rehabilitation.

Speech and language therapy (SLT) is provided for people with aphasia (PWA), focusing on improving their language and communication abilities, and helping them to participate in daily communication activities. PWA often receive SLT in the initial months following a stroke, but many discontinue therapy or receive only minimal support during the chronic phase, despite evidence supporting its benefits [6]. This gap in care is primarily due to limited and insufficient clinical resources to meet the needs of PWA [7].

Several authors have shown that the effectiveness of SLT is strongly influenced by the dosage, with the benefits of higher-intensity approaches linked to greatest overall language and functional communication gains [8,9]. A recent review concluded that the most significant improvements in language and functional communication occurred with frequent SLT (5 days/week) and were augmented when supplemented by prescribed home practice [9]. Nonetheless, providing frequent SLT during the chronic phase is often impractical due to limited access to therapists, transportation difficulties, distant geography, scheduling constraints, fatigue, and, in some countries, due to limited insurance coverage [10]. Consequently, rehabilitation for chronic PWA is often insufficient.

One approach that meets intensity requirements and supports long-term treatment is self-administered home therapy, allowing PWA to engage in therapy tasks independently. In a data analysis of over 3500 users of a therapy platform for PWA, Godlove et al. [11] compared therapy delivered at home to therapy delivered only in clinic, and found that home therapy users received more frequent sessions (at least every 2 days) and completed tasks in less time than clinic patients, who received less frequent therapy (only once every 5 days).

In addition to intensifying treatment, self-administered remote therapies for PWA offer several advantages over face-to-face therapy. They allow patients to practice in a familiar environment and eliminate the need to travel between the clinic and home, making it easier for those in rural areas or with physical limitations to continue their SLT. Breaking sessions into multiple parts throughout the day helps reduce fatigue [12]. In addition, self-administered therapies could be a lower-cost option [13,14], particularly for PWA with mild or moderate difficulties [15], potentially halving the costs [16]. From the therapists' perspective, self-administered therapies can reduce waiting lists [17], alleviate the workload, and increase the number of patients treated [18].

A scoping review focusing on self-management approaches for PWA revealed that technology-based interventions are predominantly used, enhancing autonomy and intensifying treatment [19]. In this approach, therapy is delivered on various computerized platforms, most commonly computers (51.1%) and smart tablets (40.4%), and follows the same principles as traditional face-to-face rehabilitation [20]. Computer-based tasks can be customized to suit PWA's individual needs, considering their personal context and language ability levels. This personalization can help motivate practice [21], and the resulting autonomy and opportunities are attractive to PWA [22]. Currently, the rapid development and increasing accessibility of modern technologies have led to the expansion of therapies using digital platforms in aphasia rehabilitation. There are now numerous software offering therapy activities for PWA [23,24], mostly focusing on word retrieval or a combination of language skills [23,25].

In addition to patient compliance, technology-based interventions have proven efficacy and effectiveness for chronic aphasia rehabilitation [26–28]. There is also evidence that computer-delivered therapy is effective compared to no therapy and is as effective as clinician-delivered therapy [11,29,30]. While some interventions include direct clinician involvement during therapy sessions, others allow patients to practice independently at home, suggesting a flexible approach to integrating technology in both supervised and unsupervised settings [26]. A recent randomized controlled trial compared treatment outcomes for PWA receiving self-administered computerized word-finding therapy relative to other control treatments [16]. Results showed that self-administered computerized SLT improved word-finding more than usual care did. However, there was no evidence of generalization to untrained items, functional communication, or participant perception of their own communication or participation. Another study compared the effectiveness of a self-administered speech, language, and cognitive therapy using two delivery methods: computerized or paper workbooks [10]. The results showed a greater improvement in the participant perception of quality of life and performance on language tasks in the computerized treatment group. The authors assumed that these results could be explained by the presence of immediate feedback and an automatic adaptation of the level of difficulty of the exercises for the computerized treatment group.

All these findings support the growing notion that self-administered computerized therapy can be an effective way to rehabilitate individuals with post-stroke aphasia. This approach offers a real alternative to facilitate home practice, maximize repetitive practice, and meet the dosage requirements of high-intensity SLT [25], while allowing PWA to take an active role in their rehabilitation.

Furthermore, speech–language pathologists (SLPs) show a strong interest in incorporating technology into their treatment plans. According to a questionnaire administered to SLPs in Australia and the Netherlands, 93% and 97% of respondents, respectively, expressed interest in using therapy applications with their patients [31]. In terms of the content of therapy applications, a key requirement identified by SLPs is that these tools should enable PWA to practice various language skills beyond the single word level, such as at the sentence or discourse level.

In the current clinical reality, where clinical resources are limited and insufficient to meet the needs of chronic post-stroke PWA, it is high time to consider this emerging field of computerized or smart tablet-based self-administered therapies for chronic PWA. The aims of this review are to present the treatment targets of self-administered interventions using computer or smart tablet, determine their effectiveness, and consider clinician involvement in these evolving treatments. By highlighting these issues, our aim is to provide clinicians with solid and relevant information, enabling them to make the best possible use of self-administered approaches using computers or smart tablet.

The specific aims of this systematic review are as follows:

(1) Identify the targets of computerized or smart tablet-based self-administered interventions.
(2) Investigate the effectiveness of these interventions in terms of specificity, generalization, transfer, and maintenance of targeted aphasic symptoms.
(3) Consider the extent to which clinicians are involved before and/or during the proposed interventions.

2. Materials and Methods

This systematic review was conducted following guidelines proposed by the Preferred Reporting Items for Systematic Reviews and Meta-Analyses (PRISMA, 2020) [32]. The review protocol was specified before the systematic review was conducted but was not registered.

2.1. Research Strategy and Eligibility Criteria

A comprehensive literature search was conducted using three electronic databases: SCOPUS, PubMed, and Psyc'INFO. We limited our search to studies published in English or French, with no limit for the publication years. Database searches were performed in February 2024 and were updated in August 2024. The keywords used in the search included terms related to rehabilitation, self-administration, and PWA. The following terms were entered, and the Boolean operators and truncators were adapted to each database consulted: (rehabilitati * OR readapt * OR reeducation OR training OR treatment OR therapy OR remediati * OR recovery) AND ("self-manage *" OR "self-administ *" OR "self-guided" OR "self-train *" OR "home practice" OR "home use" OR "home train *" OR remot *) AND aphasi *. The eligibility criteria shown in Table 1 were set.

Table 1. Overview of the eligibility criteria used for the selection of articles.

Eligibility Criteria	Inclusion Criteria	Exclusion Criteria
Population	People with a diagnosis of chronic aphasia following stroke (>6 months post-onset)	People with other pathologies (e.g., neurodegenerative disease, craniocerebral trauma, tumors)
Rehabilitation	Self-administered design (frequency of self-administered sessions higher than sessions in the presence of the clinician) Intervention targeting the restoration of a language disorder Computer or smart tablet-based	Augmentative and alternative communication Surgical interventions Neuromodulatory rehabilitation Pharmacological treatments
Types of studies	Peer-reviewed publication Randomized controlled trials Non-randomized controlled group trials Single subject designs Crossover design study	(Systematic) reviews or meta-analysis without original data Research protocols Preliminary Results Conference reports Book chapters

Additional references that met the inclusion and exclusion criteria were searched in the reference lists of the selected papers and were added manually.

2.2. Study Selection

The whole screening procedure was conducted independently by two examiners while recording reasons for exclusion of ineligible studies. The first selection step was based on the titles and abstracts of each record yielded by the search after all duplicates had been removed manually. The two examiners determined whether the articles were appropriate according to the eligibility criteria presented above. After the first selection step, the full text of the remaining articles was read, and the same eligibility criteria were used. During both steps of the selection process, any disagreements were resolved by discussion until consensus was reached.

2.3. Data Extraction

Three researchers independently extracted relevant data from all accepted articles using a predefined set of data fields on an a priori developed Microsoft Excel software form. When the study involved multiple conditions, only the data from participants who met the eligibility criteria were reported. Conversely, in group studies, data from all participants were included. In 50% of the studies, the first author checked the data to ensure the inter-rater reliability. Disagreements were resolved by discussion until the authors agreed on the most appropriate solution.

3. Results

Results of the literature search are reported in a PRISMA flow diagram (Figure 1). The initial search yielded 522 records. After eliminating duplicates, the remaining 286 results were reviewed to ensure they met the inclusion criteria for title and abstract screening. Agreement for inclusion of titles and abstracts was >90%. Following this step, 43 full-text articles were assessed for eligibility. An additional 39 articles, identified through a manual search of the reference lists of relevant studies, were also evaluated for eligibility through full-text review. Agreement for inclusion of full texts was >90%. Finally, a total of 39 articles were obtained from the search procedure. Most of the excluded articles did not involve an intervention with a self-administered design (the frequency of self-administered sessions was not higher than the frequency of sessions in the presence of the clinician), or the article did not provide enough information to determine this.

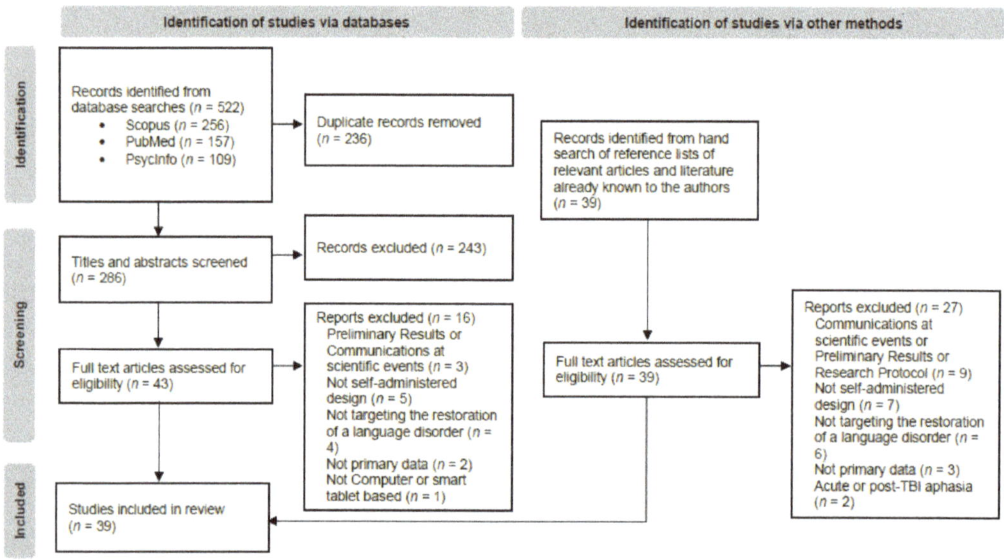

Figure 1. Flow diagram of the identification and selection records, adapted from the PRISMA flow diagram [32].

3.1. Characteristics of the Included Studies

The descriptive characteristics of the included studies are shown in Table 2, organized by treatment targets. The following information is provided for each article: author(s) and publication year, language and the number (*n*) of participants, characteristics of interventions (software/program used; type of treatment) and efficacy measures (specificity; generalization; transfer; maintenance).

3.2. Targets of Computerized or Smart-Tablet-Based Self-Administered Therapies

The primary objective of this review is to identify the treatment targets that are addressed by self-administered therapies delivered via computers or smart tablets.

Five different treatment targets were identified in this review (see Table 2). Most of the studies focused on a single specific target: word production (24 studies: 61.5%), sentence processing and/or production (7 studies: 17.9%), reading (2 studies: 5.1%), or narrative production (1 study: 2.6%). Other studies covered more than one target (5 studies: 12.8%).

Word production: Of the 24 studies targeting word production, 22 focused on spoken-word production and two on written-word production [33,34]. Most studies targeted nouns, although some studies included verbs (n = 9) [33–41], proper nouns (n = 3) [14,42,43], or adjectives (n = 1) [41]. Words were most commonly selected from vocabulary databases (n = 15) [34–37,40,43–52], although some studies focused on personally relevant words, chosen by the PWA based on their interests or communication needs (n = 9) [12,14,33,38,39,41,42, 53,54]. The therapies were delivered through a variety of programs or software (specialized or non-specialized). Specialized software, such as StepbyStep [14,38,39] or MossTalk Words [49,50], provided structured exercises with feedback and task adaptation. Non-specialized programs, such as Microsoft PowerPoint (e.g., [35,42]) or iBooks Author [36,37], presented words with varying levels of cueing through slides or interactive pages.

Sentence processing and/or production: Of the seven studies targeting sentence level, four trained personalized conversational scripts with diminishing cues, supported by a virtual SLP demonstrating correct speech movements (AphasiaScripts) [55–58]. Other treatments targeted production and comprehension of complex sentences (n = 1) [59], production of prepositions and sentences (n = 1) [60], or production of verbs and sentences (n = 1) [61].

Reading comprehension: Two studies targeted reading comprehension using an online adaptation of the Oral Reading for Language in Aphasia (ORLA) therapy [62,63]. The online system provided text and audio prompts, which mimicked SLP guidance and adjusted difficulty based on performance. One version included articulator movement models for the PWA to use as a reference [63].

Narrative production: One study targeted narrative production using an imitation-based therapy (IMITATE [64]), in which participants listened to and repeated words and phrases from different speakers.

Multiple domains: Among the five studies offering therapies targeting multiple domains, some programs focused exclusively on language tasks (n = 2), such as Tactus Therapy Solutions (reading, naming, comprehension, and writing) [65] and iAphasia (adding repetition and fluency) [66]. Others (n = 3) combined language and cognitive rehabilitation and provide access to around thirty different tasks (Constant Therapy) [10,67,68].

3.3. *Effectiveness of Technology-Based Self-Administered Therapies*

The second objective of this review was to investigate the effectiveness of these therapies for each targeted aphasic symptom. To achieve this, intervention outcomes were reported in terms of specificity (improvements on trained items), generalization (improvements on untrained items and/or standardized tests), transfer (improvements on connected speech or perceived change by PWA and/or significant other), and maintenance. None of the studies evaluated all the effectiveness measures.

Although there is considerable variability in the designs and objectives, all the selected studies unanimously highlighted the potential of self-administered, technology-based therapies to effectively reduce aphasic symptoms. Detailed data are presented in Table 2, while Figure 2 provides a summary of the effectiveness of these therapies for each targeted aphasic symptom.

When studies compared groups (e.g., self-administered vs. clinician-delivered therapy), only the results for the self-administered therapy group were reported. For information, these comparisons concerned five studies (two on word production [47,49], two on reading comprehension [62,63], and one on sentence production [59]). These studies concluded that both approaches yielded comparable outcomes in terms of specificity and generalization patterns. Similarly, two randomized controlled trials on word production treatment reported significant improvements with self-administered computer-based therapy compared to usual care [39] and an attention control group (paper-based puzzle book activities) [14].

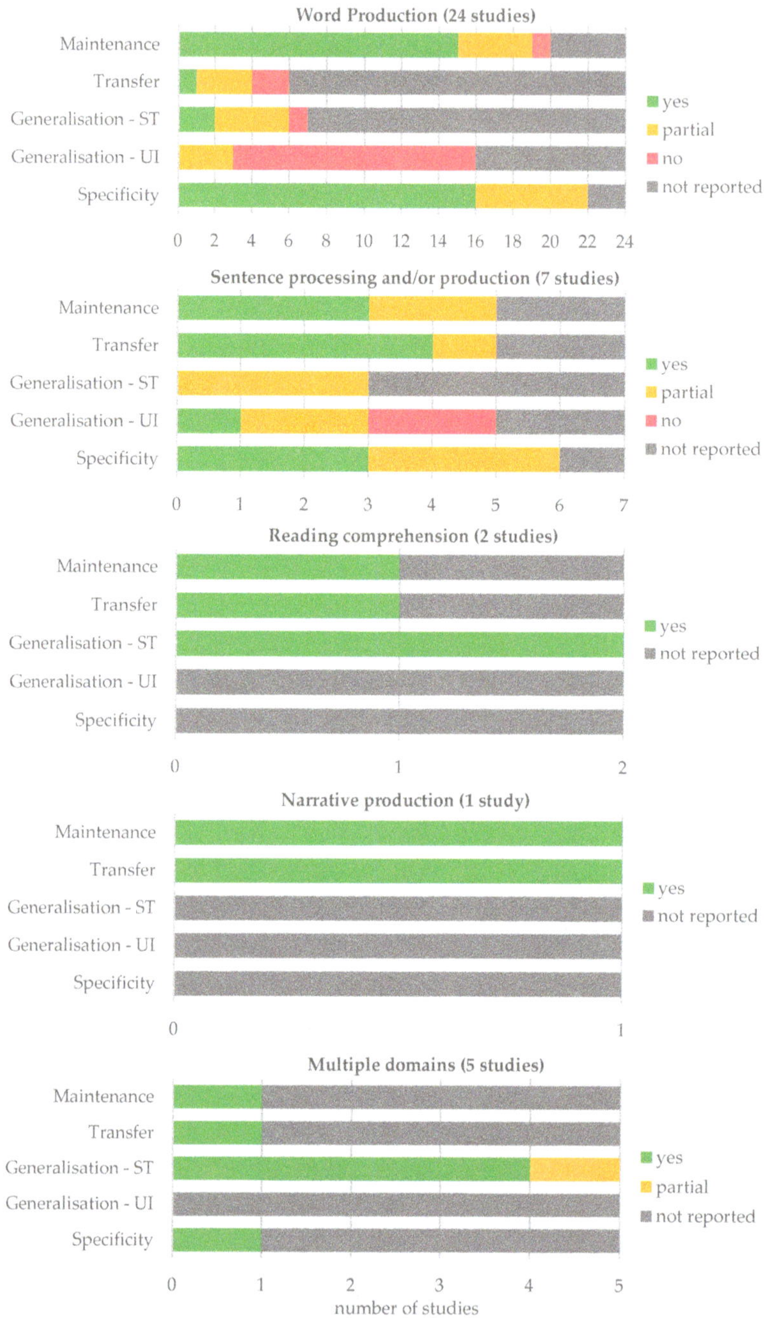

Figure 2. Summary of treatment effectiveness for targeted aphasic symptoms. Improvements are highlighted as follows: green for measures showing improvement in all groups or PWA, yellow for partial improvements (some groups or PWA), red for no improvement, and gray for unreported measures. UI, untrained items; ST, standardized tests.

Table 2. Treatment target, descriptive characteristics, and treatment effect of the included studies.

Treatment Target	Author(s), Publication Year	Participants: Language	Participants: n	Characteristics of Interventions: Treatment Type—Software/Program	Efficacy Measures: Specificity	Generalization: Untrained Items	Generalization: Standardized Tests	Transfer	Maintenance
Word production	Quique et al., 2022 [44]	Japanese	2	Cued naming with flashcards—Anki	2/2 *	0/2			2/2
	Gallée et al., 2020 [48]	English	6	Semantic and phonological naming—Constant Therapy			BNT: 4/6		
	Kurland et al., 2014 [36]	English	5	Cued naming—designed on iBooks Author	4/5	0/5			
	Kurland et al., 2018 [37]	English	21	Cued naming—designed on iBooks Author	Yes	No			Yes
	Lavoie et al., 2019 [12]	French	4	Semantic and phonological naming—iTSA	4/4	2/4		2/4	4/4
	Lavoie et al., 2016 [34]	French	1	Cued naming—designed on Keynote	1/1	0/1			1/1
	Heide et al., 2023 [41]	German	2	Cued naming—LingoTalk	2/2	0/2	WPP: 0/2	0/2 #1/2	2/2
	Choe et al., 2007 [47]	English	4	Cued naming—designed on Microsoft PowerPoint	2/4	0/4			3/4
	Choe et al., 2010 [42]	English	2	Cued naming—designed on Microsoft PowerPoint	2/2				1/2
	Choe and Stanton, 2011 [53]	English	2	Cued naming—designed on Microsoft PowerPoint	2/2	0/2			2/2
	Routhier et al., 2016 [35]	French	2	Cued naming—designed on Microsoft PowerPoint	2/2	0/2			2/2
	Mason et al., 2011 [54]	English	3	Picture-assisted repetition—designed on Microsoft PowerPoint	2/3	0/3		1/3	2/3
	Ball et al., 2011 [33]	English	3	Copy and recall therapy with repetition—Modified CART	3/3		naming WAB-R: 1/3		2/2
	Fink et al., 2002 [49]	English	3	Cued naming—MossTalk Words	3/3	0/3	PRT: 0/3; PORT: 2/3		3/3
	Ramsberger and Marie, 2007 [50]	English	4	Cued naming—MossTalk Words	4/4	2/4			4/4

Table 2. *Cont.*

Treatment Target	Author(s), Publication Year	Participants		Characteristics of Interventions: Treatment Type—Software/Program	Efficacy Measures				
		Language	n		Specificity	Generalization: Untrained Items	Generalization: Standardized Tests	Transfer	Maintenance
	Fridriksson et al., 2009 [45]	English	10	Picture/word matching—NA	1/2 group		PNT: 1/2 group		
	Laganaro et al., 2003 [46]	French	3	Metaphonological therapy—NA	3/3	0/3			3/3
	Pedersen et al., 2001 [51]	Danish	3	Semantic, phonemic and written naming—NA	2/3				2/3
	Choe et al., 2023 [43]	English	1	Spoken-naming and upper-limb—NA	1/1				0/1
	Herbert et al., 2012 [52]	English	6	Syntactic cueing—STAR	5/6	0/6		4/6	Yes
	Mortley et al., 2004 [38]	English	7	Semantic, spelling, word-to-picture matching, naming—StepbyStep	7/7	3/7		#6/6	
	Palmer et al., 2012 [39]	English	16	Listening, cued naming—StepbyStep			OANB: Yes		Yes
	Palmer et al., 2019 [14]	English	83	Listening, cued naming—StepbyStep	Yes	No		No	Yes
	Trevittaya and Chinchai, 2023 [40]	Thai	5	Cued naming—Thai Naming	5/5		Thai WAB: 5/5		5/5
Sentence process-ing/production	Cherney and Halper, 2008 [57]	English	3	Script training, reducing cues—AphasiaScripts	2/3		WAB-AQ: 1/3; CETI: 1/2; CADL: 0/3	#2/3	1/3
	Manheim et al., 2009 [58]	English	20	Script training, reducing cues—AphasiaScripts				#Yes	Yes
	Cherney et al., 2014 [56]	English	8	Script training, reducing cues—AphasiaScripts	Yes	No			Yes
	Cherney et al., 2019 [55]	English	20	Script training, reducing cues—AphasiaScripts	Yes	No		Yes	Yes
	Hickin et al., 2023 [61]	English	6	Verb and sentence therapy—designed on Microsoft PowerPoint	5/6	2/6	CETI: 4/6		2/3

Table 2. *Cont.*

Treatment Target	Author(s), Publication Year	Participants		Characteristics of Interventions: Treatment Type—Software/Program	Efficacy Measures				
						Generalization			
		Language	n		Specificity	Untrained Items	Standardized Tests	Transfer	Maintenance
	Linebarger et al., 2001 [60]	English	3	Prepositions training—Prolog and C++	2/3			1/1	
	Thompson et al., 2010 [59]	English	6	Treatment of Underlying Forms—Sentactics	6/6	6/6	SPPT: 6/6; SCT: 0/6	Yes	
Reading comprehension	Cherney, 2010 [62]	English	11	Reading sentences—Web-ORLA			WAB-AQ: Yes	Yes	
	Cherney et al., 2021 [63]	English	19	Repeating and reading sentences—Web-ORLA			WAB-R LQ: Yes		Yes
Narrative production	Duncan and Small, 2017 [64]	English	18	Intensive imitation practice—IMITATE				Yes	Yes
Multiple domains	Kiran et al., 2014 [68]	English	3	Cognitive and language tasks—Constant Therapy			WAB-R, CLQT: 3/3; BNT: 2/3		
	Des Roches et al., 2015 [67]	English	42	Cognitive and language tasks—Constant Therapy	Yes		WAB-R, CLQT: Yes		
	Braley et al., 2021 [10]	English	17	Cognitive and language tasks—Constant Therapy			WAB-R AQ: Yes		
	Choi et al., 2016 [66]	Korean	8	Language tasks—iAphasia			K-WAB: Yes		Yes
	Stark and Warburton, 2016 [65]	English	7	Language tasks—Tactus Therapy Solutions			CAT: Yes	Yes	

Note. * These fractions refer to the number of PWA who showed improvements compared to the number of PWA for whom the effectiveness measure was conducted. Binary metrics (yes/no) are reported for the outcomes of group studies. Cells are highlighted in gray if the measure was not used. When the transfer measure was preceded by a #, it refers to a perceptive measure of change by PWA and/or significant other. Data from PWA which did not meet the inclusion criteria were not reported; BNT, Boston Naming Test [69]; CADL, Communication activities of daily living [70]; CAT, Comprehensive Aphasia Test [71]; CETI, Communicative Effectiveness Index [72]; CLQT, Cognitive Linguistic Quick Test [73]; K-WAB, Korean version of the Western Aphasia Battery [74]; NAVS, Northwestern Assessment of Verbs and Sentences [75]; OANB, Object and Action Naming Battery [76]; PORT, Philadelphia Oral Reading Test (part of PNT); PNT, Philadelphia Naming Test [77]; PRT, Philadelphia Repetition Test [78]; SCT, Sentence Comprehension Test of the NAVS; SPPT, Sentence Production Priming Test of the NAVS; Thai WAB [79]; WAB, Western Aphasia Battery [80]; WAB-R, Western Aphasia Battery-Revised [81]; AQ, Aphasia Quotient, LQ, Language Quotient; WPP, Wortproduktionsprüfung [82].

3.4. Clinician Involvement Before and/or During Proposed Interventions

The third objective of this review was to consider the extent of clinician involvement before and/or during the proposed interventions. The clinician's involvement in treatment varied to some extent across studies (see Table 3). Involvement was categorized according to whether it occurred before or during the self-administered treatment period. Of note, one study did not report the SLP involvement [40].

Table 3. Clinician involvement with specification of number or intensity of sessions.

Study	Prior to Treatment		During Treatment	
	Additional SLP-Led Sessions	**Introduction and Training**	**Additional SLP-Led Sessions**	**Observation and/or Monitoring**
Pedersen et al., 2001 [51]		NR		
Linebarger et al., 2001 [60]		NR		1×/week
Fink et al., 2002 [49]				1×/week
Laganaro et al., 2003 [46]	2 weeks	1–2 sessions		
Mortley et al., 2004 [38]		1 session		< 1×/month
Choe et al., 2007 [47]		1 session	2×/week	
Ramsberger and Marie, 2007 [50]		2 sessions		
Cherney and Halper, 2008 [57]				1×/week
Fridriksson et al., 2009 [45]		1 session		
Manheim et al., 2009 [58]		5 sessions		1×/week
Choe et al., 2010 [42]		NR	2×/week	
Thompson et al., 2010 [59]				all sessions
Cherney, 2010 [62]				all sessions
Choe and Stanton, 2011 [53]		1 session		
Ball et al., 2011 [33]			1×/week	
Mason et al., 2011 [54]	4 weeks			
Palmer et al., 2012 [39]		NR		1–4×/month
Herbert et al., 2012 [52]		NR		
Kurland et al., 2014 [36]	2 weeks	NR		1×/week
Cherney et al., 2014 [56]				1×/week
Kiran et al., 2014 [68]				1×/week
Des Roches et al., 2015 [67]		NR		1×/week
Lavoie et al., 2016 [34]		2 sessions		
Routhier et al., 2016 [35]		2 sessions		
Choi et al., 2016 [66]				1×/day
Stark and Warburton, 2016 [65]		1 session		
Duncan and Small, 2017 [64]				1×/week
Kurland et al., 2018 [37]	2 weeks	2–3 sessions		1×/week
Lavoie et al., 2019 [12]		1–2 sessions		1st session
Palmer et al., 2019 [14]				NR
Cherney et al., 2019 [55]				all sessions
Gallée et al., 2020 [48]		1 session	1×/week (1/2 group)	
Braley et al., 2021 [10]				2×/week
Cherney et al., 2021 [63]		2 sessions		1×/week
Quique et al., 2022 [44]		4 sessions	1×/week	
Heide et al., 2023 [41]				2×/week
Choe et al., 2023 [43]		1 session		
Trevittaya and Chinchai, 2023 [40]				
Hickin et al., 2023 [61]		1 session	1×/week	
	4/39 *	24/39	6/39	21/39

Note. * These fractions refer to the number of studies proposing the type of clinician involvement mentioned in the column compared to the total number of studies included, highlighted in gray when not applicable. NR, number or intensity of sessions is not reported.

Prior to self-administered treatment: A large majority of studies (24/39: 61.54%) reported clinician-led introduction and training sessions just before the start of the self-administered therapy. During these sessions, the clinician provided instructions and demonstrations on how to use the computer or smart tablet. In some cases, PWA were provided with a paper guide that outlined each step of the treatment.

In a few studies (4/39: 10.26%), an SLP-led therapy preceded the self-administered treatment, with results showing that home practice programs effectively promoted continued language recovery [36]. In three of these studies, the tasks used during the SLP-led therapy differed from those used in the self-administered therapy (although the treatment goals were similar) [36,37,46], while in one study, the tasks were identical [54].

During self-administered treatment: In most studies (21/39: 53.85%), the clinician intervened weekly or even in every session during the self-administered treatment—either in person or remotely—to observe and/or monitor the therapy progress without providing any cues or reinforcement. During these sessions, the clinician initiated the administration of the protocols, adjusted the treatment plan and the difficulty of the stimuli according to the PWA's performance, guaranteed smooth computer operation, ensured appropriate and intensive practice, and/or recorded the participant's adherence. Some self-administered therapies did not require clinician monitoring; they used self-learning applications that automatically adjusted the difficulty level based on PWA performance [38,65].

In some studies (6/39: 15.38%), weekly SLP-led therapy was conducted in parallel with self-administered therapy. These sessions provided regular supervision, reinforcement, and feedback on tasks identical to those completed during self-administered therapy. In two studies, PWA also attended weekly group sessions aimed at improving their communication skills in functional settings [42,47].

4. Discussion

Acknowledging the limited and insufficient clinical resources available to meet the needs of chronic post-stroke PWA, we questioned whether self-administered computer- or smart-tablet-based therapies could effectively support SLT. Thus, the goals of this systematic review were to identify treatment targets of self-administered computer or tablet-based therapies, to assess the immediate and long-term effectiveness of these treatments as well as their transfer (to connected speech or perceived change by PWA and/or significant other), and to investigate the clinician involvement in administering such therapies for chronic post-stroke aphasia. Based on predefined selection criteria and inter-rater agreement, 39 studies were included in this literature review.

4.1. Word: A Preferred Treatment Target in Self-Administered Therapies

The review highlights a predominant focus on word-level interventions with 24 of the 39 studies reviewed addressing this target. However, other specific symptoms are also targeted, including sentence processing/production, reading comprehension, and narrative production. Additionally, some studies adopted a more holistic approach by addressing multiple domains, including language and cognitive functions.

Targeting the single word is a common outcome in standard care, where word production is a primary focus in rehabilitation [83], as word-finding difficulties are among the most prevalent and sometimes the only visible symptoms of aphasia. In addition, this emphasis on word production may reflect the relative ease of implementing word-level treatments in a self-administered setting, due to the possibility to design predictive cues and feedback. Implementing effective feedback for more complex tasks, such as narrative or sentence-level tasks, which are well-tailored to the production of the PWA, may be more challenging, making them more difficult to implement in a self-administered setting.

The papers included in this review that addressed sentence-level or narrative production focused predominantly on script training. This approach avoids the need for tailored feedback and instead relied on a gradual reduction in cues. However, advances in automatic speech recognition technologies allow for immediate and personalized feedback and have the potential to improve the implementation of self-administered therapies, extending beyond word production.

Among the word production treatments reviewed in this article, the majority primarily focus on noun retrieval, with only a few studies addressing verbs, proper nouns, or adjectives. This aligns with existing literature on anomia treatment, which most often targets concrete nouns due to their high imageability, with few studies exploring less concrete words (such as adjectives, adverbs, or abstract concepts) [84]. In addition, the prevalence of self-administered therapies focusing on nouns may be attributed to the greater ease of designing tasks with predictive feedback for concrete words compared to less tangible words, such as many verbs or adjectives.

4.2. Effectiveness of Self-Administered Therapies: Beyond Trained Items

The findings of our review confirm the effectiveness of computerized or smart-tablet-based self-administered therapies in improving word production of trained items with good maintenance over time but rare generalization and transfer. These results are consistent with a previous systematic review on technology-based therapies for anomia [28]. While transfer to connected speech is possible after anomia treatment, the functional benefit of targeting words relevant to PWA appears evident. However, most studies rely on generic vocabulary databases, with less focus on functionally relevant words for PWA. In clinical practice, personalized materials can be challenging to provide, due to time constraints. However, technology can facilitate the selection of functional items, for instance, by involving patients or families in capturing photos of meaningful objects [12].

For the first time, this review also examines the efficacy of these treatments beyond word production, extending to sentence processing and/or production, reading comprehension, and narrative production. In addition to demonstrating the effectiveness of these therapies for trained items and their maintenance over time, the findings provide promising evidence of transfer. The majority of these studies measured and demonstrated transfer to connected speech or perceived improvements reported by PWAs and/or their significant others. For instance, following self-administered script training, PWAs showed significant improvements in the accuracy of script production during conversations [55], and family members reported increased spontaneity and initiation of verbal interactions beyond the trained scripts [57,58]. In another study on sentence-level intervention [59], narrative production improved significantly, with longer utterances and a higher proportion of complex sentences. Additionally, the study specifically targeting narrative production revealed a significant increase in both the number and percentage of Correct Information Units during narrative tasks [64]. These findings are consistent with existing literature suggesting that transfer is more prevalent in therapies emphasizing verbs and sentences [85]. Interestingly, despite the absence of a real interlocutor during self-administered sessions, factors such as greater intensity, autonomy, and accessibility may contribute to successful transfer.

4.3. Clinician Involvement in Self-Administered Therapies: A Parameter Not to Be Ignored

In the majority of the studies reviewed, it is mentioned that introductory and training sessions were provided prior to the self-administered therapy. Although this was not specified in all studies, it can be assumed that explanations to PWA regarding the therapy process and the use of the computerized system are necessary before therapy to effectively carry out self-administered therapy [86].

During self-administered therapy, most of the included studies proposed weekly clinician interventions—either in person or remotely—to observe and/or monitor progress without providing any cues or reinforcement. This clinician involvement was particularly necessary when the applications used were unable to adjust the treatment plan or item difficulty based on the performance of PWAs, or to ensure compliance with the treatment protocol. As technology advances and self-learning applications become more sophisticated, the clinician's monitoring workload is likely to decrease. Some rehabilitation programs have already created a face-to-face treatment environment through virtual reality, where virtual SLP displays facial movements modeling the articulators (Web-ORLA [63]; AphasiaScripts [55–58]) or provides instructions, explanations, and modeled target sentences (Sentactics; [59]). The emerging use of virtual reality, with immersive or semi-immersive environments that simulate real-life contexts, enhances the ecological validity of language rehabilitation [87,88].

Moreover, regular contact with a clinician has been shown to increase patient adherence [86]. In their study on adherence to self-administered computer therapy, Harrison and colleagues [89] demonstrated that the time spent by the SLP on monitoring and adjusting the software for participants was directly associated with greater adherence to therapy. Regular contact can also be provided by trained volunteers. In the included study by Palmer and colleagues [39], three of the four participants who did not practice with the recommended frequency lacked support from volunteers. Less clinician contact can lead to a decline in patient motivation, as clinicians often provide essential reinforcement and encouragement [50]. These results highlight the value of regular contact with a clinician or volunteer. Such contact can be facilitated through various means, including clinic visits (e.g., [62]), home visits (e.g., [35]), video conferences (e.g., [36]), or phone calls (e.g., [39]).

A few of the studies reviewed offer therapy sessions that actively involve an SLP. In their study, Gallée and colleagues [48] examined the influence of clinician implication and demonstrated that weekly guidance from an SLP, resulted in greater engagement and greater improvement on standardized language assessments compared to independent, unsupervised therapy.

In any case, the role of the SLP remains essential in self-administered therapy. While self-administered settings offer greater intensity, autonomy, and accessibility, they may lack the personalized feedback, dynamic adjustments, and interactive support that are an integral part of therapist-led interventions. In addition, it cannot substitute for clinical expertise in areas such as selecting functionally relevant therapy targets, tailoring cues to the patient's individual profile, monitoring progress, and ensuring that treatment goals are achieved.

4.4. Limitations

It is important to acknowledge that this review presents a number of limitations and conclusions that should be interpreted in the context of these limitations. Firstly, the fact that the majority of treatments reviewed in this study resulted in improvements could be linked to publication bias, as studies without positive outcomes are typically more difficult to publish. Secondly, there is a possibility that some relevant papers may have been omitted, partly due to the lack of consensus on the terminology used to define self-administered SLT. Thirdly, the focus on a specific population—chronic PWA—limits the generalizability of the results to the acute stage. It is hypothesized, however, that computer- or smart-tablet-based therapy could also be effective in acute cases, where recovery is often more significant and rapid due to the combined effects of spontaneous recovery and therapeutic intervention. Nonetheless, we believe that the present work provides valuable insight for clinicians.

5. Conclusions

The results of this systematic review confirm that self-administered computer or smart tablet-based treatments, when guided by a SLP, are an effective rehabilitation approach for individuals with chronic post-stroke aphasia.

The large number of studies focusing on anomia have demonstrated positive outcomes for trained items with good maintenance over time but rare generalization and transfer. While fewer studies have explored self-administered therapies targeting symptoms beyond word production—such as sentence processing and/or production, reading, or narrative production—thus far, the findings are promising. Particularly, these studies highlight encouraging transfer effects to connected speech or perceived change by PWA and/or a significant other, which is often challenging to achieve in traditional SLT [85,90].

Although constant clinician supervision is not necessary, guidance sessions can significantly enhance therapy outcomes. The expert role of the SLP remains essential for defining objectives, supervising therapy, and ensuring therapeutic progress.

Finally, it is important to consider some of the challenges associated with self-administered technology-based treatments: high costs and limited access to tablets and applications, potential difficulties in using tablets for PWA, and cognitive impairments that may affect the ability to use technology independently [31].

Author Contributions: Conceptualization, C.E. and M.F.; methodology, C.E. and M.F.; eligibility criteria of articles included in systematic review's results: C.E. and A.L.; collected the data: C.E., A.L. and S.-È.P.; writing—original draft preparation, C.E.; writing—review and editing, C.E.; supervision, M.F. All authors have read and agreed to the published version of the manuscript.

Funding: This research received no external funding.

Institutional Review Board Statement: Not applicable.

Informed Consent Statement: Not applicable.

Data Availability Statement: Not applicable.

Conflicts of Interest: The authors declare no conflicts of interest.

Abbreviations

ORLA: Oral Reading for Language in Aphasia. PRISMA: Preferred Reporting Items for Systematic Reviews and Meta-Analyses. PWA: People With Aphasia. SLP: Speech and Language Pathologist. SLT: Speech and Language Therapy.

References

1. Li, L.; Scott, C.A.; Rothwell, P.M.; on behalf of the Oxford Vascular Study. Trends in Stroke Incidence in High-Income Countries in the 21st Century: Population-Based Study and Systematic Review. *Stroke* **2020**, *51*, 1372–1380. [CrossRef]
2. Grönberg, A.; Henriksson, I.; Stenman, M.; Lindgren, A.G. Incidence of Aphasia in Ischemic Stroke. *Neuroepidemiology* **2022**, *56*, 174–182. [CrossRef]
3. Engelter, S.T.; Gostynski, M.; Papa, S.; Frei, M.; Born, C.; Ajdacic-Gross, V.; Gutzwiller, F.; Lyrer, P.A. Epidemiology of Aphasia Attributable to First Ischemic Stroke: Incidence, Severity, Fluency, Etiology, and Thrombolysis. *Stroke* **2006**, *37*, 1379–1384. [CrossRef]
4. Dickey, L.; Kagan, A.; Lindsay, M.P.; Fang, J.; Rowland, A.; Black, S. Incidence and Profile of Inpatient Stroke-Induced Aphasia in Ontario, Canada. *Arch. Phys. Med. Rehabil.* **2010**, *91*, 196–202. [CrossRef] [PubMed]
5. Flowers, H.L.; Skoretz, S.A.; Silver, F.L.; Rochon, E.; Fang, J.; Flamand-Roze, C.; Martino, R. Poststroke Aphasia Frequency, Recovery, and Outcomes: A Systematic Review and Meta-Analysis. *Arch. Phys. Med. Rehabil.* **2016**, *97*, 2188–2201.e8. [CrossRef] [PubMed]
6. Allen, L.; Mehta, S.; McClure, J.A.; Teasell, R. Therapeutic Interventions for Aphasia Initiated More than Six Months Post Stroke: A Review of the Evidence. *Top. Stroke Rehabil.* **2012**, *19*, 523–535. [CrossRef] [PubMed]

7. Rose, M.; Ferguson, A.; Power, E.; Togher, L.; Worrall, L. Aphasia rehabilitation in Australia: Current practices, challenges and future directions. *Int. J. Speech-Lang. Pathol.* **2014**, *16*, 169–180. [CrossRef] [PubMed]
8. Bhogal, S.K.; Teasell, R.; Speechley, M. Intensity of Aphasia Therapy, Impact on Recovery. *Stroke* **2003**, *34*, 987–993. [CrossRef]
9. Rehabilitation and Recovery of People with Aphasia After StrokE (RELEASE) Collaborators. Dosage, Intensity, and Frequency of Language Therapy for Aphasia: A Systematic Review–Based, Individual Participant Data Network Meta-Analysis. *Stroke* **2022**, *53*, 956–967. [CrossRef] [PubMed]
10. Braley, M.; Pierce, J.S.; Saxena, S.; De Oliveira, E.; Taraboanta, L.; Anantha, V.; Lakhan, S.E.; Kiran, S. A Virtual, Randomized, Control Trial of a Digital Therapeutic for Speech, Language, and Cognitive Intervention in Post-stroke Persons with Aphasia. *Front. Neurol.* **2021**, *12*, 626780. [CrossRef] [PubMed]
11. Godlove, J.; Anantha, V.; Advani, M.; Roches, C.D.; Kiran, S. Comparison of Therapy Practice at Home and in the Clinic: A Retrospective Analysis of the Constant Therapy Platform Data Set. *Front. Neurol.* **2019**, *10*, 140. [CrossRef]
12. Lavoie, M.; Bier, N.; Macoir, J. Efficacy of a self-administered treatment using a smart tablet to improve functional vocabulary in post-stroke aphasia: A case-series study. *Int. J. Lang. Commun. Disord.* **2019**, *54*, 249–264. [CrossRef] [PubMed]
13. Latimer, N.R.; Dixon, S.; Palmer, R. Cost-utility of self-managed computer therapy for people with aphasia. *Int. J. Technol. Assess. Health Care* **2013**, *29*, 402–409. [CrossRef] [PubMed]
14. Palmer, R.; Dimairo, M.; Cooper, C.; Enderby, P.; Brady, M.; Bowen, A.; Latimer, N.; Julious, S.; Cross, E.; Alshreef, A.; et al. Self-managed, computerised speech and language therapy for patients with chronic aphasia post-stroke compared with usual care or attention control (Big CACTUS): A multicentre, single-blinded, randomised controlled trial. *Lancet Neurol.* **2019**, *18*, 821–833. [CrossRef]
15. Latimer, N.R.; Bhadhuri, A.; Alshreef, A.; Palmer, R.; Cross, E.; Dimairo, M.; Julious, S.; Cooper, C.; Enderby, P.; Brady, M.C.; et al. Self-managed, computerised word finding therapy as an add-on to usual care for chronic aphasia post-stroke: An economic evaluation. *Clin. Rehabil.* **2021**, *35*, 703–717. [CrossRef] [PubMed]
16. Palmer, R.; Dimairo, M.; Latimer, N.; Cross, E.; Brady, M.; Enderby, P.; Bowen, A.; Julious, S.; Harrison, M.; Alshreef, A.; et al. Computerised speech and language therapy or attention control added to usual care for people with long-term post-stroke aphasia: The Big CACTUS three-arm RCT. *Health Technol. Assess.* **2020**, *24*, 1–176. [CrossRef]
17. Asghar, I.; Egaji, O.A.; Griffiths, M. An overview of the digital solutions for helping people with aphasia through bibliometric analysis. *eNeurologicalSci* **2021**, *22*, 100311. [CrossRef] [PubMed]
18. Cherney, L.R.; Halper, A.S.; Kaye, R.C. Computer-based script training for aphasia: Emerging themes from post-treatment interviews. *J. Commun. Disord.* **2011**, *44*, 493–501. [CrossRef]
19. Nichol, L.; Hill, A.J.; Wallace, S.J.; Pitt, R.; Baker, C.; Rodriguez, A.D. Self-management of aphasia: A scoping review. *Aphasiology* **2019**, *33*, 903–942. [CrossRef]
20. Ben Aziza, O.; Pecci, I.; Martin, B. État de l'art sur les applications dédiées à la rééducation de personnes aphasiques: State of The Art Review on Applications Dedicated to the Rehabilitation of People with Aphasia. In Proceedings of the 32nd Conférence Francophone sur l'Interaction Homme-Machine, Virtual Event, France, 13–16 April 2021; ACM: New York, NY, USA, 2022; pp. 1–25.
21. de Luca, R.; Aragona, B.; Leonardi, S.; Torrisi, M.; Galletti, B.; Galletti, F.; Accorinti, M.; Bramanti, P.; de Cola, M.C.; Calabrò, R.S. Computerized Training in Poststroke Aphasia: What About the Long-Term Effects? A Randomized Clinical Trial. *J. Stroke Cerebrovasc. Dis.* **2018**, *27*, 2271–2276. [CrossRef] [PubMed]
22. Wade, J.; Mortley, J.; Enderby, P. Talk about IT: Views of people with aphasia and their partners on receiving remotely monitored computer-based word finding therapy. *Aphasiology* **2003**, *17*, 1031–1056. [CrossRef]
23. Cuperus, P.; De Kok, D.; De Aguiar, V.; Nickels, L. Aphasia therapy software: An investigation of the research literature and the challenges of software development. *Aphasiology* **2024**, 1–32. [CrossRef]
24. Vaezipour, A.; Campbell, J.; Theodoros, D.; Russell, T. Mobile Apps for Speech-Language Therapy in Adults with Communication Disorders: Review of Content and Quality. *JMIR mHealth uHealth* **2020**, *8*, e18858. [CrossRef] [PubMed]
25. Roches, C.A.D.; Kiran, S. Technology-Based Rehabilitation to Improve Communication after Acquired Brain Injury. *Front. Neurosci.* **2017**, *11*, 382. [CrossRef]
26. Repetto, C.; Paolillo, M.P.; Tuena, C.; Bellinzona, F.; Riva, G. Innovative technology-based interventions in aphasia rehabilitation: A systematic review. *Aphasiology* **2021**, *35*, 1623–1646. [CrossRef]
27. Cetinkaya, B.; Twomey, K.; Bullard, B.; El Kouaissi, S.; Conroy, P. Telerehabilitation of aphasia: A systematic review of the literature. *Aphasiology* **2024**, *38*, 1271–1302. [CrossRef]
28. Lavoie, M.; Macoir, J.; Bier, N. Effectiveness of technologies in the treatment of post-stroke anomia: A systematic review. *J. Commun. Disord.* **2017**, *65*, 43–53. [CrossRef] [PubMed]
29. Zheng, C.; Lynch, L.; Taylor, N. Effect of computer therapy in aphasia: A systematic review. *Aphasiology* **2016**, *30*, 211–244. [CrossRef]

30. Cacciante, L.; Kiper, P.; Garzon, M.; Baldan, F.; Federico, S.; Turolla, A.; Agostini, M. Telerehabilitation for people with aphasia: A systematic review and meta-analysis. *J. Commun. Disord.* **2021**, *92*, 106111. [CrossRef]
31. Cuperus, P.; De Kok, D.; De Aguiar, V.; Nickels, L. Understanding User Needs for Digital Aphasia Therapy: Experiences and Preferences of Speech and Language Therapists. *Aphasiology* **2023**, *37*, 1016–1038. [CrossRef]
32. Page, M.J.; McKenzie, J.E.; Bossuyt, P.M.; Boutron, I.; Hoffmann, T.C.; Mulrow, C.D.; Shamseer, L.; Tetzlaff, J.M.; Akl, E.A.; Brennan, S.E.; et al. The PRISMA 2020 statement: An updated guideline for reporting systematic reviews. *BMJ* **2021**, *372*, n71. [CrossRef] [PubMed]
33. Ball, A.L.; de Riesthal, M.; Breeding, V.E.; Mendoza, D.E. Modified ACT and CART in severe aphasia. *Aphasiology* **2011**, *25*, 836–848. [CrossRef]
34. Lavoie, M.; Routhier, S.; Légaré, A.; Macoir, J. Treatment of verb anomia in aphasia: Efficacy of self-administered therapy using a smart tablet. *Neurocase* **2016**, *22*, 109–118. [CrossRef] [PubMed]
35. Routhier, S.; Bier, N.; Macoir, J. Smart tablet for smart self-administered treatment of verb anomia: Two single-case studies in aphasia. *Aphasiology* **2016**, *30*, 269–289. [CrossRef]
36. Kurland, J.; Wilkins, A.R.; Stokes, P. IPractice: Piloting the effectiveness of a tablet-based home practice program in aphasia treatment. *Semin. Speech-Lang.* **2014**, *35*, 51–63. [CrossRef]
37. Kurland, J.; Liu, A.; Stokes, P. Effects of a tablet-based home practice program with telepractice on treatment outcomes in chronic aphasia. *J. Speech-Lang. Hear. Res.* **2018**, *61*, 1140–1156. [CrossRef]
38. Mortley, J.; Wade, J.; Enderby, P. Superhighway to promoting a client-therapist partnership? Using the internet to deliver word-retrieval computer therapy, monitored remotely with minimal speech and language therapy input. *Aphasiology* **2004**, *18*, 193–211. [CrossRef]
39. Palmer, R.; Enderby, P.; Cooper, C.; Latimer, N.; Julious, S.; Paterson, G.; Dimairo, M.; Dixon, S.; Mortley, J.; Hilton, R.; et al. Computer therapy compared with usual care for people with long-standing aphasia poststroke: A pilot randomized controlled trial. *Stroke* **2012**, *43*, 1904–1911. [CrossRef]
40. Trevittaya, T.K.P.; Chinchai, S. Development of a Thai naming application for clients with aphasia. *J. Assoc. Med. Sci.* **2023**, *56*, 152–161. [CrossRef]
41. Heide, J.; Netzebandt, J.; Ahrens, S.; Brüsch, J.; Saalfrank, T.; Schmitz-Antonischki, D. Improving lexical retrieval with LingoTalk: An app-based, self-administered treatment for clients with aphasia. *Front. Commun.* **2023**, *8*, 1210193. [CrossRef]
42. Choe, Y.; Azuma, T.; Mathy, P. The use of computers and augmentative and alternative communication devices in independent naming practice: Three single-case studies. *J. Med. Speech-Lang. Pathol.* **2010**, *18*, 12–27.
43. Choe, Y.; Asselin, A.; Foster, T.; Waymouth, T.; Van Emmerik, R. Congruent vs. incongruent tasks in interdisciplinary stroke rehabilitation: A single-case report. *Disabil. Rehabil.* **2023**, *46*, 4711–4724. [CrossRef]
44. Quique, Y.M.; Cavanaugh, R.; Lescht, E.; Evans, W.S. Applying adaptive distributed practice to self-managed computer-based anomia treatment: A single-case experimental design. *J. Commun. Disord.* **2022**, *99*, 106249. [CrossRef]
45. Fridriksson, J.; Baker, J.M.; Whiteside, J.; Eoute, D.; Moser, D.; Vesselinov, R.; Rorden, C. Treating Visual Speech Perception to Improve Speech Production in Nonfluent Aphasia. *Stroke* **2009**, *40*, 853–858. [CrossRef] [PubMed]
46. Laganaro, M.; Di Pietro, M.; Schnider, A. Computerised treatment of anomia in chronic and acute aphasia: An exploratory study. *Aphasiology* **2003**, *17*, 709–721. [CrossRef]
47. Choe, Y.-K.; Azuma, T.; Mathy, P.; Liss, J.M.; Edgar, J. The effect of home computer practice on naming in individuals with nonfluent aphasia and verbal apraxia. *J. Med. Speech-Lang. Pathol.* **2007**, *15*, 407–422.
48. Gallée, J.; Pittmann, R.; Pennington, S.; Vallila-Rohter, S. The Application of Lexical Retrieval Training in Tablet-Based Speech-Language Intervention. *Front. Neurol.* **2020**, *11*, 583246. [CrossRef] [PubMed]
49. Fink, R.B.; Brecher, A.; Schwartz, M.F.; Robey, R.R. A computer-implemented protocol for treatment of naming disorders: Evaluation of clinician-guided and partially self-guided instruction. *Aphasiology* **2002**, *16*, 1061–1086. [CrossRef]
50. Ramsberger, G.; Marie, B. Self-administered cued naming therapy: A single-participant investigation of a computer-based therapy program replicated in four cases. *Am. J. Speech-Lang. Pathol.* **2007**, *16*, 343–358. [CrossRef]
51. Pedersen, P.M.; Vinter, K.; Olsen, T.S. Improvement of oral naming by unsupervised computerised rehabilitation. *Aphasiology* **2001**, *15*, 151–169. [CrossRef]
52. Herbert, R.; Webster, D.; Dyson, L. Effects of syntactic cueing therapy on picture naming and connected speech in acquired aphasia. *Neuropsychol. Rehabil.* **2012**, *22*, 609–633. [CrossRef] [PubMed]
53. Choe, Y.; Stanton, K. The effect of visual cues provided by computerised aphasia treatment. *Aphasiology* **2011**, *25*, 983–997. [CrossRef]
54. Mason, C.; Nickels, L.; McDonald, B.; Moses, M.; Makin, K.; Taylor, C. Treatment of word retrieval impairments in aphasia: Evaluation of a self-administered home programme using personally chosen words. *Aphasiology* **2011**, *25*, 245–268. [CrossRef]
55. Cherney, L.R.; Braun, E.J.; Lee, J.B.; Kocherginsky, M.; Van Vuuren, S. Optimising recovery in aphasia: Learning following exposure to a single dose of computer-based script training. *Int. J. Speech-Lang. Pathol.* **2019**, *21*, 448–458. [CrossRef] [PubMed]

56. Cherney, L.R.; Kaye, R.C.; Van Vuuren, S. Acquisition and Maintenance of Scripts in Aphasia: A Comparison of Two Cuing Conditions. *Am. J. Speech-Lang. Pathol.* **2014**, *23*, S343–S360. [CrossRef] [PubMed]
57. Cherney, L.R.; Halper, A.S. Novel Technology for Treating Individuals with Aphasia and Concomitant Cognitive Deficits. *Top. Stroke Rehabil.* **2008**, *15*, 542–554. [CrossRef] [PubMed]
58. Manheim, L.M.; Halper, A.S.; Cherney, L. Patient-Reported Changes in Communication After Computer-Based Script Training for Aphasia. *Arch. Phys. Med. Rehabil.* **2009**, *90*, 623–627. [CrossRef] [PubMed]
59. Thompson, C.K.; Choy, J.J.; Holland, A.; Cole, R. Sentactics®: Computer-automated treatment of underlying forms. *Aphasiology* **2010**, *24*, 1242–1266. [CrossRef]
60. Linebarger, M.C.; Schwartz, M.F.; Kohn, S.E. Computer-based training of language production: An exploratory study. *Neuropsychol. Rehabil.* **2001**, *11*, 57–96. [CrossRef]
61. Hickin, J.; Cruice, M.; Dipper, L. A feasibility study of a novel computer-based treatment for sentence production deficits in aphasia, delivered by a combination of clinician-led and self-managed treatment sessions. *Aphasiology* **2023**, *37*, 1623–1645. [CrossRef]
62. Cherney, L.R. Oral Reading for Language in Aphasia (ORLA): Evaluating the Efficacy of Computer-Delivered Therapy in Chronic Nonfluent Aphasia. *Top. Stroke Rehabil.* **2010**, *17*, 423–431. [CrossRef] [PubMed]
63. Cherney, L.R.; Lee, J.B.; Kim, K.-Y.A.; van Vuuren, S. Web-based Oral Reading for Language in Aphasia (Web ORLA®): A pilot randomized control trial. *Clin. Rehabil.* **2021**, *35*, 976–987. [CrossRef] [PubMed]
64. Duncan, E.S.; Small, S.L. Imitation-based aphasia therapy increases narrative content: A case series. *Clin. Rehabil.* **2017**, *31*, 1500–1507. [CrossRef] [PubMed]
65. Stark, B.C.; Warburton, E.A. Improved language in chronic aphasia after self-delivered iPad speech therapy. *Neuropsychol. Rehabil.* **2016**, *28*, 818–831. [CrossRef]
66. Choi, Y.-H.; Park, H.K.; Paik, N.-J. A Telerehabilitation Approach for Chronic Aphasia Following Stroke. *Telemed. E-Health* **2016**, *22*, 434–440. [CrossRef]
67. Des Roches, C.A.; Balachandran, I.; Ascenso, E.M.; Tripodis, Y.; Kiran, S. Effectiveness of an impairment-based individualized rehabilitation program using an iPad-based software platform. *Front. Hum. Neurosci.* **2015**, *8*, 1015. [CrossRef]
68. Kiran, S.; Roches, C.D.; Balachandran, I.; Ascenso, E. Development of an impairment-based individualized treatment workflow using an iPad-based software platform. *Semin. Speech-Lang.* **2014**, *35*, 38–50. [CrossRef] [PubMed]
69. Goodglass, H.; Kaplan, E.; Weintraub, S. *Boston Naming Test*; Lea and Febiger: Philadelphia, PA, USA, 1983.
70. Holland, A.L.; Frattali, C.; Fromm, D. *Communication Activities of Daily Living: CADL-2*; PRO-ED Inc.: Austin, TX, USA, 1999.
71. Swinburn, K.; Porter, G.; Al, E. *Comprehensive Aphasia Test*; Psychology Press: East Sussex, UK, 2004.
72. Lomas, J.; Pickard, L.; Bester, S.; Elbard, H.; Finlayson, A.; Zoghaib, C. The Communicative Effectiveness Index: Development and Psychometric Evaluation of a Functional Communication Measure for Adult Aphasia. *J. Speech Hear. Disord.* **1989**, *54*, 113–124. [CrossRef] [PubMed]
73. Helm-Estabrooks, N. *Cognitive Linguistic Quick Test*; Harcourt Assessment: London, UK, 2001.
74. Kim, H.; Na, D.L. Normative Data on the Korean Version of the Western Aphasia Battery. *J. Clin. Exp. Neuropsychol.* **2004**, *26*, 1011–1020. [CrossRef]
75. Cho-Reyes, S.; Thompson, C.K. Verb and sentence production and comprehension in aphasia: Northwestern Assessment of Verbs and Sentences (NAVS). *Aphasiology* **2012**, *26*, 1250–1277. [CrossRef]
76. Masterson, J.; Druks, J. *An Object and Action Naming Battery*; Psychology Press: East Sussex, UK, 2000.
77. Roach, A.; Schwartz, M.F.; Martin, N.; Grewal, R.S.; Brecher, A. The Philadelphia Naming Test: Scoring and Rationale. *Clin. Aphasiology* **1996**, *24*, 121–133.
78. Dell, G.S.; Schwartz, M.F.; Martin, N.; Saffran, E.M.; Gagnon, D.A. Lexical access in aphasic and nonaphasic speakers. *Psychol. Rev.* **1997**, *104*, 801–838. [CrossRef] [PubMed]
79. Dardaranda, R.; Potisuk, S.; Gandour, J.; Holasuit, S. Thai Adaptation of The Western Aphasia Battery (WAB). *Bull. Suanprung* **1997**, *13*, 16–21.
80. Kertesz, A. *Western Aphasia Battery*; Grune & Stratton: New York, NY, USA, 1982.
81. Kertesz, A. *Western Aphasia Battery (Revised)*; PsychCorp: San Antonio, TX, USA, 2007.
82. Blanken, G.; Döppler, R.; Schlenck, K.J.; Bautz, M. *Wortproduktionsprüfung*; NAT-Verlag: Hofheim, Germany, 1999.
83. Auclair-Ouellet, N.; Tittley, L.; Root, K. Effect of an intensive comprehensive aphasia program on language and communication in chronic aphasia. *Aphasiology* **2022**, *36*, 1312–1332. [CrossRef]
84. Renvall, K.; Nickels, L.; Davidson, B. Functionally relevant items in the treatment of aphasia (part I): Challenges for current practice. *Aphasiology* **2013**, *27*, 636–650. [CrossRef]
85. Webster, J.; Whitworth, A.; Morris, J. Is it time to stop "fishing"? A review of generalisation following aphasia intervention. *Aphasiology* **2015**, *29*, 1240–1264. [CrossRef]

86. Macoir, J.; Lavoie, M.; Routhier, S.; Bier, N. Key factors for the success of self-administered treatments of poststroke aphasia using technologies. *Telemed. E-Health* **2019**, *25*, 663–670. [CrossRef] [PubMed]
87. Marshall, J.; Booth, T.; Devane, N.; Galliers, J.; Greenwood, H.; Hilari, K.; Talbot, R.; Wilson, S.; Woolf, C. Evaluating the Benefits of Aphasia Intervention Delivered in Virtual Reality: Results of a Quasi-Randomised Study. *PLoS ONE* **2016**, *11*, e0160381. [CrossRef] [PubMed]
88. Giachero, A.; Calati, M.; Pia, L.; La Vista, L.; Molo, M.; Rugiero, C.; Fornaro, C.; Marangolo, P. Conversational Therapy through Semi-Immersive Virtual Reality Environments for Language Recovery and Psychological Well-Being in Post Stroke Aphasia. *Behav. Neurol.* **2020**, *2020*, 2846046. [CrossRef]
89. Harrison, M.; Palmer, R.; Cooper, C. Factors Associated with Adherence to Self-Managed Aphasia Therapy Practice on a Computer—A Mixed Methods Study Alongside a Randomized Controlled Trial. *Front. Neurol.* **2020**, *11*, 582328. [CrossRef] [PubMed]
90. Carragher, M.; Conroy, P.; Sage, K.; Wilkinson, R. Can impairment-focused therapy change the everyday conversations of people with aphasia? A review of the literature and future directions. *Aphasiology* **2012**, *26*, 895–916. [CrossRef]

brain sciences

MDPI

Article

Adapting the Bogenhausen Dysarthria Scales (BoDyS) to Chilean Spanish Speakers: Face and Content Validation

Marcela Sanhueza-Garrido, Virginia García-Flores, Sebastián Contreras-Cubillos and Jaime Crisosto-Alarcón *

Departamento de Ciencias de la Rehabilitación en Salud, Facultad de Ciencias de la Salud y los Alimentos, Universidad del Bío-Bío, Chillán 3800708, Chile; mlsanhueza@ubiobio.cl (M.S.-G.); vgarcia@ubiobio.cl (V.G.-F.); scontreras@ubiobio.cl (S.C.-C.)

* Correspondence: jcrisosto@ubiobio.cl

Abstract: Background: Dysarthria is a neuromotor speech disorder that significantly impacts patients' quality of life. In Chile, there is a lack of culturally validated instruments for assessing dysarthria. This study aimed to cross-culturally adapt the Bogenhausen Dysarthria Scales (BoDyS) into Chilean Spanish and to conduct face and content validation. **Methods:** The adaptation process included translation and back-translation, followed by validation by a panel of experts. Clarity, format, and length were evaluated, and the Kappa index (KI), content validity index (CVI), and content validity ratio (CVR) were calculated to confirm item relevance. A pilot test was subsequently conducted with ten speech–language pathologists to apply the adapted version to patients. **Results:** The adaptation process produced a consensus version that preserved the semantic and cultural characteristics of the original scale. The statistical measures (KI = 1.00; I-CVI = 1.00; S-CVI/Ave = 1.00; S-CVI/UA = 1.00; CVR = 1.00) indicated satisfactory levels of agreement. The pilot test demonstrated the scale's appropriateness and effectiveness for assessing dysarthria within the Chilean context, although some experts recommended reducing task repetition for patients prone to fatigue. **Conclusions:** The Chilean version of the BoDyS (BoDyS-CL) is a valid and useful tool for evaluating dysarthria in Chile. This study provides a foundation for further research and the systematic implementation of this scale in local clinical practice.

Keywords: speech assessment; neuromotor speech disorders; dysarthria evaluation; cross-cultural adaptation; cross-linguistic validation

1. Introduction

Dysarthria is a group of neurogenic speech disorders characterized by abnormalities in the strength, speed, range, steadiness, tone, or accuracy of movements required for respiration, phonation, resonance, articulation, and prosody, all of which enable speech production [1]. This speech disorder is caused by damage to the areas of the central or peripheral nervous system that control the muscles responsible for speech production [1]. According to Duffy (2013), the pathogenesis of dysarthria can be classified according to the site of injury, resulting in different types of dysarthria, such as spastic, flaccid, ataxic, hypokinetic, and hyperkinetic. Damage to cerebral motor areas, such as the motor cortex, basal ganglia, cerebellum or nerve pathways, affects coordination, strength, and speech control. The severity and specific symptoms depend on the type and location of the lesion [2].

Clinical manifestations include alterations in articulation, rhythm, pitch, and intensity of speech [2]. These include slow or rapid speech, imprecision in the articulation of words, and alterations in prosody, such as monotony and increased or decreased muscle tone (spasticity or flaccidity) that affects breathing and vocal control [3]. The speech difficulties associated with dysarthria can significantly hinder communication and daily activities, impacting psychosocial dynamics and diminishing the patient's overall quality of life [4,5]. Additionally, depending on the underlying cause of dysarthria, it may coexist with other neurogenic disorders affecting language, cognition, and/or swallowing [6]. To comprehensively assess a patient's condition, it is crucial to conduct functional assessments using standardized methodologies. This approach facilitates the diagnostic process and supports the development of tailored therapeutic plans that align with the unique needs of each person.

Currently, there is no consensus on the most appropriate method for assessing dysarthria or the specific aspects that should be considered. However, a comprehensive initial speech assessment generally should include the following components: (1) a thorough clinical history, (2) an examination of the speech mechanism and oral motor skills, (3) an assessment of the speech subsystems (including respiration, phonation, articulation, resonance, and prosody), (4) a perceptual assessment, and (5) an evaluation of intelligibility [7]. The American Speech–Language–Hearing Association also suggests that functional speech characteristics, such as naturalness, efficiency, and intelligibility, should be integral to the assessment process [8]. Furthermore, it is crucial to evaluate the impact of dysarthria on the patient's activities and participation, as this helps assess the disorder's social and psycho-emotional effects on the person [5,8,9].

The evidence comparing assessment instruments for dysarthria indicates several key issues: (1) a lack of standardization regarding the aspects evaluated, scoring criteria, and administration procedures; (2) a limited scope, where some protocols concentrate solely on articulation or prosody, while others prioritize intelligibility or functional communication; (3) insufficient sensitivity, failing to account for the cultural and linguistic diversity of individuals with dysarthria; (4) notable limitations in assessing quality of life; and (5) the necessity of implementing assessment tools that extend beyond just speech components [10–14]. Furthermore, a preference for informal assessment tools over standardized ones has been noted, particularly in Spanish-speaking countries, where standardized protocols are often unavailable [14]. Additionally, Chilean Spanish's phonetic and phonological characteristics [15,16] may affect how speech intelligibility and naturalness are perceived in individuals with dysarthria. Therefore, cultural and linguistic adaptation of any assessment tool are crucial to ensure its applicability and accuracy.

Currently, the predominant framework for diagnosing dysarthria is a classification system developed at the Mayo Clinic, grounded in the perceptual characteristics of the disorder [17–19]. These characteristics aid in identifying the underlying neurological conditions associated with this speech impairment. However, it is important to note that this approach may have limitations due to the numerous variables involved [20]. Assessment protocols such as the Frenchay Dysarthria Assessment-2 [21] and the methods proposed by Duffy [2] emphasize the importance of using a broad range of assessment tasks, enabling therapists to better understand their patients' speech disorders. However, the significant inter-rater variability in clinical assessments can pose challenges in this process [20]. In this context, the Bogenhausen Dysarthria Scales (BoDyS) [20] offer a promising alternative by providing a standardized measurement of clinically relevant dimensions of speech affected by dysarthria.

The BoDyS employs speech tasks exclusively for assessment. These tasks include open-ended questions to elicit spontaneous speech, sentence repetition, reading short texts, and storytelling from images, resulting in 12 audio-visual speech samples for later analysis. The application is adaptable to meet the clinical needs of each patient, offering three versions: the complete version, which encompasses all four modalities and twelve tasks; an abbreviated version that optimizes time by reducing the number of tasks; and a version without reading, tailored for patients with visual or cognitive challenges. Assessments typically last between 15 and 30 min, with subsequent analysis taking 20 to 40 min, depending on the examiner's expertise. The BoDyS assesses nine dimensions of speech and demonstrates high inter-rater reliability, with indices exceeding 0.85, alongside empirical validity verified by factor analysis. Its practical approach facilitates treatment planning and monitoring progress or recovery, while auditory perception aids therapeutic intervention. This flexibility and clinical robustness combination makes the BoDyS a valuable tool for assessing dysarthria in clinical and research contexts [20].

In Chile, the work conducted at the Mayo Clinic has significantly contributed to the advancement of the field, leading to the development of instruments such as the Speech Assessment Protocol [22] and the Speech Assessment Protocol for Dysarthria (PEV-H-Disartria for its acronym in Spanish) [23]. These tools incorporate the assessment of neuromotor aspects of speech, as well as the assessment of various speech subsystems and their functional characteristics. However, Chile currently lacks standardized instruments for assessing speech in individuals with dysarthria, instruments that would allow for assessing their performance over time and tracking progress in treatment through speech-only tasks [14]. Based on the above, there is a need for a culturally and linguistically adapted dysarthria assessment tool adapted to Chilean Spanish, which will improve the accuracy of diagnosis, treatment planning, and progress monitoring in the rehabilitation of patients with dysarthria in Chile.

Considering the above, this study aims to carry out the cross-cultural adaptation to Chilean Spanish and the face and content validation of the Bogenhausen Dysarthria Assessment Scales (*Bogenhausener Dysarthrieskalen*, BoDyS) [20]. To our knowledge, this is the first investigation to assess the validity of the BoDyS in a language other than German. This adaptation will provide a tool tailored to the Chilean context, enabling professionals to effectively assess speech in individuals with dysarthria using stimuli, parameters, and rubric-based scoring scales relevant to the local context.

2. Materials and Methods

The methodological process was carried out in four sequential stages, encompassing the translation and adaptation of the BoDyS, expert validation, pilot testing, and final adjustments. These stages are summarized in Figure 1.

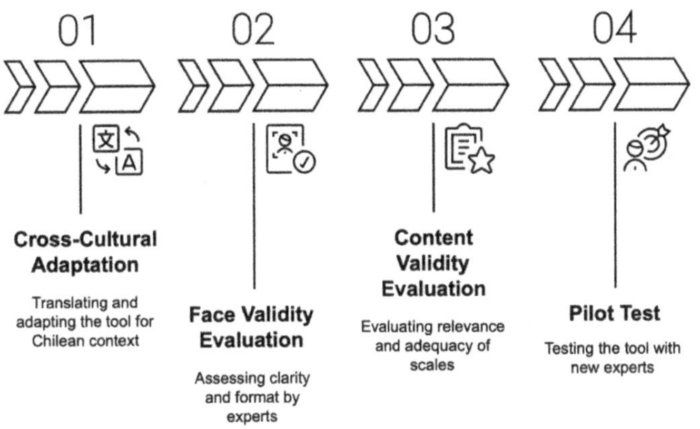

Figure 1. Flowchart of the four methodological stages: (1) translation and cultural adaptation; (2) expert assessment of face validity; (3) expert assessment of content validity; and (4) pilot testing.

In its original German version, the BoDyS is a standardized assessment tool designed to assess dysarthria in adults. It allows for assessing the overall severity of dysarthria and provides a comprehensive overview of the characteristics of the disorder by generating a detailed profile [24]. This test relies solely on speech tasks, employing auditory analysis methods that emphasize the functional aspects of motor speech performance. The BoDyS comprises four tasks: spontaneous speech, oral repetition, reading aloud, and describing sequences from images. Each task is repeated three times with different stimuli, resulting in a total of 12 speech samples to be analyzed by the evaluator. The evaluation utilizes nine scales that address the primary dimensions of dysarthric impairment: respiration, voice level, voice quality, voice stability, articulation, resonance, articulation rate, fluency, and prosodic modulation. Each speech sample is assessed on these nine scales, receiving a score from 0 to 4, where a score of 4 indicates no impairment and a score of 0 represents the most severe level of impairment [20]. Detailed characteristics of each scale are presented in Table 1.

Table 1. BoDyS and their features.

Scale	Features
RSP (respiration)	Frequent inspirations Audibly effortful inspirations Expiration beyond the resting level
VOL (voice level)	High-pitch Low-pitch Loud voice Soft voice
VOQ (voice quality)	Breathy and harsh Strained–strangled and harsh
VOS (voice stability)	Fluctuations of pitch, loudness, and/or quality Voice breaks and voice fading Voice tremor Involuntary vocalizations
ART (articulation)	Reduced Overshooting Wide Narrow Variable
RSN (resonance)	Hypernasality and nasal emission Hyponasality Intermittent hyper- or hyponasality
TEM (articulation rate)	Decreased articulation rate Increased articulation rate
FLU (fluency)	Pauses Iterations
MOD (prosodic modulation)	Reduced prosodic modulation of pitch and/or loudness Excessive prosodic modulation of pitch and/or loudness Conspicuous rhythm and/or stress patterns

Note: Referenced based on the proposal by Ziegler et al. (2017) [20].

2.1. Stage 1: Cross-Cultural Adaptation Procedure

To ensure the cross-cultural adaptation of the instrument in its complete version for this stage of work, we employed the back-translation methods proposed by Ramada-Rodilla [25] and Wild et al. [26]. Initially, two certified native Chilean translators independently translated the scales and the stimuli manual from German into Spanish. In collaboration with field experts, the research team then developed a consensus version based on these translations. This preliminary consensus version was subsequently presented to three Chilean speech–language pathologists, who evaluated its comprehensibility, the appropriateness of the vocabulary, and the relevance of the translated material. The professionals selected for this task met the following inclusion criteria: (a) they were licensed speech–language pathologists, (b) they had a minimum of five years of professional experience, and (c) they possessed at least three years of experience in speech assessment and intervention. Following this evaluation, the consensus version underwent back-translation into the original language conducted by a professional Spanish–German translator who was a native German speaker. This step aimed to compare the original version of the BoDyS protocol with the back-translated version, identifying any similarities or potential discrepancies in content or vocabulary arising from the modifications.

Specifically, to evaluate the translation of the stimuli manual, the translation was submitted to a professional linguist, who had to adapt the stimuli of the assessment tasks, considering semantic, phonological, and lexical aspects pertinent to Chilean Spanish. It is important to mention that the stimuli used in the BoDyS-CL incorporated a syllabic metric for the stimulus words, like the German version. Likewise, the length and types of sentences (affirmative, exclamatory, and interrogative) have the same characteristics as the original version. Regarding the texts created for the BoDyS-CL, similar themes to those addressed in the German version were considered. For example, one of the texts in the original version is about a traditional German character, the "carpet cleaner." In contrast, a culturally equivalent character was considered in the Spanish version: the "knife sharpener." The entire process of adapting the stimuli was evaluated and guided by a team of speech–language pathologists, a translator, and a linguist.

Additionally, it is important to note that new images were created, distinct from the original German version, to align with the Chilean cultural context while preserving the number of scenes in each sequence.

At this stage of the study, the potential inter-dialectal applicability of the instrument across Spanish-speaking populations has not yet been examined. This is because the phrases, texts, and images were developed with reference to the cultural context of Chile. Nevertheless, future stages may involve the adaptation of stimuli to ensure their suitability and relevance for other varieties of Spanish.

2.2. Stage 2: Procedure for Face Validity

Both the face and content validity were evaluated by a panel of experts consisting of ten Chilean speech–language pathologists (eight women and two men), who were selected through convenience sampling and in accordance with the criteria set forth by Cabero and Barroso [27]. These criteria allow for the calculation of the so-called 'expert competence coefficient' or 'K coefficient', which is derived from a self-assessment conducted by the individual to determine their expert competence in the research subject. It is calculated using the following formula: $K = 1/2\ (Kc + Ka)$, where Kc is the 'knowledge coefficient', representing the expert's knowledge about the given topic or problem, and Ka is the 'argumentation coefficient', which refers to the reasoning behind the expert's criteria [27].

All panel members had at least five years of experience and worked in different health centers in Chile to care for patients with neuromotor speech disorders.

The BoDyS underwent a face validity evaluation process. These scales consist of the stimuli manual, a description of the scales and their features, and the recording sheet used to note the frequency of occurrence of dysarthria characteristics, also referred to as features, for each scale. Each component of the BoDyS was assessed using three criteria: (1) clarity, which examines whether the text is clear and possesses appropriate semantic and syntactic qualities; (2) format, which evaluates whether the presentation of information, images, font, and font size are suitable; and (3) length, which considers whether the length of the texts is appropriate and facilitates the reader's understanding. These criteria were rated on a 5-point Likert scale, with 1 indicating 'strongly disagree' and 5 signifying 'strongly agree'.

2.3. Stage 3: Procedure for Content Validity

Content validity refers to the extent to which a measurement instrument encompasses the various dimensions of the concept being assessed. An instrument is deemed content-valid if it effectively addresses all relevant aspects of the concept under investigation [28]. Both qualitative and quantitative approaches were used to evaluate the content validity of the instrument [29,30].

In this stage, the same ten expert speech–language pathologists who participated in the face validity assessment were involved. Participants were tasked with completing a rubric to assess the clarity and relevance of each item, as well as the adequacy of the scales. They used Likert-type indicators rated from 1 to 4 to assess each criterion. The scale corresponded to the following meanings: 1 indicated that the item does not meet the criterion, 2 signified a low level of compliance, 3 represented a moderate level of compliance, and 4 indicated a high level of compliance [31]. Empirical techniques were then applied to calculate the content validity index (CVI), the content validity ratio (CVR), and the modified Kappa agreement index (K) [32,33].

The content validity index (CVI) quantifies the content validity of multiple-item scales based on expert ratings regarding the relevance of each item. This indicator is determined by the proportion of experts who rated an item with a score of 3 or 4 compared to the total number of experts who assessed that item in terms of clarity and relevance [32]. The scale content validity index (S-CVI) is used to assess validity by scale, which is derived from the proportion of items that received a score of 3 or 4 in terms of clarity and relevance from at least 80% of the experts [34]. The item content validity index (I-CVI) is employed to evaluate each feature within the scales, calculated as the proportion of experts who rated 3 or 4 divided by the total number of experts who rated that item [32]. An I-CVI value of $\geq$0.83 is deemed acceptable. Additionally, universal agreement among experts (S-CVI/UA) and average agreement (S-CVI/Ave) were computed, with values of S-CVI/UA $\geq$ 0.83 and S-CVI/Ave $\geq$ 0.9 considered satisfactory [29]. To calculate the modified Kappa statistic, first, the probability of chance agreement is calculated for each element through the following formula: $K = (\text{I-CVI} - Pc)/(1 - Pc)$, where $Pc = [N!/A!\,(N - A)!] * 0.5\,N$. In this formula, Pc is the probability of chance agreement, N is the number of experts, and A is the number of experts who agree that the item is relevant. Kappa values above 0.74 are considered excellent [32].

The qualitative data for content validity were obtained from all the comments provided by the experts in the observations section of the evaluation rubric.

The entire validation process was carried out in two stages. In the first stage, the experts evaluated the material in terms of face and content, after which the research team made the changes suggested based on their observations. Subsequently, in the second stage, the experts were asked to confirm whether these modifications appropriately addressed the recommended adjustments.

2.4. Stage 4: Pilot Test Procedure

Following the cross-cultural adaptation and validation, both for face and content, a pilot test was conducted to obtain a final version. For the pilot test, 10 new experts were selected through convenience sampling (7 women and 3 men), all working in different healthcare centers in Chile, providing care to patients with speech neuromotor disorders.

In this case, the K coefficient was not calculated, as the primary interest at this stage was the participants' clinical experience in assessing subjects with dysarthria rather than their status as experts from an academic perspective. For this reason, the language-speech pathologists who participated in this phase were not the same as those in the previous stages.

Once the instrument was adapted and validated, each expert was required to assess a patient with dysarthria. Following this, the participating speech–language pathologists had to complete a cognitive interview. Through this interview, the following aspects were explored: (1) their overall impression of the BoDyS, (2) their opinion regarding the instructions of the BoDyS-CL, (3) their insights on the feasibility of applying the BoDyS-CL, (4) their perspectives on the content of the BoDyS-CL, and (5) their evaluation of the tasks in the BoDyS-CL. Based on the analysis of the experts' feedback, final modifications were made to the instrument's final version.

3. Results

3.1. Stage 1: Findings from Cross-Cultural Adaptation

In this stage, based on the translation and back-translation procedures, some technical terms used in Chile to describe dysarthria characteristics were adjusted in the 'Respiration during Speech' scale ('respiration' in the original version) and the 'Vocal Quality' scale ('vocal quality' in the original version). Additionally, a clinical description was incorporated for each feature to guide clinicians and provide greater clarity on how to identify them during clinical assessment. This process resulted in the agreed-upon and adapted version of the description of the scales and their features.

Regarding the linguistic adaptations of the stimuli in the Chilean Spanish version, the following changes were made. (1) In the "Spontaneous Language" task, which consists of 12 questions to request a speech sample, the questions were adapted following the rules of the original scale in terms of semantic accuracy, length, and complexity. (2) In the "Sentence Repetition" task, which consists of 15 sentences, these were adapted following the rules of the original scale regarding lexical, phonological, length, and complexity aspects. (3) An example of the sentences in German and Spanish, respectively, is "What is your profession?" (translated from German), with the Chilean adaptation being "What is your occupation?". The "Reading" task, consisting of three new texts, was created following the rules of the original scale regarding semantic accuracy, length, and complexity. (4) In the "Picture Description" task, which consists of three new sequences of four images each, they were created following the rules of the original scale regarding the number of images and complexity (Figure 2).

Figure 2. Picture No. 2 for the description of BoDyS-CL.

3.2. Stage 2: Face Validation Results

Experts for this study stage were selected through a questionnaire that assessed their expertise in the relevant area, employing the K coefficient for expert selection [27]. Those experts who achieved a score higher than 0.8 were included, as their level of knowledge was deemed sufficient. A group of 10 experts were engaged in this process, which was conducted in two rounds to allow the experts to review the changes made after the first round and to facilitate a new validation process.

The face validation focused on specific sections of the BoDyS: the stimuli manual, the descriptions of the scales and features, and the recording sheet. Each section was evaluated based on clarity, length, and format using a 5-point Likert scale. In the first round, the experts evaluated the face of the instrument as follows:

In the evaluation of the stimuli manual, regarding the criterion of clarity, 30% of the experts suggested adjusting the instruction for the task "Spontaneous Language Instruction," while the remaining 70% fully agreed with the clarity of the instructions. Regarding the format criterion, 90% of the experts were completely satisfied with the presentation of the information, images, font, and/or font size. However, one expert suggested increasing the font and image sizes to improve visibility for potential patients. Regarding the length of the manual, 90% of the experts indicated that the length of the stimuli was appropriate and allowed for reading comprehension. However, one expert mentioned that the number of tasks might be excessive for patients who experience higher fatigability. It is important to note that the original version of BoDyS includes a reduced version with fewer tasks to mitigate fatigue effects. Nevertheless, this study was conducted using the full version.

Regarding the scales' description and features, 80% of the experts stated that the text is clear and contains appropriate semantic and syntactic aspects. Similarly, 80% of the experts indicated that the format, in terms of information presentation, images, font, and/or font size, is adequate. Regarding the length of this section, 90% of the experts indicated that the "length of the texts is appropriate and allows for reading comprehension." However, one expert mentioned that the description of the features might be too extensive to cover within a single assessment session.

Regarding the recording sheet, where the frequency of occurrence of dysarthria characteristics—or features—is recorded for each scale, the entire group of experts fully agreed that the text contains appropriate semantic, syntactic, and length-related aspects concerning the criteria of clarity and length. Concerning the format criterion, 20% of the experts indicated that it would be necessary to include a summary table of performance scores to guide the process of constructing the dysarthria profile.

Based on the experts' evaluation, the following changes were introduced in the recording sheet: (1) the instruction on how to record the features was incorporated, (2) a table summarizing the total task scores for the scales was added, and (3) a table for constructing the dysarthria profile was included. Similarly, based on the experts' feedback, the following modifications were made to the stimuli manual: (1) the font size of the stimuli was increased, and (2) numbering was added to the image sequences to facilitate their description.

During the second round of face validation, unanimous agreement was achieved for clarity, length, and format.

3.3. Stage 3: Content Validation Results

This process involved validating the description section of the scales and their corresponding features to assess the clarity and relevance of each feature, as well as the adequacy of the scales. Both qualitative and quantitative analyses of the results were conducted. For the qualitative analysis, all expert comments were critically examined. Based on this analysis, the experts suggested the only modification to the 'Vocal Characteristics During Speech (REG)' scale, changing feature 3 and feature 4 from 'very loud volume' and 'very soft volume' to 'increased volume' and 'decreased volume', respectively.

On the quantitative side, regarding the I-CVI indicator, which was applied to evaluate the clarity and relevance of the scale items, there was absolute agreement among the evaluators (I-CVI = 1.00) for each feature in every scale. This indicates that all the evaluated features were considered clear and relevant.

For the S-CVI indicator, which was applied to quantify the clarity and relevance criteria of the scale in general and used to evaluate each of the scales, the values obtained were 1.00 for both the average agreement (S-CVI/Ave) and the universal agreement among experts (S-CVI/UA) for the two evaluated criteria of clarity and relevance. In this case, following Zamanzadeh's criteria [29], high levels of agreement among experts were observed in the evaluation of these criteria.

The Kappa index was calculated based on the clarity and relevance criteria. In both review rounds, all Kappa indexes were equal to 1, indicating an excellent level of agreement.

The CVR index was calculated to determine the relevance of each item. In both rounds, values of 1.00 were obtained, exceeding the previously established thresholds [29,30,35] for acceptability. As a result, all items were considered essential for the instrument.

After completing the cross-cultural adaptation and the face and content validation processes, a pilot test was conducted to finalize the Chilean Spanish-adapted version of the BoDyS.

3.4. Results from the Pilot Test

Ten expert speech–language pathologists assessed a patient with dysarthria using the adapted and validated version of the BoDyS. Following this, each professional participated in a cognitive interview to identify potential difficulties in applying the instrument and to understand the clinician's experience during the process.

The information gathered from the interviews was systematically organized, yielding the following results:

Concerning the general impression of the cross-culturally adapted Chilean Spanish version of the BoDyS (BoDyS-CL), the interviewees noted that the test provides accurate information about the patient's performance and is easy to apply. However, they pointed out that the application time could be lengthy, which may cause patients with more severe impairments to fail to complete the test due to fatigue.

Regarding the instructions of the BoDyS-CL, the interviewees remarked that the guidelines for each task are clear, and all experts unanimously agreed that no changes are necessary.

In terms of the feasibility of applying the BoDyS-CL, the interviewees indicated that the organization, instructions, and format of the assessment tasks are appropriate. However, they suggested improving the clarity and size of images used in the picture sequence description task.

Concerning the content of the BoDyS-CL, interviewees reported that the instrument is highly comprehensible, and the instructions are appropriate, with no need for rewording. They highlighted that the open-ended question task to elicit spontaneous speech is the most effective way of assessing the patient's speech performance. However, they also pointed out that the complete application of the test is lengthy for patients, and they did not see the need to repeat the tasks three times, as the original full version of the test requires.

The opinions regarding the tasks of the BoDyS-CL were generally positive, with the instrument regarded as sufficient. However, there is a highlighted need for tasks that allow for the determination of the neuromotor performance of speech structures and the necessity of evaluating the fundamental motor processes of articulation in more detail. This approach would enable a more accurate characterization of altered speech sounds.

Based on the evaluation, it was decided to include more detailed information regarding test scoring in the instruction manual. With all the processes outlined here, the Chilean version of the BoDyS is considered both adapted and validated.

4. Discussion

The cultural adaptation and content validation process for the BoDyS, which evaluates speech in individuals with dysarthria and was initially developed in German, was carried out using exhaustive and iterative processes. The final version includes 9 scales describing different dimensions of speech, along with a stimuli manual containing 12 speech tasks. The Chilean Spanish adapted scales were critically evaluated for face and content validity through a judgment from 10 experts. Following this, they were subjected to a pilot test by 10 professionals with experience working with individuals with dysarthria.

While the direct translation process for the cross-cultural adaptation progressed without issues, a "clinical description" section was added to the scale descriptions of the original version for each of the features per scale. This addition aimed to assist evaluators and provide more precise guidance on identifying speech impairments during assessments. As a result, a consensus was reached on the adapted versions of the scale descriptions and their associated features.

As a result of the observations made by the experts regarding face validity, modifications were made to the instrument format, including an increase in the size of the stimuli and the font. These changes were implemented due to concerns about their potential impact on patients' performance. This adjustment considers that dysarthria is more prevalent among older adults, a demographic that is at a higher risk of visual impairments, such as age-related macular degeneration (AMD). This condition significantly decreases quality of life, particularly affecting the ability to read, recognize faces, and drive [36]. The most recent National Disability and Dependency Survey in Chile [37] reports that 44% of individuals over 60 years old experience some degree of visual impairment, with 42% experiencing partial vision loss and 2% being entirely blind. Additionally, dysarthria is commonly associated with Parkinson's disease, which can also lead to visual impairments, such as reduced eye movement and difficulty distinguishing visuospatial aspects. Even with visual stimuli, Parkinson's patients may have difficulty processing certain information [38]. It is estimated that 82% of individuals with Parkinson's disease experience at least one ophthalmological symptom [39]. It is worth noting that the original BoDyS version provides standard norms for a test variant that excludes reading tasks, thus supporting the assessment of patients with visual (and other) reading difficulties [20].

Another modification implemented was the addition of numbering to the images used in storytelling. This measure was introduced to reduce cognitive load, particularly considering some patients may experience dysarthria alongside language or cognitive impairments, which could impact assessment outcomes [40,41]. The intention behind this scale is not to increase the cognitive effort needed to follow the sequence of images but rather to ensure that patients can comprehend the images and narrate a story fluently, thereby enabling the collection of a meaningful speech sample. In older adults, there is a decrease in cognitive and motor processing speed [42]. Therefore, it is crucial to consider the effects of this slowdown on performance in various cognitive tasks when interpreting impairment profiles in this population, particularly by distinguishing between the speed and quality of performance across different activities. Additionally, it is essential to note that patients who have suffered a stroke frequently experience an immediate decline in their cognitive abilities, including thinking and reasoning skills, as well as a gradual deterioration in mental processes over time. These patients have a higher likelihood of experiencing accelerated deterioration in their cognitive and planning abilities for at least six years following the medical incident [43,44]. This aspect must be considered when assessing patients with neuromotor disorders, such as dysarthria, and their performance in speech-related tasks.

On the other hand, modifications aimed at improving clarity focused on how to track results. Experts noted that the original instructions were insufficiently precise, potentially leading to errors in analysis and incorrect diagnoses. Proper record-keeping enables more accurate result analysis, thereby enhancing decision-making for more effective therapeutic interventions [45]. Furthermore, accurate records facilitate data collection that can be compared with future assessments, supporting more rigorous monitoring of the patient's progress.

Regarding task length, some experts have noted that repeating tasks three times may be excessively tiring for specific patients with dysarthria. This insight stems from the understanding that, depending on the underlying cause of dysarthria, a range of cognitive or motor characteristics may be present. These characteristics can vary based on the level and severity of the lesion, potentially affecting the central nervous system, the peripheral nervous system, or both [22]. For instance, patients may experience cognitive fatigue, particularly those with hypokinetic dysarthria associated with Parkinson's disease. Cognitive impairments in this type of dysarthria are primarily evident in executive functions and visuospatial skills, accompanied by a marked decline in information processing speed, which can hinder performance during extended assessments [46]. On the other hand, in other cases, fatigue can have a significant impact at the motor level, as seen in myasthenia gravis, which affects the neuromuscular junction and causes an inability of muscles to maintain the strength of their contractions over time, leading to rapid fatigue and muscle weakness [47]. However, it is important to highlight that the BoDyS have abbreviated versions, consisting of eight speech samples—that is, two rounds of tasks—specifically designed for patients with greater fatigability and attentional difficulties. Additionally, there is a non-reading version that includes a total of nine speech samples and excludes all tasks involving reading.

These versions which will be evaluated and validated in a subsequent stage. The availability of abbreviated formats will facilitate more appropriate assessments for patients experiencing either motor or cognitive fatigue.

Regarding the content validity process, the scale that received the most suggestions from the expert panel was the 'vocal characteristics during speech' scale, particularly the 'volume' feature, where the terms 'very soft' and 'very loud' were replaced with 'increased' and 'decreased'. This consensus and subsequent modification were based on the technical language used in the existing dysarthria literature, where the concepts of increase and decrease describe variability in tone and volume [1]. In terms of the quantitative content validity results, both the first and second rounds were deemed adequate in clarity, relevance, and sufficiency, meeting the necessary criteria to proceed to the pilot phase.

The pilot phase allowed for identifying potential new difficulties during the application of the instrument and for gaining insights into the clinician's experience throughout the process. The experts indicated that the test provides accurate information about the patient's performance, is easy to apply, and provides clear instructions. This aspect is particularly significant, as it enhances the validity and reliability of the results, minimizing the risk of errors that could compromise diagnostic accuracy [45]. It also reduces inter-observer variability, which is crucial for maintaining assessment consistency and improving clinical outcomes [48]. Furthermore, the clarity of the instructions contributes to operational efficiency, allowing procedures to be carried out more smoothly with fewer clarifications or repetitions. Evidence suggests that clear protocols reduce the time required to complete assessments without compromising their accuracy [49].

Among the various tasks in the assessment, the "Spontaneous Speech" task stands out as it offers a more comprehensive understanding of speech performance. This task provides valuable insights into how patients communicate in real-life situations, an aspect that structured or repetition tasks cannot fully capture. The inherent variability of spontaneous tasks more accurately reflects the challenges patients face daily, leading to a more representative assessment [50]. Furthermore, it is important to note that spontaneous speech tasks require the integration of the subcomponents of speech, allowing the clinician to observe how these systems interact during natural speech production. This perspective is supported by studies such as those conducted by Dos Santos et al. [51] or Brendel et al. [52], indicating that spontaneous speech tasks deliver a broader view of a patient's neuromotor speech capabilities. However, experts generally mentioned that repeating the tasks three times results in an overly long application time, during which older patients and/or those with neurological impairments may experience cognitive and motor fatigue, thus preventing them from completing the test. It is worth noting that these tasks do not necessarily need to take much time. Each task consists of only one topic/question, and the examiner may decide to end the interaction as soon as she/he has picked up enough of the patient's speech symptoms in her/his response. This characteristic is essential to prevent patient fatigue when taking the test; however, due to the nature of the validation process, the judges had to familiarize themselves with the full instrument rather than its shortened versions to ensure the availability of the complete Spanish version.

Additionally, experts highlighted the need for this scale to include tasks assessing the neuromotor performance of the anatomical structures involved in speech. Duffy [2] emphasizes the importance of a detailed understanding of anatomical alterations when accurately identifying the source of speech difficulties in patients with dysarthria. These alterations can impact speech production in various ways, ranging from muscular weakness to coordination issues in speech movements. Conversely, Feenaughty (2021) [41] points out that utilizing detailed clinical assessments allows for a more precise assessment of the orofacial structures, which in turn facilitates the customization of treatments to meet each patient's specific needs.

Given the importance of accurate diagnosis and precise monitoring during the therapeutic management of dysarthria, the BoDyS can potentially improve clinical decision-making, therapeutic planning, and outcome monitoring in the Chilean healthcare context by providing comparable measures of patient performance over time [53].

While this study presents valuable contributions, several limitations must be acknowledged. The sample size was small, which may affect the representativeness of the results. However, a notable degree of consensus was observed among the judges during both rounds of the Delphi process. It is important to recognize that since the adaptation is specifically for Chilean Spanish, all judges were Chilean, which may restrict its applicability to other Spanish variants. Additionally, it is essential to note that content and face validation do not assess the instrument's performance in actual clinical practice, thus not ensuring its effectiveness in assessing patients with dysarthria. Therefore, future research should complement this phase with reliability studies that examine the scales' performance in pilot tests with patients, including test–retest reliability and inter-rater consistency, to ensure the instrument's stability over time and across evaluators. Also, comparisons of the instrument's results with objective measures of speech and across different populations with dysarthria should also be included to verify its effectiveness in varied clinical and linguistic contexts.

Overall, the BoDyS should be supplemented to ensure a comprehensive assessment of a patient's performance, as it currently overlooks aspects related to the patient's quality of life and does not include neuromotor or orofacial assessments. These elements are essential for providing more precise insights into the dysarthria level and type, as recommended by ASHA [8]. Therefore, it is suggested that the BoDyS be associated with additional instruments, such as orofacial motor examinations and the QOL-DyS, a self-assessment questionnaire that measures health-related quality of life in individuals with dysarthria [54]. Including such complementary assessments would enhance the clinical utility of the BoDyS and provide a more holistic understanding of the patient's communicative functioning.

5. Conclusions

The cross-cultural adaptation and validation of the BoDyS to Chilean Spanish provides a valuable tool for assessing dysarthria within the Chilean context. The findings indicate that this adapted version is clear and relevant in its instructions, content, and format, making it suitable for assessing the functional aspects of speech. However, experts identified areas for improvement, such as task duration and the need for supplementary assessment of neuromotor and orofacial functions.

This study provides a foundation for the systematic implementation of the adapted BoDyS in clinical practice in Chile. It paves the way for future research that may include the assessment of other relevant dimensions in dysarthria treatment. Ultimately, this tool aims to facilitate the development of more effective and personalized therapeutic interventions, thereby enhancing the quality of life for patients with dysarthria.

Author Contributions: Conceptualization, M.S.-G., V.G.-F. and S.C.-C.; methodology, M.S.-G. and J.C.-A.; software, M.S.-G., V.G.-F. and J.C.-A.; validation, M.S.-G. and J.C.-A.; formal analysis, M.S.-G. and J.C.-A.; investigation, M.S.-G., V.G.-F., S.C.-C. and J.C.-A.; resources, M.S.-G. and J.C.-A.; data curation, M.S.-G. and J.C.-A.; writing—original draft preparation, M.S.-G., V.G.-F., S.C.-C. and J.C.-A.; writing—review and editing, M.S.-G., V.G.-F., S.C.-C. and J.C.-A.; visualization, M.S.-G. and J.C.-A.; supervision, J.C.-A.; project administration, M.S.-G.; funding acquisition, M.S.-G. and J.C.-A. All authors have read and agreed to the published version of the manuscript.

Funding: This research was funded by Universidad del Bío-Bío, Chile, grant number GI2309834 and 2040443IF/R.

Institutional Review Board Statement: The study was conducted in accordance with the Declaration of Helsinki and approved by the Ethics Committee of Hospital San Juan de Dios of Santiago, Chile (n 225/2023, approved on 6 July 2023).

Informed Consent Statement: Informed consent was obtained from all subjects involved in the study.

Data Availability Statement: The data supporting the conclusions of this article cannot be provided due to the sensitivity of the data collected during the study.

Conflicts of Interest: The authors declare no conflicts of interest.

References

1. Duffy, J.R. *Motor Speech Disorders*, 4th ed.; Mosby: London, UK, 2019.
2. Duffy, J.R. *Motor Speech Disorders*, 3rd ed.; Mosby: London, UK, 2013.
3. Freed, D.B. *Motor Speech Disorders: Diagnosis and Treatment*; Plural Publishing: San Diego, CA, USA, 2018.
4. Ball, L.J.; Beukelman, D.R.; Pattee, G.L. Communication effectiveness of individuals with amyotrophic lateral sclerosis. *J. Commun. Disord.* **2004**, *37*, 197–215. [CrossRef] [PubMed]
5. Aceituno, C.; Aranda, S.; Palma, G.; Pino, C.; Villegas, P.; Quezada, C.; Toledo-Rodríguez, L. Calidad de vida en personas con disartria. *Rev. Logop. Foniatría Audiol.* **2019**, *39*, 27–31. [CrossRef]

6. Flowers, H.L.; Silver, F.L.; Fang, J.; Rochon, E.; Martino, R. The incidence, co-occurrence, and predictors of dysphagia, dysarthria, and aphasia after first-ever acute ischemic stroke. *J. Commun. Disord.* **2013**, *46*, 238–248. [CrossRef] [PubMed]
7. Spencer, K.A.; Brown, K.A. Dysarthria following Stroke. *Semin. Speech Lang.* **2018**, *39*, 15–24. [CrossRef]
8. American Speech-Language-Hearing Association. Dysarthria in Adults. Available online: https://www.asha.org/Practice-Portal/Clinical-Topics/Dysarthria-in-Adults/ (accessed on 13 August 2024).
9. World Health Organization. *Clasificación Internacional del Funcionamiento, de la Discapacidad y de la Salud*; World Health Organization: Geneva, Switzerland, 2001.
10. Altaher, A.M.; Chu, S.Y.; Kam, R.B.M.; Razak, R.A. A Report of Assessment Tools for Individuals with Dysarthria. *Open Public Health J.* **2019**, *12*, 384–386. [CrossRef]
11. Ziegler, W.; Schölderle, T.; Brendel, B.; Risch, V.; Felber, S.; Ott, K.; Goldenberg, G.; Vogel, M.; Bötzel, K.; Zettl, L.; et al. Speech and Nonspeech Parameters in the Clinical Assessment of Dysarthria: A Dimensional Analysis. *Brain Sci.* **2023**, *13*, 113. [CrossRef]
12. Pommée, T.; Balaguer, M.; Mauclair, J.; Pinquier, J.; Woisard, V. Assessment of adult speech disorders: Current situation and needs in French-speaking clinical practice. *Logop. Phoniatr. Vocology* **2022**, *47*, 92–108. [CrossRef]
13. Rumbach, A.F.; Finch, E.; Stevenson, G. What are the usual assessment practices in adult non-progressive dysarthria rehabilitation? A survey of Australian dysarthria practice patterns. *J. Commun. Disord.* **2019**, *79*, 46–57. [CrossRef]
14. Sanhueza-Garrido, M.; García-Flores, V.; Rojas-Zepeda, C.; Crisosto-Alarcón, J. Comparative analysis of speech assessment tools for individuals with dysarthria in Chile and worldwide. *Salud Cienc. Y Tecnol.* **2024**, *4*, 626. [CrossRef]
15. Salinas-Marchant, C.; Pérez Herrera, D. Dialectal Variation: Relevant Aspects in Speech Therapy Research and Practice. *Rev. CUHSO* **2024**, *34*, 998–1009. [CrossRef]
16. Sadowsky, S.; Salamanca, G. El inventario fonético del español de Chile: Principios orientadores, inventario provisorio de consonantes y sistema de representación (AFI-CL). *Onomázein* **2011**, *24*, 61–84. [CrossRef]
17. Darley, F.L.; Aronson, A.E.; Brown, J.R. Differential Diagnostic Patterns of Dysarthria. *J. Speech Hear. Res.* **1969**, *12*, 246–269. [CrossRef] [PubMed]
18. Darley, F.L.; Aronson, A.E.; Brown, J.R. Clusters of Deviant Speech Dimensions in the Dysarthrias. *J. Speech Hear. Res.* **1969**, *12*, 462–496. [CrossRef] [PubMed]
19. Darley, F.L.; Aronson, A.E.; Brown, J. *Motor Speech Disorders*; W.B. Saunders: London, UK, 1975.
20. Ziegler, W.; Staiger, A.; Schölderle, T.; Vogel, M. Gauging the Auditory Dimensions of Dysarthric Impairment: Reliability and Construct Validity of the Bogenhausen Dysarthria Scales (BoDyS). *J. Speech Lang. Hear. Res.* **2017**, *60*, 1516–1534. [CrossRef]
21. Enderby, P.; Palmer, R. *FDA-2: Frenchay Dysarthria Assessment*, 2nd ed.; Pro Ed: Austin, TX, USA, 2008.
22. González, R.; Bevilacqua, J. Las Disartrias. *Rev. Hosp. Clínico Univ. Chile* **2012**, *23*, 299–309.
23. Toledo, L. Evaluación de las funciones orofaciales en la persona con disartria. In *Motricidad Orofacial II: Evaluación y Diagnóstico*; Villanueva, P., Ed.; Editorial Universitaria: Santiago, Chile, 2017; pp. 131–140.
24. Ziegler, W.; Schölderle, T.; Staiger, A.; Vogel, M. Die Bogenhausener Dysarthrieskalen (BoDyS): Ein standardisierter Test für die Dysarthriediagnostik bei Erwachsenen. *Sprache · Stimme · Gehör* **2015**, *39*, 171–175. [CrossRef]
25. Ramada-Rodilla, J.M.; Serra-Pujadas, C.; Delclós-Clanchet, G.L. Adaptación cultural y validación de cuestionarios de salud: Revisión y recomendaciones metodológicas. *Salud Pública México* **2013**, *55*, 57–66. [CrossRef]
26. Wild, D.; Grove, A.; Martin, M.; Eremenco, S.; McElroy, S.; Verjee-Lorenz, A.; Erikson, P. Principles of Good Practice for the Translation and Cultural Adaptation Process for Patient-Reported Outcomes (PRO) Measures: Report of the ISPOR Task Force for Translation and Cultural Adaptation. *Value Health* **2005**, *8*, 94–104. [CrossRef]
27. Almenara, J.C.; Osuna, J.M.B. La Utilización del Juicio de Experto para la Evaluación de TIC: El Coeficiente de Competencia Experta. *Bordón Rev. Pedagog.* **2013**, *65*, 25–38. [CrossRef]
28. García de Yébenes Prous, M.J.; Rodríguez Salvanés, F.; Carmona Ortells, L. Validación de cuestionarios. *Reumatol. Clínica* **2009**, *5*, 171–177. [CrossRef]
29. Zamanzadeh, V.; Ghahramanian, A.; Rassouli, M.; Abbaszadeh, A.; Alavi-Majd, H.; Nikanfar, A.R. Design and Implementation Content Validity Study: Development of an instrument for measuring Patient-Centered Communication. *J. Caring Sci.* **2015**, *4*, 165–178. [CrossRef] [PubMed]
30. Achury Saldaña, D.M.; Sepúlveda, G.J.; Rodríguez Colmenares, S.M. Validez de apariencia y de contenido de un instrumento para evaluar la capacidad de agencia de autocuidado en el paciente con hipertensión arterial. *Investig. Enfermería: Imagen Desarro.* **2011**, *10*, 93–111.
31. Galicia Alarcón, L.A.; Balderrama Trápaga, J.A.; Edel Navarro, R. Content validity by experts judgment: Proposal for a virtual tool. *Apertura* **2017**, *9*, 42–53. [CrossRef]
32. Andreu, M.F.; Policastro, P.; Terrasa, S.A.; Rossettini, G.; Testa, M.; Dias de Carvalho, T.; Pardo, Y. Adaptación transcultural y validación de contenido de un cuestionario para explorar las perspectivas de fisioterapeutas de Argentina con respecto al uso de factores contextuales en el dolor musculoesquelético. *Argent. J. Respir. Phys. Ther.* **2022**, *4*, 20–31. [CrossRef]

33. Rodrigues, I.B.; Adachi, J.D.; Beattie, K.A.; MacDermid, J.C. Development and validation of a new tool to measure the facilitators, barriers and preferences to exercise in people with osteoporosis. *BMC Musculoskelet. Disord.* **2017**, *18*, 540. [CrossRef]
34. Polit, D.F.; Beck, C.T.; Owen, S.V. Is the CVI an acceptable indicator of content validity? Appraisal and recommendations. *Res. Nurs. Health* **2007**, *30*, 459–467. [CrossRef]
35. Lawshe, C.H. A quantitative approach to content validity. *Pers. Psychol.* **1975**, *28*, 563–575. [CrossRef]
36. Díaz, J.I.V. Tratamiento actual de la degeneración macular relacionada con la edad. *Rev. Médica Clínica Las Condes* **2023**, *34*, 335–343. [CrossRef]
37. Ministerio de Desarrollo Social y Familia de Chile. *Encuesta Nacional de Discapacidad y Dependencia*; Ministerio de Desarrollo Social y Familia de Chile: Santiago, Chile, 2022.
38. Del Pino, R.; Acera, M.; Murueta-Goyena, A.; Lucas-Jiménez, O.; Ojeda, N.; Ibarretxe-Bilbao, N.; Peña, J.; Reyero, P.; Cortés, J.; Tijero, B.; et al. Visual dysfunction is associated with cognitive impairment in Parkinson's disease. *Park. Relat. Disord.* **2021**, *92*, 22–25. [CrossRef]
39. Borm, C.D.J.M.; Visser, F.; Werkmann, M.; de Graaf, D.; Putz, D.; Seppi, K.; Poewe, W.; Vlaar, A.M.M.; Hoyng, C.; Bloem, B.R.; et al. Seeing ophthalmologic problems in Parkinson disease. *Neurology* **2020**, *94*, e1539–e1547. [CrossRef]
40. Sterling, L.E.; Jawaid, A.; Salamone, A.R.; Murthy, S.B.; Mosnik, D.M.; McDowell, E.; Wheaton, M.; Strutt, A.M.; Simpson, E.; Appel, S.; et al. Association between dysarthria and cognitive impairment in ALS: A prospective study. *Amyotroph. Lateral Scler.* **2010**, *11*, 46–51. [CrossRef] [PubMed]
41. Feenaughty, L.; Guo, L.-Y.; Weinstock-Guttman, B.; Ray, M.; Benedict, R.H.B.; Tjaden, K. Impact of Cognitive Impairment and Dysarthria on Spoken Language in Multiple Sclerosis. *J. Int. Neuropsychol. Soc.* **2021**, *27*, 450–460. [CrossRef] [PubMed]
42. Oveisgharan, S.; Wang, T.; Barnes, L.L.; Schneider, J.A.; Bennett, D.A.; Buchman, A.S. The time course of motor and cognitive decline in older adults and their associations with brain pathologies: A multicohort study. *Lancet Healthy Longev.* **2024**, *5*, e336–e345. [CrossRef] [PubMed]
43. Einstad, M.S.; Saltvedt, I.; Lydersen, S.; Ursin, M.H.; Munthe-Kaas, R.; Ihle-Hansen, H.; Knapskog, A.-B.; Askim, T.; Beyer, M.K.; Næss, H.; et al. Associations between post-stroke motor and cognitive function: A cross-sectional study. *BMC Geriatr.* **2021**, *21*, 103. [CrossRef]
44. Talarico, T.R.; Venegas, M.J.; Ortiz, K.Z. Perfil populacional de pacientes com distúrbios da comunicação humana decorrentes de lesão cerebral, assistidos em hospital terciário. *Rev. CEFAC* **2010**, *13*, 330–339. [CrossRef]
45. Cabrera-Miguieles, M.G.; Sepúlveda-Ibarra, C.; Almeida-Marcos, N.; Crisosto-Alarcón, J. Planificación terapéutica fonoaudiológica para el abordaje de usuarios adultos con dificultades comunicativas y/o lingüísticas de origen neurológico. *Rev. Chil. Fonoaudiol.* **2023**, *22*, 1–16. [CrossRef]
46. Picó Berenguer, M.; Yévenes Briones, H.A. Trastornos del habla en Enfermedad de Parkinson. Revisión. *Rev. Científica Cienc. Médica* **2019**, *22*, 36–42. [CrossRef]
47. Pasnoor, M.; Wolfe, G.I.; Barohn, R.J. Chapter 12—Myasthenia gravis. In *Handbook of Clinical Neurology*; Hanna, M.G., Ed.; Elsevier: Amsterdam, The Netherlands, 2024; Volume 203, pp. 185–203.
48. Fonville, S.; van der Worp, H.B.; Maat, P.; Aldenhoven, M.; Algra, A.; van Gijn, J. Accuracy and inter-observer variation in the classification of dysarthria from speech recordings. *J. Neurol.* **2008**, *255*, 1545–1548. [CrossRef]
49. Ramos, R.A.; de los Reyes, V.C.; Sucaldito, M.N.; Tayag, E. Rapid health assessments of evacuation centres in areas affected by Typhoon Haiyan. *West. Pac. Surveill. Response J.* **2015**, *6* (Suppl. 1), 39–43. [CrossRef]
50. Ma, W.; Xu, L.; Zhang, H.; Zhang, S. Can Natural Speech Prosody Distinguish Autism Spectrum Disorders? A Meta-Analysis. *Behav. Sci.* **2024**, *14*, 90. [CrossRef]
51. Dos Santos, V.B.; Ayres, A.; Kieling, M.L.M.; Miglorini, E.C.; Jardim, L.B.; Schumacher-Schuh, A.F.; Rieder, C.R.D.M.; Castilhos, R.M.D.; Spencer, K.; Rothe-Neves, R.; et al. Differences in spontaneous speech fluency between Parkinson's disease and spinocerebellar ataxia type 3. *Front. Neurol.* **2023**, *14*, 1179287. [CrossRef] [PubMed]
52. Brendel, B.; Synofzik, M.; Ackermann, H.; Lindig, T.; Schölderle, T.; Schöls, L.; Ziegler, W. Comparing speech characteristics in spinocerebellar ataxias type 3 and type 6 with Friedreich ataxia. *J. Neurol.* **2015**, *262*, 21–26. [CrossRef] [PubMed]
53. Crisosto-Alarcón, J. How to build measurable treatment goals? A taxonomy of outcome criteria for vocal intervention monitoring. *Rev. Investig. E Innovación Cienc. Salud* **2023**, *5*, 205–231. [CrossRef]
54. Martínez-Cifuentes, R.; Pierart-Vásquez, T.; Castro-Mellado, J.I.; Lara-Salas, C.; Fuentes-Parra, C.; Martínez-González, F.; Montecinos-Madrid, N.; Quilodrán-Anaba, F. Traducción y adaptación del "Quality of Life in the Dysarthric Speaker" (QOL-DYS) para su uso en usuarios chilenos con disartria. *Rehabil. Integral* **2022**, *16*, 27–38. [CrossRef]

brain sciences

MDPI

Review

Communication Abilities, Assessment Procedures, and Intervention Approaches in Rett Syndrome: A Narrative Review

Louiza Voniati [1,*], Angelos Papadopoulos [2], Nafsika Ziavra [2] and Dionysios Tafiadis [2]

1 Department of Health Sciences, Speech and Language Therapy, European University, Nicosia 22006, Cyprus
2 Department of Speech and Language Therapy, School of Health Sciences, University of Ioannina, 45 500 Ioannina, Greece; angelos@uoi.gr (A.P.); nziavra@uoi.gr (N.Z.); tafiadis@uoi.gr (D.T.)
* Correspondence: l.voniati@euc.ac.cy

Abstract: Background/Objectives: Rett syndrome (RTT) is a rare neurodevelopmental disorder that affects movement and communication skills primarily in females. This study aimed to synthesize the research from the last two decades regarding the verbal and nonverbal communication abilities, assessment procedures, and intervention approaches for individuals with RTT. **Methods:** A structured literature search was conducted using the Embase, Scopus, and PubMed databases. Fifty-seven studies were selected and analyzed based on inclusion criteria. The data were categorized into four domains (verbal communication skills, nonverbal communication skills, assessment procedures, and intervention approaches). **Results:** The findings indicated a wide variety of communicative behaviors across the RTT population, including prelinguistic signals, regression in verbal output, and preserved nonverbal communicative intent. Moreover, the results highlighted the importance of tailored assessments (Inventory of Potential Communicative Acts, eye tracking tools, and Augmentative and Alternative Communication) to facilitate functional communication. The individualized intervention approaches were found to be the most effective in improving communicative participation. **Conclusions:** The current review provides an overview of the current evidence with an emphasis on the need for personalized and evidence-based clinical practices. Additionally, it provided guidance for professionals, clinicians, and researchers seeking to improve the quality of life for individuals with RTT.

Keywords: Rett syndrome (RTT); communicative abilities; assessment; intervention; speech-pathology; narrative review

1. Introduction

Rett syndrome (RTT) is a rare neurodevelopmental disorder caused by a genetically based mutation of the methyl-CpG-binding protein 2 (MECP2) gene on the X-chromosome that occurs almost entirely in females in early childhood [1–3]. Mutations in the gene encoding MECP2 are related to rare familial cases of RTT and typical RTT [2]. By employing a variety of advanced mutation detection tests, it is possible to identify mutations in the MECP2 gene in around 95 to 97% of patients who exhibit typical symptoms of RTT [3–6]. However, even with the most effective techniques, 3 to 5% of patients who meet all the clinical criteria for RTT are found to be missing a known mutation in the MECP2 gene. This suggests that a mutation in this gene is unnecessary for diagnosing typical RTT [3,4]. Of the diagnosed cases, 88.4% showed a positive mutation of the MECP2 gene upon testing. In unusual cases, recognized mutations in the MECP2 gene are present in only 50 to 70%

of affected individuals. Additionally, MECP2 gene mutations have been identified in individuals who do not display the clinical features associated with RTT [3–6].

1.1. Diagnostic Criteria for Rett Syndrome

In 2010, Neul et al. [3] updated the 2002 diagnostic clinical criteria for typical Rett syndrome to enhance clarity and streamline the diagnostic process. The revised criteria encompassed the key criterion of regression, along with four key signs important for diagnosing typical Rett syndrome (RTT). The characteristic symptoms of typical RTT include a decline in purposeful hand movements and verbal communication, accompanied by the development of atypical walking patterns and repetitive hand stereotypies [3,7]. The median age for diagnosing classic Rett syndrome is typically 3 years, with a diagnostic range spanning from 2 to 4 years. Andreas Rett reported that all children were healthy until 9 months but experienced some feeding issues. In addition, reports revealed the first neurological dysfunctions at 8–10 months with stagnated development and diminished social interest by 15 months. According to a review study [8] detailing the acquired speech and language and socio-communicative abilities of females with RTT before the onset of regression, the authors reported that all infants who were later diagnosed with RTT engaged in eye contact before regression. However, the examination of videos from children during the initial 6 months of their lives revealed that many infants displayed frozen, developmentally unusual smiles at 2.5 months and again at 6 months of age [8]. The authors of the review noted that during the regression phase, children exhibited reduced interest in objects and social interactions and did not respond when their name was called, despite sustaining eye contact [8]. For typically developing infants, the act of producing soft, prelinguistic vocal sounds progressively declines, eventually ceasing between 6 and 10 months of age, and is replaced by the early babbling stage. The early stage of babbling is then followed by the production of more complex syllables and babbling noises, including a mixture of different sounds [8]. After a phase of development that appeared typical but showed subtle impairments, recent evaluations of receptive language skills indicate that some individuals with Rett syndrome may exhibit cognitive abilities within the below-average or mild cognitive impairment range [9]. However, as late infancy progresses, affected individuals can experience a decline in functioning, leading to severe intellectual disability and serious neurological complications [10,11].

Between the ages of 5 and 18 months, individuals with Rett syndrome may experience either progressive deterioration or a relatively slow improvement in motor and adaptive functioning, affecting cognitive and communicative abilities [9]. Additionally, they frequently exhibit impaired communication skills [12,13], sensory processing challenges [14], stereotyped hand movements [15], unexplained respiratory irregularities such as hyperventilation [8], seizures [16], scoliosis [17], and motor disorders like spasticity [18], which may eventually lead to wheelchair dependence [19]. Reduced ambulation may lead to dystonia, which can result in foot and hand deformities [20].

As RTT progresses, affected individuals may develop commonly co-occurring conditions, including episodes resembling panic attacks, bruxism, ataxia and apraxia, tremors, and acquired microcephaly. People with RTT generally live into adulthood, with an average life expectancy of around 50 years, underscoring the long-term need for interdisciplinary care and tailored communication interventions throughout the lifespan. In this context, understanding the evolving nature of communication impairments becomes critical to enhancing the quality of life for both individuals with RTT and their caregivers.

1.2. Rett Syndrome and Communication Characteristics

One of the key diagnostic criteria suggesting the possibility of RTT in a child is deterioration and significant weaknesses in their communicative repertoire [12,13,21]. Several studies have indicated that individuals with RTT often experience a range of prelinguistic behaviors needed for communicative purposes (basic oral, verbal, and gestural expressions and nonverbal communication with signing, gestures, nods, sounds, touch, and eye interaction) [8,21], but their speech and language development conspicuously include at most only a few single words [17,22,23]. However, across the literature, there is notable variability in how communication abilities are defined, measured, and interpreted. Studies differ in their use of terminology, assessment procedures, and outcome reporting, which contributes to a fragmented and inconsistent knowledge base on communication in RTT. This makes it difficult for researchers and clinicians to draw definitive conclusions or compare findings across studies.

In addition, autistic behaviors observed in individuals with RTT can implicate the ability to participate in joint social interaction, impaired verbal and nonverbal shared communication, and limited imagination with repetitive and fixed behaviors [24,25]. The presence of symptoms that overlap with autism spectrum disorders, such as stereotyped behaviors, impaired social reciprocity, and restricted communication, can obscure the accurate diagnosis and categorization of communicative deficits specific to RTT. This overlap complicates both clinical decision-making and the selection of appropriate assessment instruments, which are often not normed for RTT-specific profiles. Moreover, the potential for misdiagnosis or diagnostic overshadowing may result in interventions that are not adequately tailored to the nuanced needs of the individual. These communication deficits influence the quality of life of an individual with RTT [26]. Thus, individuals with RTT cannot express their feelings or personal needs to live a quality life [26]. Effective communication skills are essential for achieving a high quality of life, as they play a crucial role in building relationships, shaping identity, fostering personal development, and engaging socially.

Communication difficulties also place a considerable emotional and logistical burden on caregivers, who are often responsible for interpreting subtle cues and advocating for appropriate services. Additionally, cultural and linguistic contexts may influence how symptoms are interpreted and the accessibility of augmentative strategies, thereby limiting the generalizability of some findings to diverse populations. Diverse cultural attitudes toward disability, gender, and nonverbal expression may influence how symptoms are identified, reported, and addressed. For example, in some cultural settings, limited verbal output might be perceived as shyness or developmental delay rather than a potential marker of a neurodevelopmental disorder, thereby delaying diagnosis and intervention [27].

Access to communication assessments and interventions—especially those involving advanced technologies like AAC or eye-tracking—also varies significantly across geographical regions due to disparities in healthcare infrastructure, clinician training, funding, and the language availability of tools. While advances in AAC technology, such as eye-tracking systems, tablet-based applications, and even neuroimaging tools, have opened new pathways to support communication, these resources remain underutilized or insufficiently addressed in many studies. Their integration into assessment and intervention models warrants further exploration [28].

Numerous studies and reviews have explored the communication abilities of children with RTT [12,13,29,30]. Additionally, some research has concentrated solely on assessment methods for communication abilities [8,31], while others have focused on therapeutic strategies [12,13,32,33]. Yet, despite this growing body of literature, there remains no unified synthesis that integrates communication characteristics, assessment procedures, and evidence-based intervention strategies into a coherent framework. This critical gap underscores the urgent need for a comprehensive, up-to-date narrative review that not only summarizes findings but also highlights practical considerations for clinical application.

Although numerous studies have been conducted, the past two decades have lacked a comprehensive review that systematically consolidates all available information on the communication difficulties experienced by individuals with Rett syndrome. Such a review is crucial for improving the identification and management of these challenges. There is a critical need for research to consolidate all this information, enabling SLPs and other professionals to gain an extensive understanding of the communication abilities of individuals with RTT, along with detailed knowledge of assessment and intervention methods [26]. By consolidating existing knowledge, SLPs and other professionals would gain a more comprehensive perspective on the unique communication challenges faced by this population. Additionally, the research would enhance the precision and effectiveness of assessment and intervention strategies, allowing clinicians to tailor their approaches based on a deeper, evidence-based understanding. This, in turn, could lead to improved interaction opportunities, greater social engagement, and ultimately a higher quality of life for individuals with RTT [12,13,22,29].

1.3. Purpose of the Present Review

In response to the critical need for a comprehensive understanding of the communication abilities of individuals with Rett syndrome, along with effective assessment and intervention methods, a narrative review protocol was chosen. This approach enables the systematic integration of extensive data on both verbal and nonverbal communication skills, while also providing valuable insights into evidence-based strategies for improving communication outcomes in individuals with RTT.

An integrative narrative review offers a unique advantage by bridging diverse research methodologies and translating complex evidence into accessible guidance for clinical and educational settings. This synthesis not only maps the landscape of current knowledge but also identifies areas for innovation in diagnostic practices and therapeutic support [34]. Moreover, this type of narrative review critically evaluates and integrates publications from evidence-based studies employing diverse methodologies, ensuring a comprehensive synthesis of existing research [35]. By identifying patterns, inconsistencies, and emerging trends in the literature, this review aims to inform future research priorities and enhance multidisciplinary collaboration for improved communication support across the lifespan of affected individuals.

This narrative review was conducted to address the research question: What are the communication abilities of individuals with Rett syndrome, and how can professionals effectively assess their communication repertoire and implement appropriate intervention strategies? The primary objective of the review was to identify and synthesize existing research on communication skills, assessment methodologies, and intervention approaches for individuals with RTT. Specific objectives included:

1. Categorize the communication abilities of individuals with RTT participating in research into verbal and nonverbal skills.
2. Indicate possible speech and language assessment procedures for identifying the communication abilities of research participants and clinical clients with RTT.
3. Identify potential intervention approaches for SLPs and other professionals involved in assessing, planning, and implementing interventions to address communication deficits in individuals with RTT.

2. Materials and Methods

2.1. Search Strategy

The present review adhered to the structured 12-step approach outlined by Kable et al. (2012) [36] to ensure a systematic and comprehensive evaluation of the relevant literature. Each step was carefully implemented, beginning with the formulation of the research question, which aimed to examine the communication abilities of individuals with Rett syndrome, as well as the assessment and intervention strategies employed by professionals.

A thorough literature search was conducted in June 2024 using three well-established electronic databases: Embase, Scopus, and PubMed. We used the terms and the Boolean operators as below: ("Rett Syndrome" OR "Rett disorder" OR "MECP2") AND ("communication" OR "communicative abilities" OR "language" OR "speech" OR "verbal" OR "nonverbal") AND ("assessment" OR "evaluation" OR "diagnosis" OR "screening") AND ("intervention" OR "therapy" OR "treatment" OR "rehabilitation" OR "speech therapy" OR "language intervention"). Regarding the filters, to ensure the inclusion of high-quality and relevant studies, clear search parameters were applied, focusing on published articles from the last two decades (2004–2024). The search terms, as described before, allowed for a targeted and comprehensive exploration of available research. Moreover, backward citation tracking was performed to identify additional relevant studies beyond those retrieved directly from the databases. The Mendeley Reference Manager was used to help us organize citations and remove duplicate studies. The data were extracted using an Excel matrix for (a) authors, (b) sample, (c) sample's age, (d) purpose of the study, (e) study design, (f) communication skills, (g) assessment of communication skills, and (h) intervention of communication skills. To reduce bias and increase methodological transparency regarding interrater agreement, two independent researchers conducted the screening process. In the event of any disagreement, a consensus was reached.

The retrieved studies were systematically screened based on predefined inclusion and exclusion criteria, ensuring relevance and reliability. Titles and abstracts were reviewed to verify their alignment with the objectives of the present study. To synthesize existing knowledge effectively, the selected articles were categorized into four primary domains: (a) verbal competencies, (b) nonverbal competencies, (c) speech and language assessment, and (d) intervention strategies. This classification facilitated a structured analysis of communication characteristics in individuals with Rett syndrome.

Furthermore, an expert consensus on the current context and potential future developments in the field was examined across each of these four areas. The findings from this process were systematically documented, providing a synthesized overview of key insights. Additionally, the study process is visually represented in the following flow diagram, illustrating the methodological framework applied throughout the review.

By adhering to these guidelines, the present review ensured a robust methodological foundation, maintaining reliability and relevance in its findings. The inclusion of specific search parameters, keywords, and time frames enabled a comprehensive synthesis of the most pertinent literature on communication abilities in individuals with Rett syndrome, ultimately supporting the development of evidence-based assessment and intervention strategies.

2.2. Study Selection Criteria

A comprehensive literature search was carried out utilizing major academic databases, including Embase, Scopus, and PubMed, to ensure a thorough evaluation of relevant studies. Given the narrative nature of this review, PRISMA guidelines were not applied, as the focus was on synthesizing the existing literature rather than conducting a systematic review. Research articles were screened, duplicates were disregarded, and 64 full papers were found for review. Then, abstracts of all articles were examined, and 57 studies were selected and retained for screening and reviewing against the inclusion criteria (see Figure 1). Inclusion criteria further focused on recognizing the adequacy of the information presented in the research. Two [2] independent researchers evaluated the eligibility of the studies. Following this assessment, the full texts of the selected studies were retrieved for further analysis. The two researchers worked collaboratively to ensure that all studies met the established eligibility criteria, resolving any potential discrepancies to maintain consistency in the review process. The selected papers met the established criteria, offering relevant information that aligned with the objectives of the review. Additionally, the chronological distribution of the included studies was documented to ensure a comprehensive assessment of the evolving research landscape.

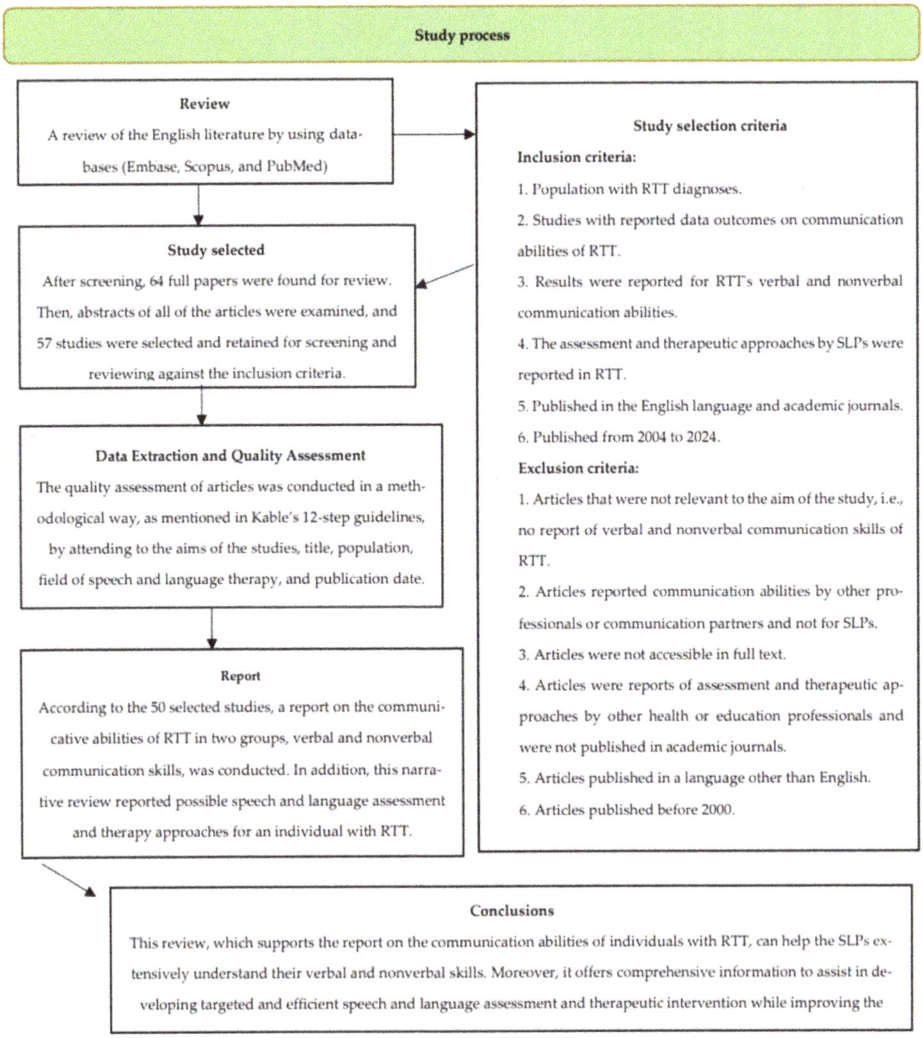

Figure 1. Flow diagram of the description of the study process.

The inclusion criteria were as follows:

1. Population with RTT diagnosis;

2. All the types of studies with reported data outcomes on the communication abilities of RTT;
3. The findings focused exclusively on the verbal and nonverbal communication abilities of individuals with Rett syndrome, without imposing restrictions based on participant characteristics;
4. The assessment and intervention approaches implemented by SLPs and other healthcare professionals in individuals with RTT were documented;
5. Published in peer-reviewed academic journals available in the English language;
6. They were published from 2004 to 2024.

This review encompasses studies that present data on the verbal and nonverbal communication abilities of individuals with Rett syndrome, along with approaches for their assessment and intervention.

The exclusion criteria were as follows:

1. Articles irrelevant to the study's aim that did not report the verbal and nonverbal communication skills of individuals with RTT;
2. Articles reported communication abilities assessed by other non-SLP professionals or communication partners;
3. Articles were inaccessible to the authors in full text;
4. Articles reported assessment and intervention approaches by other health or education professionals or were not published in peer-reviewed journals;
5. Articles were published in a language other than English;
6. Articles published before 2004.

2.3. Data Extraction and Quality Assessment

As previously outlined in this review, the study adhered to a structured 12-step framework to document the search strategy before conducting a critical analysis and synthesis of the retrieved literature. Following Kable's 12-step guidelines (2012) [36], the review systematically examined the studies' objectives, titles, target populations, relevance to speech and language therapy, and publication dates. Currently, there are no universally recognized guidelines for formulating narrative reviews, which present two primary limitations: (a) the inherent biases related to assumptions, planning, selection, and evaluation are often not explicitly identified, and (b) the lack of reproducibility in narrative studies. However, to enhance the methodological rigor and ensure a higher quality standard in this review, the researchers employed the Scale for the Assessment of Narrative Review Articles (SANRA) as a guiding framework during manuscript preparation [37,38].

The Scale for the Assessment of Narrative Review Articles (SANRA) was developed between 2010 and 2017 by a team of editors as a concise and practical tool for evaluating the quality of narrative review articles. It serves as a guideline for editors, reviewers, and readers in assessing both the rigor of academic papers and the effectiveness of authors in structuring narrative reviews. In 2014, the authors refined SANRA to enhance its clarity and reliability, streamlining the wording of its items and eliminating a criterion related to manuscript writing and accessibility due to inconsistent ratings [37,38].

The revised SANRA scale consists of six criteria, each rated on a scale from 0 (low quality) to 2 (high quality), with a maximum possible score of 12. The six assessment components include (1) the explanation of the review's significance, (2) the articulation of the review's objectives, (3) the description of the literature search process, (4) the accuracy and adequacy of referencing, (5) the strength of the scientific reasoning, and (6) the presentation of relevant endpoint data [37].

In the present review, SANRA was employed as part of the self-assessment process for the current study's data extraction and quality assessment to ensure methodological rigor and enhance the reliability of the findings. The review was systematically assessed according to the SANRA criteria, allowing for a structured evaluation of the comprehensiveness, relevance, and scientific validity of the selected literature. By integrating SANRA into the quality assessment framework, this review aimed to maintain high academic standards while synthesizing the existing knowledge related to communication abilities, assessment, and intervention strategies for RTT. Appendix B presents the SANRA evaluation of the current study.

3. Results

Based on the analysis of 57 selected studies, the communicative abilities of individuals with Rett syndrome have been examined within two primary domains: verbal and nonverbal communication skills. Furthermore, a range of speech and language assessment methods and intervention strategies have been employed across research studies to evaluate and support the communication development of individuals with RTT (see Table A1 in the Appendix A).

3.1. Sample Characteristics

To reduce potential sample overlap in commonly conducted research, the sample characteristics were systematically classified according to study type. Specifically, the studies were organized into three distinct categories: empirical research studies, literature review studies, and studies in which the sample size was not explicitly reported. This classification ensured a structured approach to data organization, facilitating a clearer understanding of the participant demographics and methodological variations across the included studies. By distinguishing between these categories, the review aimed to enhance the accuracy of data interpretation, prevent redundancy, and provide a more comprehensive synthesis of communication-related findings in individuals with Rett syndrome. The narrative review incorporated 30 selected research studies, collectively analyzing data from 1052 individuals diagnosed with RTT.

Additionally, six of these research studies examined the perspectives and experiences of 1764 parents, caregivers, SLPs, and other professionals working with individuals with RTT. Further, four of the selected research studies included comparative analyses between 93 individuals with RTT and 96 typically developing females, contributing to a broader understanding of communication differences in this population. In addition to empirical research, seven of the 57 selected papers were literature reviews focused on individuals with RTT, increasing the total number of examined studies to 187. Lastly, the remaining ten studies did not specify the exact sample size, though they provided relevant insights into communication assessment and intervention strategies for individuals with RTT.

3.2. Verbal Communication Abilities

Based on the extracted data from the included studies examining verbal communication abilities, individuals with RTT demonstrated multiple deficits. Attention-related limitations appeared to hinder their learning capacity, as supported by findings in the literature [15,19,39]. The findings indicate that enhancing cognitive functions in individuals with RTT may contribute to more purposeful and effective communication [40]. However, this is often constrained by motor impairments, such as difficulties in coordinating hand movements and oral motor control required for speech production [17,28], which were observed in several of the reviewed studies.

Although there is a notable accomplishment in speech and language milestones, such as cooing, babbling, first words, and even word combinations, during the first two years of life, longitudinal analyses indicate a subsequent regression characterized by reduced verbal output and diminished social engagement in RTT [8,21,31,39,41]. This developmental regression has been linked to a deterioration in speech, language, and overall communicative abilities, alongside a reduced inclination for social interaction and impairments in purposeful hand use [8,34,42].

Several studies documented the emergence of idiosyncratic verbalizations and limited topic maintenance despite gradual expansion in articulation, phonological awareness, and mental vocabulary [31,41] Behaviors such as finger-pointing [43] and repetition of known common words, along with the refusal to follow directions or respond to their name, can indicate a decrease in their mental lexicon development [8,15,22]. They can often lose proto-conventional words but build on their passive lexicon skills [8,21,33].

Additionally, a significant phonological deficiency has been detected, along with poor precision of the articulation mode, thus inhibiting overall speech intelligibility [44,45]. Verbal abilities are further hindered by morphosyntactic and pragmatic errors and disfluency, possibly resulting from immediate echolalia or questioning [22].

Multiple studies also reported a progressive decline in vocalization ability, even when expiratory airflow was present, coinciding with increased motor stereotypies—such as repetitive hand-to-mouth movements—which may serve as diagnostic indicators for RTT and its variants [8,46,47]. Individuals with RTT who can no longer use speech to communicate also display inconsistent imitation and social skills [9]. Despite this, observational studies reveal that many maintain expressive language at a pre-intentional level, evidenced by the use of specific nonverbal cues with potential communicative intent. These include vocalizations [9,35,48], laughter and crying [9], facial expressions [49,50], touch, and gazing [51]. The inability of individuals with Rett syndrome to regain speech skills can be primarily attributed to two key factors: the presence of severe dyspraxia, which significantly impairs motor planning for speech production, and, in some cases, cognitive abilities that may fall within the below-average to mild range [52].

3.3. Non Verbal Communication Abilities

The disengagement of verbal communication and the regression of language abilities often occur after reaching Stage I. At the same time, some children display some form of nonverbal prelinguistic communication, which is also seen in later stages [45]. Empirical observations confirm that many individuals with RTT continue to use nonverbal channels well into later stages of development. Several studies investigating developmental trajectories in typically developing infants demonstrate that early preverbal communication emerges through shared social engagement [8,41]. The child uses communicative acts such as pointing or reaching out to show or request objects [8,53–55]. The action of more than two persons directing their attention to a common drive of interest—'joint attention'—is an important part of this informative communicative process [20,55,56]. Making eye contact and following the gaze of others while attending to their focus of interest (eye-direction detection') are behaviors necessary for developing joint attention. Such behaviors also play a vital role in identifying movements as a means of intentional communication [21].

Nonverbal communication can be exhibited through eye gaze/eye contact to request a non-expressive interaction [34,46,49]. Caregiver and clinician reports highlight the use of body positioning, gestures, and affective signals such as frowning or withdrawing to express preferences or discomfort [8,25,48]. Eye pointing is used clearly for communicative intentions and gestural attention [52,57–59].

Research examining language comprehension in individuals with RTT, including both classical and forme fruste (a partial symptom of the syndrome) presentations, has indicated that they primarily recognize their names, as well as key words or simple sentences. These identifications are largely influenced by visual or contextual cues [29,53].

However, receptive abilities are difficult to assess consistently, as moderate to severe motor impairments often limit active response options. Many studies, therefore, evaluated comprehension relative to estimated mental age or through proxy measures, such as gaze preference or behavioral responses [45]. These challenges underscore the need for individualized, multimodal approaches to assessing and interpreting receptive communication in RTT.

3.4. Speech and Language Assessment Procedure

Considering the need for accurate assessment, numerous sources of information, such as medical records, parent and staff reports, video recordings, and clinical observations, were used to explain the participants' developmental history and current functioning [43]. Standardized methods are often found to be limited, largely due to the lack of assessment tools specifically tailored to the unique communicative and cognitive profiles of individuals with RTT [60].

To assess the potential communicative forms and functions of existing prelinguistic behaviors in individuals with severe disabilities such as RTT, the Inventory of Potential Communicative Acts (IPCA) was used in several studies. These studies reported that, despite extremely limited behavioral repertoires—such as eye gazing or simple body movements—children with RTT exhibited consistent patterns interpreted by caregivers and educators as communicative acts [30,47].

The communicative functions inferred from these behaviors typically included acknowledgment, initiating dialogue, requesting objects, and expressing rejection [29]. However, these interpretations were based primarily on informant-reported responses via the IPCA, and some studies cautioned that clinical observation alone did not always verify these functions conclusively, reflecting the need for complementary assessment methods to validate communicative intent [61].

Receptive language abilities were assessed in multiple studies using both technology-based and traditional instruments. These included eye-tracking technology (ETT) and standardized tools such as the Peabody Picture Vocabulary Test (PPVT) [62]. The Assessment of Visual Attention in Interaction (AVAI) tool, developed to assess visual attention operationalized as concentrated gazes at the communication partner, an object, and a symbol set, might also be beneficial throughout the receptive and expressive language skills assessment procedure. The AVAI tool enables the assessment of visual attention in more realistic interactions with communication partners [13].

Other tools for the identification of the girls' communication levels include: the Early Language Milestones (ELM) Scale, the Vineland Adaptive Behavior Scales (VABSs) [53,63–65], structured descriptive assessment (SDA) [66], and the parent–interview portions of the Sequenced Inventory of Communicative Development (SICD), along with medical records [30]. These tools were often used in conjunction with clinical records to provide a more holistic picture of communicative functioning, motor skills, and associated medical conditions.

Some studies also described scoring approaches to quantify total communication abilities—for example, by assigning weighted values to behaviors such as eye pointing, cause/effect signaling, and the ability to make choices—providing a structured means of summarizing observed communicative intent [58,61].

The caregiver-completed Rett Syndrome behavior Questionnaire (RSBQ) is one of the most extensively used effectiveness measures in the clinical evaluation of RTT, due to its specificity for essential RTT symptoms, including eye gazing and expressionless facial features [57]. Moreover, the Rett Assessment Rating Scale (RARS) enables a complete and reliable evaluation of individuals with RTT by individually analyzing and rating key features, including the cognitive area, which encompasses communication skills [49,55]. Another useful tool is the Responsive Augmentative and Alternative Communication Style Scale (RAACS), which was created to measure parents' communication styles with children who have difficulties with communication [46]. Furthermore, for the Communication and Symbolic Behavior Scales Developmental Profile Infant–Toddler Checklist (CSBS-DP-IT) [43,67], the social composite score generated from the CSBS-DP which was designed to measure communication and social interaction abilities in young children (ages 12 to 24 months), could also be considering in assessment of affected individuals. It may also be used with older children with developmental delays, including those with RTT [67]. RTT-specific clinician rating scales related to discrete aspects of communication (RTT-COMC) are a clinician-completed assessment of the individual's ability to communicate their choices or preferences, including the use of nonverbal means such as eye contact or gestures, using an eight-point Likert scale [67,68].

Across the studies included, findings suggest that a subset of individuals with RTT retain communicative potential identifiable through structured assessment of prelinguistic behaviors. These findings reinforce the importance of early, individualized, and multimodal communication assessments. They also support therapeutic approaches that actively build upon observable communicative intent rather than relying solely on verbal performance as a metric of ability [34].

3.5. Speech and Language Intervention Approaches

Based on the assessment findings, four distinct modes of communication have been identified as key components of the intervention process. These include speech or vocalization, gestures, graphic symbols and/or written language, and electronic or computer-based systems such as Assistive Augmentative Communication (AAC). The reviewed studies indicate that these modalities support various levels of communicative engagement, particularly among individuals with limited verbal output. AAC methods can involve speech-generating devices, computers with speech output, graphic symbols [13], and printed vocabulary systems.

Most individuals with RTT demonstrate significant limitations in expressive language, while their receptive comprehension is often relatively preserved [33,60]. Studies have shown that integrating AAC strategies can help bridge this expressive–receptive discrepancy, enabling greater participation in communication exchanges. The role of the SLP is to identify appropriate modalities and tailor intervention plans accordingly. AAC devices are divided into high, mid, and low-tech. High-tech AAC encompasses speech-output systems such as touch-screen iPads or tablets that are manually activated, as well as Tobii devices that operate through eye gaze. These tools have demonstrated effectiveness in controlled settings for facilitating sentence construction and initiating communication. High-tech systems use a digital or synthesized voice, enable computer interfacing, and have features like text-to-speech and sentence creation. Mid-tech AAC (e.g., microswitches) lacks some speech-output technologies. There are no electronics in low-tech AAC. A few instances are choice boards and communication books [28,66]. Among specific approaches reported in intervention studies, 'mind training'—where AAC is used to request reinforcers with single-word prompts—has been cited as effective when implemented through graduated cueing and prompt-fading procedures [69]. In addition to AAC-based techniques, some studies have explored the use of Transcranial Direct Current Stimulation (tDCS) as a therapeutic modality. While research is preliminary, positive outcomes were noted in speech production and comprehension, suggesting potential as an adjunct to behavioral interventions [64]. Other methods for intervention are to teach speech or communicative intent through communicative gestures (for example, crossing arms, eye gaze, and pointing) [65,66], graphic symbols [13], or using a communication board while enabling the individuals with RTT to make choices and interact [43]. The reviewed literature emphasizes that the use of these strategies should be grounded in individualized assessment results and responsive to the child's developmental stage.

Numerous researchers vouch for the need for intensive, continued treatment to improve the long-term outcome in syndromes, and specifically in neurodevelopmental disorders. Intervention planning should, on the one hand, deal with the present functional and medical actuality, but it should also focus on the long-term prognosis of the potential of an individual with RTT [64]. This dual focus requires alignment between therapeutic goals, educational programming, residential supports, and future vocational considerations [30,34].

A successful intervention program must also include the adequate participation and cooperation of the individual's family. Empirical studies suggest that individualized, multi-domain intervention programs—targeting communication, cognitive functioning, and motor abilities—yield the most beneficial outcomes [67,70]. Regardless of the individual's age, treatment planning should establish precise learning opportunities and generalization of what is learned [71]. Developing an effective IEP requires goal-oriented planning that incorporates communication-specific objectives, AAC strategies, and partner training [72]. Furthermore, to establish the IEP, a structured, intervention-oriented procedure is necessary. It is suggested that the communicative functions and partner strategies be targeted in general communication intervention and communication aids (CAs) [13,33,34,46,72,73].

The communication skills targeted for intervention were coded into pragmatic functions based on the classification system described by Sigafoos, Arthur-Kelly, and Butterfield [74]: imitative speech, joint attention, turn-taking, asking, requesting access to preferred stimuli (e.g., food, drinks, toys, or songs), requesting social interaction/topic initiation, naming objects or commenting, narrating, following instructions, and receptive language. Partner strategies that support these functions involve: modeling communication behaviors, allowing extra response time, recognizing and expanding on interests, regulating stimulation, gaining attention, and respecting communicative intent through supported choice-making [9,55].

When conventional approaches yield limited results, identifying an effective Communication Aid (CA) can guide the next phase of intervention. Reported examples include single-message speech-generating devices [69,75], visual communication charts [44,46,60], picture exchange cards [54], and both low-tech (books, boards) and high-tech (eye-gaze tablets, iPads) systems [57,69,74]. Communication passports, a PECS (Picture Exchange Communication System), talking mats, multi-message AAC systems, and custom apps were also identified in the literature as tools that support individual needs and enhance communication independence [15,75].

4. Discussion

This review aimed to provide further knowledge and report on the communicative intent of individuals with RTT regarding communication abilities, assessment procedures, and intervention approaches. The study's strength lies in its synthesis of findings from the past two decades, offering a comprehensive understanding of communication challenges and support strategies that can inform clinical practice and enhance multidisciplinary approaches.

4.1. Communication Abilities of Individuals with RTT (Verbal and Nonverbal)

In terms of communication abilities in individuals with RTT, studies by Cianfaglione et al. [45], Roche et al. [15], and Townend et al. [19] indicate that their verbal communication skills are closely tied to their cognitive abilities, which are difficult to evaluate. Fabio et al. [20] and Elefant and Wigram [39] agreed with previous researchers. They noted that improving cognitive abilities can help in deliberate communication. Still, these skills can be misunderstood due to their physical limitations in coordinating hand movements and the oral motor skills required to produce speech. All the researchers who analyzed RTT's verbal communication skills noted that, gradually, vocalization skills decreased, and hand-to-mouth stereotypes increased, as individuals with RTT could no longer use speech to communicate and showed inconsistencies in imitation and social skills [8,61].

In addition, several studies concluded that verbal communication disengagement and language delay [3,17] typically occur after Stage I, with the children displaying nonverbal prelinguistic communication at later stages [8,41,45,53,54]. As it was indicated, individuals with RTT can engage in nonverbal dialogue through bodily expressions; their ability to convey engagement, reluctance, and displeasure is often manifested through gestures and other nonverbal cues, serving as an essential means of interaction and expression [21,56,65]. However, it is essential to note that many of these findings stem from studies with small sample sizes and heterogeneous RTT phenotypes, which limit generalizability. Additionally, much of the data rely on caregiver reports, which, while valuable, may introduce subjectivity or bias. These methodological limitations should be considered when interpreting conclusions.

Longitudinal research on communication development in RTT remains limited but essential. Understanding how verbal and nonverbal skills evolve across life stages—from early regression to adulthood—could guide more adaptive, age-appropriate interventions and monitoring strategies. These findings regarding the communication abilities of individuals with Rett syndrome, encompassing both verbal and nonverbal aspects, offer significant insights into how affected individuals express themselves. The existing literature emphasizes substantial variability in communication skills, highlighting differences in expressive capabilities and receptive processing.

Furthermore, the findings highlight the multifaceted challenges that individuals with RTT face in conveying their thoughts, emotions, and needs. Factors such as motor impairments, cognitive limitations, and speech-related deficits contribute to the complexity of their communication difficulties. Understanding these variations is crucial for developing effective intervention strategies, including tailored speech-language therapies and AAC systems [66,69]. Additionally, cultural, linguistic, and socioeconomic factors may significantly influence the interpretation of communicative behaviors, access to intervention tools (such as AAC devices), and caregiver engagement. These contextual influences are often underreported and should be addressed in future, more inclusive research.

4.2. Speech and Language Assessment Procedures

To accurately evaluate RTT's verbal and nonverbal communication skills, an assessment of their developmental history and current functioning in persons with severe impairments, such as RTT, was necessary [47,60]. The IPCA was used to examine the possible communication forms and roles of current prelinguistic behaviors [42,60]. As an interview protocol, the IPCA aimed to identify the communicative responses of people with a developmental disability [76]. A recent narrative review discussed the findings of numerous studies that support the positive effects on the reliability and validity of the IPCA, as well as its utility for communication acts, profiles, and interventions for individuals with developmental disabilities [77]. Other research has found that the ELM Scale, VABSs, and the parent–interview sections of the SICD are often utilized during RTT communication evaluation [52]. The ELM scale [78] was used as a brief screening (parental/caregiver report, examiner observation, and direct testing) of the language abilities of children under the age of 3 years (how understandable the child's speech is). This assessment consists of three sections: auditory expressive, auditory receptive, and visual [78,79]. Furthermore, as indicated in the literature [71–83], it was crucial to use a VABS as a standardized assessment tool that utilizes semi-structured interviews to measure adaptive behavior and assist in the diagnosis of developmental disabilities. Regarding the utilization of the SICD [84], this tool assesses the language development of children by evaluating expressive communication (including imitation and generation of sounds and words) and comprehension of commands [84,85].

Despite the utility of these instruments, many were not designed initially or normed for the RTT population. As such, their sensitivity and specificity in capturing nuanced communicative intent, particularly nonverbal behaviors, are limited. This further underscores the need for RTT-specific assessment tools that encompass the full range of communicative capacities. Emerging technologies such as eye-tracking systems, brain–computer interfaces (BCIs), and AI-assisted AAC platforms offer promising advances in assessing and supporting communication in nonverbal or minimally verbal individuals with RTT.

These approaches aim to identify strengths and limitations in verbal and nonverbal expression, facilitate alternative communication methods, and enhance overall interaction abilities. The findings contribute to a broader understanding of evidence-based practices for assessing communication in individuals with RTT, informing future clinical applications and research directions.

4.3. Intervention Approaches for SLPs and Other Professionals

Some researchers have provided a precise picture of communication forms and goals in individuals with RTT, suggesting that some children may retain communicative skills detected by prelinguistic behavior [30,34].

Five different communication forms are inferred from assessment results: speech, gesture, graphic symbols, written lexicon, and electronic or computer-based systems (AAC) [15,86]. Thus, interventions to enhance speech or communicative intent are attempted through gestures, graphic symbols, or a communication board. Intervention planning should consider both the present functional and medical actuality and the long-term prognosis of a child's potential with the adequate involvement of the family [33,75].

However, intervention planning must extend beyond speech–language pathology. The coordination of care with occupational therapists, psychologists, educators, and neurologists is vital to addressing the multifactorial needs of individuals with RTT. Multidisciplinary collaboration ensures that motor, cognitive, emotional, and sensory components of communication are not overlooked. An IEP for individuals with RTT should be directed by identifying intervention methods used to assess and evaluate communication skills and the overall aims and estimated outcome of a general communication intervention [30,34,64]. Communication skills targeted for intervention are coded into pragmatic functions based on the classification system described by Sigafoos, Arthur-Kelly, and Butterfield. Partner communication strategies' target skills are imitative speech, joint attention, turn-taking, asking, requesting access to preferred stimuli, requesting social interaction, naming objects or commenting, narrating, following instructions, and receptive language [70]. Recognizing adequate and efficient communication aids facilitate further intervention when other means have been unsuccessful [44,53,73,74].

While families are acknowledged in intervention planning, their ongoing role in training, emotional support, and implementation is equally crucial. Sustained caregiver engagement—bolstered through accessible resources and coaching—can greatly influence outcomes and reinforce skills across home and community environments.

Ethical considerations must also be integrated into communication intervention, including the individual's right to express themselves meaningfully, the need for informed consent in research or clinical settings involving nonverbal populations, and broader advocacy for inclusive access to communication support tools and services.

Although past research and case studies have provided us with a wealth of data and facts on the communication abilities of individuals with RTT, many of these findings are based on small samples, including diverse RTT phenotypes, and frequently rely on caregiver-reported outcomes. These methodological limitations must be acknowledged when interpreting communication profiles and assessing intervention effectiveness [15,44,45]. This presents a critical opportunity to conduct an in-depth examination of the existing intervention methodologies and explore innovative approaches for enhancing support for individuals with RTT who experience challenges with both verbal and nonverbal communication. Notably, longitudinal and developmental perspectives are underrepresented in the current research. The evolution of communication abilities across different life stages—from early childhood to adulthood—warrants a more systematic study to tailor interventions to the changing needs of individuals over time.

The ongoing refinement of current practice techniques enables the advancement of evidence-based, effective strategies specifically tailored to the unique communication needs of individuals with RTT. Emerging technologies, including eye-tracking systems, brain–computer interfaces, and AI-supported AAC tools, have the potential to transform assessment and intervention frameworks. Although still underutilized in many clinical settings, these technologies offer promising avenues for accessing and interpreting communicative intent, particularly for individuals with limited expressive language.

While strategic planning remains an essential aspect of intervention, typically formulated before or during its implementation, the insights provided by this study offer valuable supplementary perspectives. These additional considerations contribute to a more comprehensive understanding of the communication-related challenges in RTT, informing practitioners and researchers as they strive to optimize intervention strategies.

A comprehensive understanding of communication in individuals with RTT must account for cultural, linguistic, and socioeconomic factors that shape access to services, caregiver involvement, and intervention outcomes. Inclusive methodologies are essential to ensure equitable support. Beyond initial assessment planning, families play a central role in daily implementation and emotional support, warranting greater emphasis. Effective intervention also requires a multidisciplinary approach, engaging professionals from various fields, including speech–language pathology, occupational therapy, psychology, education, and neurology, to meet diverse needs. Ethical dimensions, including the right to communication, informed consent for nonverbal individuals, and advocacy for inclusive services, must be actively integrated into both practice and research. This review highlights a pressing need for more rigorous, longitudinal studies that examine the trajectory of communication development across the lifespan in individuals with Rett syndrome (RTT). Future research should prioritize identifying predictors of communication success and intervention responsiveness, as well as conducting randomized controlled trials to rigorously evaluate the effectiveness of AAC, behavioral, and neuromodulatory interventions. There is a clear necessity to develop RTT-specific assessment instruments capable of capturing subtle communicative behaviors, including those conveyed through gaze or gestures, and to investigate how emerging technologies—such as AI-driven AAC systems and brain–computer interfaces—can be integrated into both assessment and intervention practices. Moreover, embracing cultural and linguistic diversity in study samples will be crucial to improving the global relevance and applicability of findings. A transparent and critical appraisal of the existing evidence base, combined with methodological innovation and inclusive research approaches, will be pivotal in shaping best practices and advancing equitable, person-centered communication support for individuals with RTT.

5. Conclusions

This narrative review integrated evidence from 57 studies published in the past two decades (2004–2024), providing a comprehensive overview of the verbal and nonverbal communication profiles of individuals with Rett syndrome (RTT). The findings indicate that, despite marked regression in verbal abilities common among this population, many individuals retain a repertoire of prelinguistic and nonverbal behaviors, such as eye-pointing, facial expressions, and gestural communication. When appropriately identified and supported, these preserved communicative behaviors can serve as functional and meaningful channels for expression, underscoring the importance of tailored intervention strategies to maximize communicative potential.

Regarding the assessment strategies, they frequently relied on several tools (e.g., Inventory of Potential Communicative Acts, Vineland Adaptive Behavior Scales, and eye tracking technology). However, there is a lack of standardized communication tools exclusively for individuals with RTT.

Assessment approaches frequently utilized instruments such as the Inventory of Potential Communicative Acts, the Vineland Adaptive Behavior Scales, and eye-tracking technology to evaluate communicative abilities in individuals with Rett syndrome. However, a persistent gap in the literature is the absence of standardized assessment tools specifically developed and validated for the RTT population, which limits the precision and consistency of communication profiling in both clinical and research contexts.

In relation to clinical practice, this narrative review highlights the importance of delivering interventions that are tailored to the individualized and responsive needs of each person with Rett syndrome, taking into account their distinct communicative and developmental profiles. It is recommended that professionals across various disciplines, including speech–language pathologists, occupational therapists, educators, and neurologists, collaborate with caregivers to implement personalized strategies. Crucially, active family involvement throughout the intervention process remains a central objective, as it contributes significantly to the consistency, relevance, and overall effectiveness of communication support.

In conclusion, future research should prioritize the development of standardized assessment instruments specifically designed for individuals with Rett syndrome, alongside rigorous investigations into the efficacy and applicability of technology-based intervention approaches.

Author Contributions: L.V., A.P., N.Z. and D.T., writing—original draft preparation. L.V., A.P., N.Z. and D.T., writing—review and editing. All authors have read and agreed to the published version of the manuscript.

Funding: This research received no external funding.

Institutional Review Board Statement: Not applicable.

Informed Consent Statement: Not applicable.

Data Availability Statement: Data sharing is not applicable to this article.

Acknowledgments: This review was supported and sustained by the I Communicate Research Centre (iCRC), European University Cyprus.

Conflicts of Interest: The authors declare no conflicts of interest.

Appendix A

Table A1. Studies characteristics involved in this review.

Authors	Sample [Age]	Purpose of the Study	Study Design	Results		
				Communication Skills	Assessment of Communication Skills	Intervention of Communication Skills
Cianfaglione et al. [45]	42 individuals with RTT [4 to 47 years, and at follow-up was 7 to 48 years]	Gained a UK national sample of people with RTT across the age range and (1) conduct a cross-sectional comparison of age groups and (2) undertake a longitudinal follow-up.	Research study	Preverbal and pre-intentional level matches their cognitive skill bundle, inhibiting speech intelligibility, body and gestural engagements and expressing unwillingness and displeasure can often happen nonverbally, and receptive abilities are judged alongside mental age	VABS and RSBQ	NA
Roche et al. [15]	23 studies: 13 studies on ASD, 8 studies on RTT, and 2 studies on FXS [0–24 months]	Evaluated early vocal behaviors present in the first two years of life for individuals later diagnosed with ASD, RTT, or FXS.	Systematic analysis on the early development of vocalizations in children later diagnosed with ASD, RTT, or FXS	Inhibited learning ability, decreased mental lexicon development, and verbal communication skills being closely linked to RTTs' cognitive skills	IPCA	NA
Townend et al. [19]	67 families who have individuals with RTT [3 years 6 months to 60 years 6 months]	A brief report on families' experiences of eye gaze technology as one form of AAC for individuals with RTT and the advice, training, and support they receive about this.	Research study/online survey relating to communication skills and families' experiences of communication support	Using eye gaze technology to express their communication abilities	Eye gaze technology	NA
Sigafoos et al. [30]	Eight studies that involved 41 individuals with RTT [2.7 years to 36 years]	Reviewed studies that aimed to determine whether behaviors, such as body movements, vocalizations, eye gaze, and facial expressions, served a communicative function for individuals with RTT.	Systematic analysis	Eye pointing/eye gaze, body movements, leading, clapping, reaching, pushing away items, tantrums/screaming	VABS, PPVT-3, Ordinal Scales of Psychological Development, behavioral observations, and interviews	NA
Fabio et al. [20]	20 children (10 individuals with RTT and 10 control girls, matched on mental age) [6 and 21 years]	Analyzed attention and communication abilities in RTT.	Research study	Joint attention, requesting, pointing at the object, and reaching for toys that were out of reach	VABS and behavioral observations	NA
Elefant and Wigram [39]	Seven girls with RTT [from 4 to 10 years]	Presented results of a research study examining learning ability in individuals with RTT.	Research study	Choice making and learning ability	NA	Choice-making through eye gaze, nose pointing, or with their hand, according to their preference song and ability

Table A1. *Cont.*

Authors	Sample [Age]	Purpose of the Study	Study Design	Results		
				Communication Skills	Assessment of Communication Skills	Intervention of Communication Skills
Einspieler and Marschik [8]	NA [infants during their first 2 months of life up to 3 years]	This review sheds light on atypical neurofunctions and potential behavioral biomarkers before the onset of the regression of RTT.	Review study	Eye contact, responsive smile, reaction to being called by name, cooing, vocalizations, babbling, index finger pointing, gestures, body movement, proto-words, and two-word combinations	NA	NA
Segawa and Nomura [42]	51 articles with individuals with RTT [NA]	NA	Review study	Avoidance of social interaction	NA	NA
Urbanowicz et al. [61]	16 parents who have individuals with RTT [from 2.3 to 33.7]	Investigated the communicative use of eye gaze and gestures in females with RTT.	Research study	Eye pointing, assigning a point to each feature, the ability to make choices, requesting attention, facial expressions, and social interactions	Interview/several questions about expressive and receptive communication	NA
Fabio et al. [22]	A girl with RTT [21 years old]	Investigated the possibility of training communication abilities in people with RTT.	Case study	Cognitive and communicative delays	NA	Training reading–writing abilities
Wandin et al. [44]	Nine individuals with RTT [15 to 52 years]	Aimed to (a) develop a tool for assessing visual attention in individuals with RTT using AAC with a communication partner during naturalistic interactions in clinical settings and (b) explored aspects of the tool's reliability, validity, and utility.	Research study	Limited speech, no intelligible words, and word approximations used inconsistently	AVAI tool	NA
Marschik et al. [47]	An individual with RTT [The participant was longitudinally observed from birth to 11 years of age]	Assessed various aspects of speech and language and communicative functions of an individual with the preserved speech variant of RTT to describe her developmental profile over 11 years.	Case study	Hand stereotypies (such as rubbing and washing-like hand movements), as well as hand-to-mouth and hand-to-tongue stereotypies, morphosyntactic and socio-pragmatic limitations, and dysfluency of speech, caused mainly by immediate echolalia	Audio–video recordings of spontaneous speech, picture book reading, storytelling (five stories) and daily routines, parental diaries, and checklists: (i) the Austrian Rett survey; (ii) the Austrian adaptation of the MB-CDIs, a checklist to assess early socio-communicative functions, early gestures, vocabulary, and grammar; (iii) IPCA, AWST-R, TROG-D, SET-K, and the PADLD.	NA

Table A1. *Cont.*

Authors	Sample [Age]	Purpose of the Study	Study Design	Results		
				Communication Skills	Assessment of Communication Skills	Intervention of Communication Skills
Ahonniska-Assa et al. [9]	17 girls with RTT [3 years and 4 months to 12 years and 2 months]	Investigated whether eye-tracking technology in forced-choice tasks can enable children with very severe motor and speech impairments to respond consistently, allowing a more reliable evaluation of their language comprehension.	Research study	Inconsistent imitation and social skills	Eye-tracking technology and PPVT-4	NA
Castelli et al. [51]	Nine individuals with RTT [5 to 19 years old]	Investigated Theory of Mind capacities with RTT, which is similar to autism yet with more significant impairments.	Research study	Prelinguistic behaviors	Non-verbal false-belief task	NA
Sigafoos et al. [29]	Nine studies, 31 participants with RTT [2 years and 7 months to 17 years]	Reviewed communication intervention studies involving people with RTT.	Systematic analysis	Imitative speech, requesting, naming/commenting, and various receptive language skills	NA	Early intensive behavioral intervention
Hagberg [50]	NA	Reviewed presentation and clinical diagnosis of RTT at various ages and stages.	Review study	Eye communication or facial expressions as a prominent feature in most individuals with RTT	NA	NA
Didden et al. [34]	120 children and adults (girls) with RTT [5 to 55 years]	Assessed the forms and functions of prelinguistic communicative behaviors for 120 children and adults with RTT using the IPCA.	Research study	Eye gaze, laughing, smiling, vocalization, approaching a person, and distancing from a person	IPCA	NA
Marschik et al. [41]	Six individuals with RTT [aged 7 to 24 months]	Contributed new findings related to the pre-regression verbal development of females with a variant of RTT.	Research study	Babbling, (proto-)words and (proto-)word combinations, and nonverbal skills	Spontaneous speech analyses and additional parental report	NA
Djukic et al. [53]	37 Individuals with RTT and 34 typical development girls [2 to 31 years]	Examined the use of eye tracking technology, which is a technique uniquely suited for studying cognition in RTT.	Research study	Using eye gaze to communicate their desires, greet, point, request, and refuse	NA	NA
Kaufmann et al. [54]	80 individuals with RTT [1.6 to 14.9 years]	Examined the social impairments in RTT, the characteristics, and the relationship with clinical severity.	Research study	Pointing at or reaching for an object and requesting	SSI, a measure of autistic behavior suited for individuals with severe communication and motor impairment, RSBQ, VABS, and RSSS	NA
Hetzroni et al. [56]	Three individuals with RTT [8 to 10 years]	Investigated whether the use of assistive technology would assist in the ability to identify symbols of individuals with RTT.	Research study	Body movements for making choices, attention, and eye gaze movement	Observations of classroom activities and informal teacher and parent interviews	NA

Table A1. *Cont.*

Authors	Sample [Age]	Purpose of the Study	Study Design	Results		
				Communication Skills	**Assessment of Communication Skills**	**Intervention of Communication Skills**
Marschik et al. [21]	15 young individuals with RTT [first 2 years]	Aimed at shedding light on pre-regression development in RTT by focusing on early speech and language development using a video database.	Research study	Cooing, babbling, (proto-)words, word combinations and vocalization atypicalities	NA	NA
Fabio et al. [59]	28 Participants with RTT [4 to 22 years]	Examined the effects of cognitive rehabilitation with eye-tracker technology on attention, choice behaviors, and language over 2 years in patients with RTT.	Research study	Eye pointing, choice making, gestures, and body movements	NA	Cognitive rehabilitation
Skotko et al. [60]	Four individuals with RTT and their mothers [3.6 to 7 years old]	Described evidence and intervention strategies for parents, educators, and researchers who seek to enhance communication and literacy in individuals with RTT.	Research study	Attention and vocalizations	NA	Parent–child storybook interactions
Smeets et al. [52]	Two girls with RTT [9 and 11.5 years]	Discussed the clinical aspects with special emphasis on the behavioral phenotype and reviewed current perspectives in clinical management alongside perspectives on altering gene expression.	Review study	Dyspraxia, attention, and eye pointing	NA	NA
Barnes et al. [63]	74 individuals with RTT [2 and 11 years]	Examined the profiles of anxious behavior in individuals with RTT, with a focus on identifying the instrument with the best psychometric properties in this population.	Research study	NA	RSBQ, ADAMS, ABC-C, VABS-II, and CHQ	NA
Byiers and Symons [62]	NA [NA]	NA	Commentary study	NA	Eye-tracking technology	NA
Cass et al. [58]	87 individuals with RTT [2 years 1 month to 44 years 10 months]	Presented systematic data from a multidisciplinary clinical assessment of a large series of females with RTT.	Research study	Eye pointing, understanding cause and effect, making choices, and using words with and without meaning	NA	NA
Sigafoos et al. [13]	16 studies that involved 100 participants with RTT [NA]	Identified and summarized 16 communication intervention studies for individuals with RTT.	Systematic–narrative review	NA	NA	AAC device, graphic symbols, or activated microswitches/speech-generating devices to request preferred items, music therapy, eye tracking technology, and transcranial stimulation

Table A1. *Cont.*

Authors	Sample [Age]	Purpose of the Study	Study Design	Communication Skills	Results: Assessment of Communication Skills	Results: Intervention of Communication Skills
Lotan and Ben-Zeev [33]	NA [NA]	Presented the basic understanding of common characteristics typical of this disorder and the variants from the classical expression of RTT.	Review study	Express a few words or partial vocabulary	NA	AAC
Koppenhaver et al. [75]	6 individuals with RTT [6 to 7 years old]	Explored the impact of resting hand splints, light-tech augmentative communication systems, such as voice-output devices and symbols, and very basic parent training on the symbolic communication and labelling behaviors of six individuals with RTT.	Research study	NA	NA	Picture communication symbols, single-message Big-Mack, a multi-message four in-line cheap talk, a variety of stands made from PVC pipe
Ibrahim et al. [86]	Five children with speech and physical impairments [6 to 9 years]	Examined how children and school staff interact in task-oriented events when speech SGDs are not present or focal.	Research study	NA	NA	AAC
Fabio et al. [64]	Three individuals with RTT [aged 29, 30, and 31 years old]	Examined the neurophysiological and cognitive effects of tDCS in three individuals with RTT with chronic language impairments.	Research study	NA	VABS and RARS	tDCS and cognitive empowerment
Lamb et al. [70]	396 caregivers who have individuals with RTT [1.6 to 50 years]	Investigated factors related to family functioning and adaptation in caregivers of individuals with RTT.	Research study	NA	PSOC, WCC-R, FAM-III, and PAS	PMT and CET intervention
Byiers et al. [71]	Three individuals with RTT [ages 15 to 47 years]	Examined the effectiveness of functional assessment and functional communication training methods for teaching three children with classic RTT novel communicative behaviors.	Research study	NA	Functional assessment	Functional communication training
Stahlhut et al. [74]	40 participants (parents, care assistants, professionals from schools, and professionals at day centers of RTT) [NA]	Explored facilitators and barriers to "uptime" (non-sedentary) activities in Danish girls and women with RTT as perceived by parents and professionals using focus groups.	Research study	NA	NA	Eye gaze computer

Table A1. *Cont.*

Authors	Sample [Age]	Purpose of the Study	Study Design	Results		
				Communication Skills	Assessment of Communication Skills	Intervention of Communication Skills
Lim et al. [72]	62 articles with communication skills of children' RTT [from 9 to 60 years]	Summarized existing interventions and their outcomes in RTT rehabilitation and identified gaps in the literature.	Scoping review	NA	NA	Communication interventions to improve choice-making, communicative language, and social communication abilities
Wandin, Lindberg and Sonnander [87]	Three women with RTT [aged 27, 29, and 31 years]	Examined the effect of a communication intervention package on expressive communication and visual attention in individuals with RTT.	Case studies	Visual attention	NA	ALM using responsive partner strategies and a gaze-controlled device
Wandin et al., 2015 [73]	320 SLPs of RTT [NA]	Investigated communication intervention that SLPs provide to people with RTT.	Research study	NA	NA	AAC and communication aids
Loffler and Gordon [80]	NA [NA]	NA	Review study	NA	NA	Eye-tracking technology
Sandberg et al. [40]	Eight individuals with RTT [11 to 36 years]	Investigated the interrelationships between communication, cognition, and autistic features in young women with one of two variants of RTT complex disorders.	Research study	Loss and reappearance of verbal speech abilities, joint attention, and intentional communication	Video recordings, observations, and VABS	NA
Kolb et al. [69]	Three individuals with RTT [1.75 to 29 years old]	Systematically replicated the training procedures to teach three individuals with RTT to use AAC to make requests through caregiver coaching by researchers via telehealth.	Research study	NA	SDA	AAC in the form of manually activated single-message microswitches, eye-gaze device, mind training
Neul et al. [88]	925 caregivers who have individuals with RTT [From less than 1 year to over 40 years]	Determined the top concerns in RTT and RTT-related disorders.	Research study	Communication difficulties and physical inability to coordinate hand movements for communication	CGI-S and the RTT CSS	NA
Percy et al. [57]	NA [from 2 to 47 years for clinical studies using the Rett Syndrome Behavior Questionnaire, and the phase 3 LAVENDER study, aged 5 to 20 years]	Reviewed the RSBQ and its utilization in the assessment of symptoms associated with RTT.	Review study	Eye pointing and vocalizations	RSBQ	Eye gaze device, switches, and iPad

Table A1. *Cont.*

Authors	Sample [Age]	Purpose of the Study	Study Design	Communication Skills	Results Assessment of Communication Skills	Intervention of Communication Skills
Vilvarajan et al. [48]	103 girls with RTT [aged less than 18 years]	Comprehensive review of the clinical features, comorbidities, and multidisciplinary management of a well-characterized cohort of females with classical RTT.	Research study	Single words or short phrases, body language, eye gaze, gestures, and vocalizations	NA	NA
Romano, Lotan and Fabio [49]	195 girls and women with RTT [3 to 40 years]	Evaluated the severity level of girls and women with RTT living in Italy and Israel, two countries with different approaches to caring for people with complex disabilities.	Research study	Spatial and temporal orientation, memory, verbal communication skills, non-verbal communication through facial expressions, ability to maintain eye contact and shared attention, and the presence of responsive smiling	RARS	NA
Portnova et al. [65]	32 Individuals with RTT and 41 typically developing girls [1.9 to 17.1 and 2.58 to 17.98]	Combined clinical, qualitative, and experimental/quantitative approaches to the resting EEG analysis in a search for the EEG characteristics that are linked with RTT symptoms.	Research study	Pleasure and displeasure sounds, and representational gestures	VABS, PDMS-2, and PLS-5	NA
Fabio et al. [55]	28 young and women with RTT [4 to 22 years]	Examined the effects of motor training on attention, reaching skills, and stereotypies in patients with RTT.	Research study	Attention, reaching skills, and stereotypies	VABS and RARS	Intervention in motor abilities and cognitive intervention
Wandin, Lindberg and Sonnander [46]	Three adults with RTT [27, 29 and 31 years old]	Explored and describe a trained communication partner's use of responsive strategies in dyadic interaction with adults with RTT.	Research study	Eye gaze, body movements, and simple hand movements	RAACS	Gaze-controlled device
Unholz-Bowden et al. [66]	Three individuals with RTT [3, 5 and 19 years old]	Investigated the use of high- and low-tech AAC modalities by three individuals with RTT given similar instruction for using both modalities.	Research study	Gestures, requesting, pointing, and eye gaze	NA	High- and low-tech AAC instruments
Girtler et al. [89]	Three individuals with RTT [3, 5 and 19 years old]	Evaluated the effects of systematic individualized instruction procedures on the page-linking skills of individuals with RTT.	Research study	Gestures, laughing, crying, and vocalizations	IPCA and SDA	AAC

Table A1. *Cont.*

Authors	Sample [Age]	Purpose of the Study	Study Design	Results: Communication Skills	Results: Assessment of Communication Skills	Results: Intervention of Communication Skills
Xavier et al. [90]	14 girls with RTT and 11 girls with typical development [7–13 years and 3–4 years]	Evaluated the receptive vocabulary of individuals with RTT using eye-tracking technology and examined how these objective measures compared with parents' perceptions of their daughters' language abilities.	Research study	NA	Kerr Scale, a parental questionnaire on communication skills, the Peabody Picture Vocabulary Test (PPVT-4), and eye-tracking equipment	NA
Passaro et al. [91]	A narrative review NA [NA]	Evaluated studies conducted on the use of ETT to improve cognitive abilities in girls with RTT and to examine its potential application.	Review	NA	Eye tracking technology	NA
Neul et al. [67]	187 females with RTT [5 to 20 years old]	Evaluated the efficacy of trofinetide in communication of individuals with RTT.	Research study	Inability to communicate or make choices	The caregiver-rated CSBS-DP-IT social composite score and RTT-specific clinician rating scales related to discrete aspects of communication (RTT-COMC and RTT-VCOM)	NA
Lee et al., 2024 [28]	A case study with a female and a Narrative review with RTT [22 years old]	Evaluated the impact of AAC devices on communication outcomes and quality of life in individuals with RTT through a case report and narrative review.	Research study	NA	AAC	AAC
Voniati et al., 2024 [92]	Scoping Review with 7 articles with a total of 381 participants with RTT [NA]	Aimed to provide an overview of the identification of communication abilities in children with RTT in their daily routine from the perspectives of communication partners who interact with individuals diagnosed with RTT.	Scoping review	Eye gaze, expressing pleasure, requesting desired items, making choices, vocalizations, pointing, interaction, and strong receptive skills	Functional Ability Checklist, Functional Independence Measure for Children Communication and Symbolic Behavior Scales Developmental Profile Infant–Toddler Checklist, MacArthur–Bates Communicative Development Inventories and sections of the Vineland Adaptive Behavior Scales, Second Edition, the Assessment of Visual Attention in Interaction tool, the Child Participation in Family Activities questionnaire, the Clinical Severity Score and Hoffer Ambulation Scale	NA

Table A1. *Cont.*

Authors	Sample [Age]	Purpose of the Study	Study Design	Results		
				Communication Skills	Assessment of Communication Skills	Intervention of Communication Skills
Voniati et al., 2023 [43]	Narrative Review: 19 articles with individuals with RTT [NA]	Aimed to support the clinical work of speech–language pathologists (SLPs) while assessing the communication aptitudes of children with RTT.	Review	NA	Medical and developmental intake, functional assessment, Mullen scales, IPCA, PPVT, Vineland Adaptive Behavior Scales, a communication matrix, and AAC assessment	NA

RTT: individuals with Rett syndrome, ASD: Autism Spectrum Disorder, NA: non-applicable, FXS: fragile X syndrome, VABS: Vineland Adaptive Behavior Scale, RSBQ: Rett Syndrome Behavioral Questionnaire, PDMS-2: Peabody Developmental Motor Scale 2, IPCA: Inventory of Potential Communicative Acts, AVAI Tool: Assess Visual Attention in Rett syndrome, PSOC: Parenting Sense of Competence Scale, WCC-R: ways of coping checklist, revised, FAM-III: Family Assessment Measure III, PAS: psychological adaptation scale, tDCS: Transcranial Direct Current Stimulation, MB-CDIs: The MacArthur–Bates Communicative Development Inventories, AWST-R: Aktiver Wortschatztest für 3- bis 5-jährige Kinder–Revision–(German vocabulary Test), TROG-D: Test zur Überprüfung des Grammatikverständnisses (Test for the Reception of Grammar), SET-K: SprachEntwicklungsTest für Kinder (speech and language development test), PADLD: Patholinguistic Assessment of Developmental Language Disorders, SSI: Screen for Social Interaction, ADAMS: Anxiety, Depression, and Mood Scale, ABC-C: Aberrant Behavior Checklist-Community, CHQ: child health questionnaire, PMT: parent management training, CET: coping effectiveness training, SGDs: speech generating devices, RARS: Rett Assessment Rating Scale, RAACS: Responsive Augmentative and Alternative Communication Style Scale, PLS-5: preschool language scale, CGIS-S: Clinical Global Impression-Severity, CSS: clinical severity score, SDA: structural descriptive assessment, ALM: aided language modelling, AAC: Augmentative and Alternative Communication, EEG: Electroencephalography.

B.

Table 2. SANRA Brief Scale for the quality assessment of the current narrative review.

SANRA Criterion	Score (0–2)	Justification for Scoring
1. Importance for the readership	2	The introduction contains all the elements to explain the significance of the communication deficits of individuals with RTT.
2. Stated aims or research questions	2	The aims of the study are stated clearly in the abstract and Section 1.2 with main research questions and subobjectives.
3. Description of the literature search	2	A transparent, structured search is described. The databases, keywords, Boolean, date range, inclusion/exclusion criteria, and the use of Kable's 12-step method are noted.
4. Referencing	2	Up-to-date, relevant, and peer-reviewed references (n = 91) were used from Databases.
5. Scientific reasoning	2	The discussion was developed by making a synthesis of the findings across all studies and discussed supported with citations and any possible evidence.
6. Appropriate presentation of data	2	The results were described by domain (verbal, nonverbal, assessment, intervention) and presented by a table in the Appendix A.
Total Score	12	

References

1. Chahil, G.; Bollu, P.C. Rett Syndrome. 2018. Available online: https://europepmc.org/books/NBK482252/figure/article-28490.image.f1/?report=objectonly (accessed on 23 June 2025).
2. Amir, R.E.; Van den Veyver, I.B.; Wan, M.; Tran, C.Q.; Francke, U.; Zoghbi, H.Y. Rett syndrome is caused by mutations in X-linked MECP2, encoding methyl-CpG-binding protein 2. *Nat. Genet.* **1999**, *23*, 185–188. [CrossRef] [PubMed]
3. Neul, J.L.; Kaufmann, W.E.; Glaze, D.G.; Christodoulou, J.; Clarke, A.J.; Bahi-Buisson, N.; Leonard, H.; Bailey, M.E.S.; Schanen, N.C.; Zappella, M.; et al. Rett syndrome: Revised diagnostic criteria and nomenclature. *Ann. Neurol.* **2010**, *68*, 944–950. [CrossRef] [PubMed]
4. Poleg, T.; Hadar, N.; Heimer, G.; Dolgin, V.; Aminov, I.; Safran, A.; Agam, N.; Jean, M.M.; Freund, O.; Kaur, S.; et al. Unraveling MECP2 structural variants in previously elusive Rett syndrome cases through IGV interpretation. *NPJ Genom. Med.* **2025**, *10*, 23. [CrossRef] [PubMed]
5. Byun, C.K.; Lee, J.S.; Lim, B.C.; Kim, K.J.; Hwang, Y.S.; Chae, J.H. FOXG1 mutation is a low-incidence genetic cause in atypical Rett syndrome. *Child Neurol. Open* **2015**, *2*, 2329048X14568151. [CrossRef] [PubMed]
6. Weaving, L.S.; Christodoulou, J.; Williamson, S.L.; Friend, K.L.; McKenzie, O.L.D.; Archer, H.; Evans, J.; Clarke, A.; Pelka, G.J.; Tam, P.P.; et al. Mutations of CDKL5 cause a severe neurodevelopmental disorder with infantile spasms and mental retardation. *Am. J. Hum. Genet.* **2004**, *75*, 1079–1093. [CrossRef] [PubMed]
7. Asuncion, R.; Ramani, P. *Rett Syndrome;* [Updated 27 March 2025]; StatPearls Publishing: Treasure Island, FL, USA, 2025. Available online: https://www.ncbi.nlm.nih.gov/books/NBK482252/ (accessed on 23 June 2025).
8. Einspieler, C.; Marschik, P.B. Regression in Rett syndrome: Developmental pathways to its onset. *Neurosci. Biobehav. Rev.* **2019**, *98*, 320–332. [CrossRef] [PubMed]
9. Ahonniska-Assa, J.; Polack, O.; Saraf, E.; Wine, J.; Silberg, T.; Nissenkorn, A.; Ben-Zeev, B. Assessing cognitive functioning in females with Rett syndrome by eye-tracking methodology. *Eur. J. Paediatr. Neurol.* **2018**, *22*, 39–45. [CrossRef] [PubMed]
10. Zhang, X.; Bao, X.; Zhang, J.; Zhao, Y.; Cao, G.; Pan, H.; Zhang, J.; Wei, L.; Wu, X. Molecular characteristics of Chinese patients with Rett syndrome. *Eur. J. Med. Genet.* **2012**, *55*, 677–681. [CrossRef] [PubMed]
11. Tarquinio, D.C.; Hou, W.; Neul, J.L.; Berkmen, G.K.; Drummond, J.; Aronoff, E.; Harris, J.; Lane, J.B.; Kaufmann, W.E.; Motil, K.J.; et al. The course of awake breathing disturbances across the lifespan in Rett syndrome. *Brain Dev.* **2018**, *40*, 515–529. [CrossRef] [PubMed]

12. Sigafoos, J.; O'Reilly, M.F.; Ledbetter-Cho, K.; Lim, N.; Lancioni, G.E.; Marschik, P.B. Addressing sequelae of developmental regression associated with developmental disabilities: A systematic review of behavioral and educational intervention studies. *Neurosci. Biobehav. Rev.* **2019**, *96*, 56–71. [CrossRef] [PubMed]
13. Sigafoos, J.; Roche, L.; O'Reilly, M.F.; Lancioni, G.E.; Marschik, P.B. Updated systematic-narrative review on communication intervention in Rett Syndrome: 2010–2022. *Augment. Altern. Commun.* **2023**, *39*, 241–255. [CrossRef] [PubMed]
14. Hornof, A.; Whitman, H.; Sutherland, M.; Gerendasy, S.; McGrenere, J. Designing for the" Universe of One" Personalized Interactive Media Systems for People with the Severe Cognitive Impairment Associated with Rett Syndrome. In Proceedings of the 2017 CHI Conference on Human Factors in Computing Systems, Denver, CO, USA, 6–11 May 2017; pp. 2137–2148.
15. Roche, L.; Zhang, D.; Bartl-Pokorny, K.D.; Pokorny, F.B.; Schuller, B.W.; Esposito, G.; Bölte, S.; Roeyers, H.; Poustka, L.; Gugatschka, M.; et al. Early vocal development in autism spectrum disorder, Rett syndrome, and fragile X syndrome: Insights from studies using retrospective video analysis. *Adv. Neurodev. Disord.* **2018**, *2*, 49–61. [CrossRef] [PubMed]
16. Jian, L.; Nagarajan, L.; De Klerk, N.; Ravine, D.; Bower, C.; Anderson, A.; Williamson, S.; Christodoulou, J.; Leonard, H. Predictors of seizure onset in Rett syndrome. *J. Pediatr.* **2006**, *149*, 542–547. [CrossRef] [PubMed]
17. Neul, J.L.; Lane, J.B.; Lee, H.S.; Geerts, S.; Barrish, J.O.; Annese, F.; Baggett, L.M.; Barnes, K.; ASkinner, S.; Motil, K.J.; et al. Developmental delay in Rett syndrome: Data from the natural history study. *J. Neurodev. Disord.* **2014**, *6*, 20. [CrossRef] [PubMed]
18. Brunetti, S.; Lumsden, D.E. Rett Syndrome as a movement and motor disorder—A narrative review. *Eur. J. Paediatr. Neurol.* **2020**, *28*, 29–37. [CrossRef] [PubMed]
19. Townend, G.S.; Marschik, P.B.; Smeets, E.; Van de Berg, R.; van den Berg, M.; Curfs, L.M.G. Eye gaze technology as a form of augmentative and alternative communication for individuals with Rett syndrome: Experiences of families in The Netherlands. *J. Dev. Phys. Disabil.* **2016**, *28*, 101–112. [CrossRef] [PubMed]
20. Fabio, R.A.; Giannatiempo, S.; Caprì, T. Attention and identification of the same and the similar visual stimuli in Rett Syndrome. *Life Span. Disabil.* **2019**, *22*, 113–127.
21. Marschik, P.B.; Kaufmann, W.E.; Sigafoos, J.; Wolin, T.; Zhang, D.; Bartl-Pokorny, K.D.; Pini, G.; Zappella, M.; Tager-Flusberg, H.; Einspieler, C.; et al. Changing the perspective on early development of Rett syndrome. *Res. Dev. Disabil.* **2013**, *34*, 1236–1239. [CrossRef] [PubMed]
22. Fabio, R.A.; Castelli, I.; Marchetti, A.; Antonietti, A. Training communication abilities in Rett Syndrome through reading and writing. *Front. Psychol.* **2013**, *4*, 66589. [CrossRef] [PubMed]
23. Bartolotta, T.E.; Zipp, G.P.; Simpkins, S.D.; Glazewski, B. Communication skills in girls with Rett syndrome. *Focus Autism Other Dev. Disabl.* **2011**, *26*, 15–24. [CrossRef]
24. Bartl-Pokorny, K.D.; Marschik, P.B.; Sigafoos, J.; Tager-Flusberg, H.; Kaufmann, W.E.; Grossmann, T.; Einspieler, C. Early socio-communicative forms and functions in typical Rett syndrome. *Res. Dev. Disabil.* **2013**, *34*, 3133–3138. [CrossRef] [PubMed]
25. Cianfaglione, R.; Clarke, A.; Kerr, M.; Hastings, R.P.; Oliver, C.; Moss, J.; Heald, M.; Felce, D. A national survey of Rett syndrome: Behavioural characteristics. *J. Neurodev. Disord.* **2015**, *7*, 11. [CrossRef] [PubMed]
26. Rozensztrauch, A.; Sebzda, A.; Śmigiel, R. Clinical presentation of Rett syndrome in relation to quality of life and family functioning. *J. Int. Med. Res.* **2021**, *49*, 03000605211007714. [CrossRef] [PubMed]
27. Townend, G.S.; Bartolotta, T.E.; Urbanowicz, A.; Wandin, H.; Curfs, L.M.G. Development of consensus-based guidelines for managing communication of individuals with Rett syndrome. *Augment. Altern. Commun.* **2020**, *36*, 71–81. [CrossRef] [PubMed]
28. Lee, J.; Lee, J.; Abu-Daoud, M.; Al-Ajlouni, Y.A. Assistive Communication Devices in Rett Syndrome: A Case Report and Narrative Review. *BioMed* **2024**, *4*, 146–155. [CrossRef]
29. Sigafoos, J.; Green, V.A.; Schlosser, R.; O'eilly, M.F.; Lancioni, G.E.; Rispoli, M.; Lang, R. Communication intervention in Rett syndrome: A systematic review. *Res. Autism Spectr. Disord.* **2009**, *3*, 304–318. [CrossRef]
30. Sigafoos, J.; Kagohara, D.; van der Meer, L.; Green, V.A.; O'Reilly, M.F.; Lancioni, G.E.; Lang, R.; Rispoli, M.; Zisimopoulos, D. Communication assessment for individuals with Rett syndrome: A systematic review. *Res. Autism Spectr. Disord.* **2011**, *5*, 692–700. [CrossRef]
31. Einspieler, C.; Kerr, A.M.; Prechtl, H.F.R. Is the early development of girls with Rett disorder really normal? *Pediatr. Res.* **2005**, *57*, 696–700. [CrossRef] [PubMed]
32. Lotan, M. Rett syndrome. Guidelines for individual intervention. *Sci. World J.* **2006**, *6*, 1504–1516. [CrossRef] [PubMed]
33. Lotan, M.; Ben-Zeev, B. Rett syndrome. A review with emphasis on clinical characteristics and intervention. *Sci. World J.* **2006**, *6*, 1517–1541. [CrossRef] [PubMed]
34. Didden, R.; Korzilius, H.; Smeets, E.; Green, V.A.; Lang, R.; Lancioni, G.E.; Curfs, L.M. Communication in individuals with Rett syndrome: An assessment of forms and functions. *J. Dev. Phys. Disabil.* **2010**, *22*, 105–118. [CrossRef] [PubMed]
35. Sukhera, J. Narrative reviews: Flexible, rigorous, and practical. *J. Grad. Med. Educ.* **2022**, *14*, 414–417. [CrossRef] [PubMed]
36. Kable, A.K.; Pich, J.; Maslin-Prothero, S.E. A structured approach to documenting a search strategy for publication: A 12 step guideline for authors. *Nurse Educ. Today* **2012**, *32*, 878–886. [CrossRef] [PubMed]

37. Baethge, C.; Goldbeck-Wood, S.; Mertens, S. SANRA—A scale for the quality assessment of narrative review articles. *Res. Integr. Peer Rev.* **2019**, *4*, 5. [CrossRef] [PubMed]
38. Baethge, C.; Mertens, S.; Goldbeck-Wood, S. Development of a quality score for the assessment of non-systematic review articles (SANRA). In Proceedings of the Poster, Seventh International Congress on Peer Review and Biomedical Publication, Chicago, IL, USA, 8–10 September 2013.
39. Elefant, C.; Wigram, T. Learning ability in children with Rett syndrome. *Brain Dev.* **2005**, *27*, S97–S101. [CrossRef] [PubMed]
40. Sandberg, A.D.; Ehlers, S.; Hagberg, B.; Gillberg, C. The Rett syndrome complex: Communicative functions in relation to developmental level and autistic features. *Autism* **2000**, *4*, 249–267. [CrossRef]
41. Marschik, P.B.; Pini, G.; Bartl-Pokorny, K.D.; Duckworth, M.; Gugatschka, M.; Vollmann, R.; Zappella, M.; Einspieler, C. Early speech–language development in females with Rett syndrome: Focusing on the preserved speech variant. *Dev. Med. Child. Neurol.* **2012**, *54*, 451–456. [CrossRef] [PubMed]
42. Segawa, M.; Nomura, Y. Rett syndrome. *Curr. Opin. Neurol.* **2005**, *18*, 97–104. [CrossRef] [PubMed]
43. Voniati, L.; Papaleontiou, A.; Georgiou, R.; Tafiadis, D. Identifying key aspects of the communicative capabilities of children with Rett syndrome. *Neurosci. Res. Notes* **2023**, *6*, 175. [CrossRef]
44. Wandin, H.; Lindberg, P.; Sonnander, K. Development of a tool to assess visual attention in Rett syn-drome: A pilot study. *Augment. Altern. Commun.* **2020**, *36*, 118–127. [CrossRef] [PubMed]
45. Cianfaglione, R.; Clarke, A.; Kerr, M.; Hastings, R.P.; Oliver, C.; Felce, D. Ageing in Rett syndrome. *J. Intellect. Disabil. Res.* **2016**, *60*, 182–190. [CrossRef] [PubMed]
46. Wandin, H.; Lindberg, P.; Sonnander, K. A trained communication partner's use of responsive strategies in aided communication with three adults with Rett syndrome: A case report. *Front. Psychol.* **2022**, *13*, 989319. [CrossRef] [PubMed]
47. Marschik, P.B.; Vollmann, R.; Bartl-Pokorny, K.D.; Green, V.A.; van der Meer, L.; Wolin, T.; Einspieler, C. Develop-mental profile of speech-language and communicative functions in an individual with the Preserved Speech Variant of Rett syndrome. *Dev. Neurorehabil* **2014**, *17*, 284–290. [CrossRef] [PubMed]
48. Vilvarajan, S.; McDonald, M.; Douglas, L.; Newham, J.; Kirkland, R.; Tzannes, G.; Tay, D.; Christodoulou, J.; Thompson, S.; Ellaway, C. Multidisciplinary Management of Rett Syndrome: Twenty Years' Experience. *Genes* **2023**, *14*, 1607. [CrossRef] [PubMed]
49. Romano, A.; Lotan, M.; Fabio, R.A. A Severity Comparison between Italian and Israeli Rett Syndrome Cohorts. *Diagnostics* **2023**, *13*, 3390. [CrossRef] [PubMed]
50. Hagberg, B. Clinical manifestations and stages of Rett syndrome. *Ment. Retard. Dev. Disabil. Res. Rev.* **2002**, *8*, 61–65. [CrossRef] [PubMed]
51. Castelli, I.; Antonietti, A.; Fabio, R.A.; Lucchini, B.; Marchetti, A. Do rett syndrome persons possess theory of mind? Some evidence from not-treated girls. *Life Span. Disabil.* **2013**, *16*, 157–168.
52. Smeets, E.E.J.; Pelc, K.; Dan, B. Rett syndrome. *Mol. Syndromol.* **2012**, *2*, 113–127. [CrossRef] [PubMed]
53. Djukic, A.; Rose, S.A.; Jankowski, J.J.; Feldman, J.F. Rett syndrome: Recognition of facial expression and its relation to scanning patterns. *Pediatr. Neurol.* **2014**, *51*, 650–656. [CrossRef] [PubMed]
54. Kaufmann, W.E.; Tierney, E.; Rohde, C.A.; Suarez-Pedraza, M.C.; Clarke, M.A.; Salorio, C.F.; Bibat, G.; Bukelis, I.; Naram, D.; Lanham, D.C.; et al. Social impairments in Rett syndrome: Characteristics and relationship with clinical severity. *J. Intellect. Disabil. Res.* **2012**, *56*, 233–247. [CrossRef] [PubMed]
55. Fabio, R.A.; Giannatiempo, S.; Caprì, T.; Semino, M. Repeated Motor Training on Attention Reaching Skills and Stereotypies in Rett Syndrome. *Mov. Disord. Clin. Pract.* **2022**, *9*, 637–646. [CrossRef] [PubMed]
56. Hetzroni, O.; Rubin, C.; Konkol, O. The use of assistive technology for symbol identification by children with Rett syndrome. *J. Intellect. Dev. Disabil.* **2002**, *27*, 57–71. [CrossRef]
57. Percy, A.K.; Neul, J.L.; Benke, T.A.; Marsh, E.D.; Glaze, D.G. A review of the Rett Syndrome Behaviour Questionnaire and its utilization in the assessment of symptoms associated with Rett syndrome. *Front. Pediatr.* **2023**, *11*, 1229553. [CrossRef] [PubMed]
58. Cass, H.; Reilly, S.; Owen, L.; Wisbeach, A.; Weekes, L.; Slonims, V.; Wigram, T.; Charman, T. Findings from a multidisciplinary clinical case series of females with Rett syndrome. *Dev. Med. Child. Neurol.* **2003**, *45*, 325–337. [CrossRef] [PubMed]
59. Fabio, R.A.; Giannatiempo, S.; Semino, M.; Caprì, T. Longitudinal cognitive rehabilitation applied with eye-tracker for patients with Rett Syndrome. *Res. Dev. Disabil.* **2021**, *111*, 103891. [CrossRef] [PubMed]
60. Skotko, B.G.; Koppenhaver, D.A.; Erickson, K.A. Parent reading behaviors and communication outcomes in girls with Rett syndrome. *Except. Child.* **2004**, *70*, 145–166. [CrossRef]
61. Urbanowicz, A.; Downs, J.; Girdler, S.; Ciccone, N.; Leonard, H. An exploration of the use of eye gaze and gestures in females with Rett syndrome. *J. Speech Lang. Hear. Res.* **2016**, *59*, 1373–1383. [CrossRef] [PubMed]
62. Byiers, B.; Symons, F. The need for unbiased cognitive assessment in Rett syndrome: Is eye tracking the answer? *Dev. Med. Child. Neurol.* **2013**, *55*, 301. [CrossRef] [PubMed]

63. Barnes, K.V.; Coughlin, F.R.; O'Leary, H.M.; Bruck, N.; Bazin, G.A.; Beinecke, E.B.; Walco, A.C.; Cantwell, N.G.; Kaufmann, W.E. Anxiety-like behavior in Rett syndrome: Characteristics and assessment by anxiety scales. *J. Neurodev. Disord.* **2015**, *7*, 30. [CrossRef] [PubMed]
64. Fabio, R.A.; Gangemi, A.; Capri, T.; Budden, S.; Falzone, A. Neurophysiological and cognitive effects of Transcranial Direct Current Stimulation in three girls with Rett Syndrome with chronic language impairments. *Res. Dev. Disabil.* **2018**, *76*, 76–87. [CrossRef] [PubMed]
65. Portnova, G.; Neklyudova, A.; Voinova, V.; Sysoeva, O. Clinical EEG of Rett syndrome: Group analysis supplemented with longitudinal case report. *J. Pers. Med.* **2022**, *12*, 1973. [CrossRef] [PubMed]
66. Unholz-Bowden, E.K.; Girtler, S.N.; Shipchandler, A.; Kolb, R.L.; McComas, J.J. Use of augmentative and alternative communication by individuals with rett syndrome part 2: High-tech and low-tech modalities. *J. Dev. Phys. Disabil.* **2024**, *36*, 147–167. [CrossRef] [PubMed]
67. Neul, J.L.; Percy, A.K.; Benke, T.A.; Berry-Kravis, E.M.; Glaze, D.G.; Peters, S.U.; Marsh, E.D.; An, D.; Bishop, K.M.; Youakim, J.M. Trofinetide treatment demonstrates a benefit over placebo for the ability to communicate in Rett syndrome. *Pediatr. Neurol.* **2024**, *152*, 63–72. [CrossRef] [PubMed]
68. Urbanowicz, A.; Ciccone, N.; Girdler, S.; Leonard, H.; Downs, J. Choice making in Rett syndrome: A descriptive study using video data. *Disabil. Rehabil.* **2018**, *40*, 813–819. [CrossRef] [PubMed]
69. Kolb, R.L.; McComas, J.J.; Girtler, S.N.; Simacek, J.; Dimian, A.F.; Unholz-Bowden, E.K.; Shipchandler, A.K. Teaching requesting to individuals with Rett syndrome using alternative augmentative communication (AAC) through caregiver coaching via telehealth. *J. Dev. Phys. Disabil.* **2023**, *35*, 1063–1090. [CrossRef] [PubMed]
70. Lamb, A.E.; Biesecker, B.B.; Umstead, K.L.; Muratori, M.; Biesecker, L.G.; Erby, L.H. Family functioning mediates adaptation in caregivers of individuals with Rett syndrome. *Patient Educ. Couns.* **2016**, *99*, 1873–1879. [CrossRef] [PubMed]
71. Byiers, B.J.; Dimian, A.; Symons, F.J. Functional communication training in Rett syndrome: A preliminary study. *Am. J. Intellect. Dev. Disabil.* **2014**, *119*, 340–350. [CrossRef] [PubMed]
72. Lim, J.; Greenspoon, D.; Hunt, A.; McAdam, L. Rehabilitation interventions in Rett syndrome: A scoping review. *Dev. Med. Child. Neurol.* **2020**, *62*, 906–916. [CrossRef] [PubMed]
73. Wandin, H.; Lindberg, P.; Sonnander, K. Communication intervention in Rett syndrome: A survey of speech language pathologists in Swedish health services. *Disabil. Rehabil.* **2015**, *37*, 1324–1333. [CrossRef] [PubMed]
74. Stahlhut, M.; Esbensen, B.A.; Larsen, J.L.; Bisgaard, A.M.; Downs, J.; Nordmark, E. Facilitators and Barriers of Participation in "Uptime" Activities in Girls and Women with Rett Syndrome: Perspectives from Parents and Professionals. *Qual. Health Res.* **2019**, *29*, 609–619. [CrossRef] [PubMed]
75. Koppenhaver, D.; Erickson, K.; Harris, B.; McLellan, J.; Skotko, B.; Newton, R. Storybook-based communication intervention for girls with Rett syndrome and their mothers. *Disabil. Rehabil.* **2001**, *23*, 149–159. [PubMed]
76. Sigafoos, J.; Woodyatt, G.; Keen, D.; Tait, K.; Tucker, M.; Roberts-Pennell, D.; Pittendreigh, N. Identifying potential communicative acts in children with developmental and physical disabilities. *Commun. Disord. Q.* **2000**, *21*, 77–86. [CrossRef]
77. Sigafoos, J.; O'Reilly, M.F.; Lancioni, G.E.; Zhang, D.; Marschik, P.B. The Inventory of Potential Communicative Acts (IPCA): Narrative review of reliability, validity, and clinical utility. *Evid. Based Commun. Assess. Interv.* **2022**, *16*, 105–122. [CrossRef]
78. Coplan, J. Early Language Milestone Scale: Examiner's Manual. PRO-ED. 1993. Available online: https://www.aucd.org/uploads/itacresourcefiles/ELM.PDF (accessed on 23 June 2025).
79. Lewis, M. Early Language Milestone Scale. In *Encyclopedia of Autism Spectrum Disorders*; Springer: New York, NY, USA, 2013; pp. 1032–1033.
80. Loffler, G.; Gordon, G.E. Cognitive function in Rett syndrome: Profoundly impaired or near normal? *Eur. J. Paediatr. Neurol.* **2018**, *22*, 2–3. [CrossRef] [PubMed]
81. Levine, J.; Hakim, F.; Kooy, R.F.; Gozes, I. Vineland adaptive behavior scale in a cohort of four ADNP syndrome patients implicates age-dependent developmental delays with increased impact of activities of daily living. *J. Mol. Neurosci.* **2022**, *72*, 1531–1546. [CrossRef] [PubMed]
82. Perry, A.; Flanagan, H.E.; Dunn Geier, J.; Freeman, N.L. Brief report: The Vineland Adaptive Behavior Scales in young children with autism spectrum disorders at different cognitive levels. *J. Autism Dev. Disord.* **2009**, *39*, 1066–1078. [CrossRef] [PubMed]
83. Salomon-Estebanez, M.; Mohamed, Z.; Michaelidou, M.; Collins, H.; Rigby, L.; Skae, M.; Padidela, R.; Rust, S.; Dunne, M.; Cosgrove, K. Vineland adaptive behavior scales to identify neurodevelopmental problems in children with Congenital Hyperinsulinism (CHI). *Orphanet J. Rare Dis.* **2017**, *12*, 96. [CrossRef] [PubMed]
84. Hedrick, D.L.; Prather, E.M.; Tobin, A.R. *Sequenced Inventory of Communication Development*; University of Washington Press: Washington, DC, USA, 1975.
85. Thompson, L.A.; Plomin, R. The sequenced inventory of communication development: An adoption study of two and three-year olds. *Int. J. Behav. Dev.* **1988**, *11*, 219–231. [CrossRef]

86. Ibrahim, S.B.; Vasalou, A.; Clarke, M. Design opportunities for AAC and children with severe speech and physical impairments. In Proceedings of the 2018 CHI Conference on Human Factors in Computing Systems, Montreal, QC, Canada, 21–26 April 2018; pp. 1–13.
87. Wandin, H.; Lindberg, P.; Sonnander, K. Aided language modelling, responsive communication and eye-gaze technology as communication intervention for adults with Rett syndrome: Three experimental single case studies. *Disabil. Rehabil. Assist. Technol.* **2023**, *18*, 1011–1025. [CrossRef] [PubMed]
88. Neul, J.L.; Benke, T.A.; Marsh, E.D.; Suter, B.; Silveira, L.; Fu, C.; Peters, S.U.; Percy, A.K.; for the Rett syndrome Natural History Study Group; Skinner, S.A.; et al. Top caregiver concerns in Rett syndrome and related disorders: Data from the US natural history study. *J. Neurodev. Disord.* **2023**, *15*, 33. [CrossRef] [PubMed]
89. Girtler, S.N.; Unholz-Bowden, E.K.; Shipchandler, A.; Kolb, R.L.; McComas, J.J. Use of Augmentative and Alternative Communication by Individuals with Rett Syndrome Part 1: Page-Linking. *J. Dev. Phys. Disabil.* **2024**, *36*, 125–145. [CrossRef] [PubMed]
90. Xavier, J.S.; Orsati, F.T.; Laurence, P.G.; Barbosa, L.K.; Gouveia, A.S.; D'ANtino, M.E.F.; Lukasova, K.; Macedo, E.C.; Schwartzman, J.S. Insights into Rett Syndrome girls' receptive vocabulary through eye movements and parental perception. *Eur. J. Paediatr. Neurol.* **2024**, *51*, 1–8. [CrossRef] [PubMed]
91. Passaro, A.; Zullo, A.; Di Gioia, M.; Curcio, E.; Stasolla, F. A Narrative Review on the Use of Eye-Tracking in Rett Syndrome: Implications for Diagnosis and Treatment. *OBM Genet.* **2024**, *8*, 250. [CrossRef]
92. Voniati, L.; Georgiou, R.; Papaleontiou, A.; Tsapara, A.; Papadopoulos, A.; Tafiadis, D. Scoping Review of Communication Abilities of Children with Rett Syndrome in Daily Routine: A Communication Partners' Perspectives. *Adv. Neurodev. Disord.* **2024**, *9*, 185–194. [CrossRef]

MDPI AG
Grosspeteranlage 5
4052 Basel
Switzerland
Tel.: +41 61 683 77 34

Brain Sciences Editorial Office
E-mail: brainsci@mdpi.com
www.mdpi.com/journal/brainsci

www.ingramcontent.com/pod-product-compliance
Lightning Source LLC
Chambersburg PA
CBHW040823250926
51850CB00007BA/1246
* 9 7 8 3 7 2 5 8 5 9 4 1 2 *